ALLE · ZEIT · WACH · 1842

SJ

Werner Brilon (Editor)

Intersections without Traffic Signals II

Proceedings of an International Workshop
18-19 July, 1991 in Bochum, Germany

With 163 Figures

Springer-Verlag
Berlin Heidelberg NewYork
London Paris Tokyo
Hong Kong Barcelona Budapest

Prof. Dr.-Ing. Werner Brilon

Lehrstuhl für Verkehrswesen
Ruhr-Universität Bochum
Universitätsstraße 150
4630 Bochum

This workshop was made possible by the generous support of the following organisations:

Deutsche Forschungsgemeinschaft (German Council for Research)

Minister für Stadtentwicklung und Verkehr
des Landes Nordrhein-Westfalen

ISBN-13:978-3-642-84539-0 e-ISBN-13:978-3-642-84537-6
DOI: 10.1007/978-3-642-84537-6

Softcover reprint of the hardcover 1st edition 1991

Typesetting: Camera ready by author

61/3020-543210 – Printed on acid-free paper

Preface

In March 1988, an international workshop on intersections without traffic signals was held at the Ruhr-University in Bochum, Germany. The proceedings of this workshop were published by Springer [1]. The workshop was performed in a very harmonious atmosphere, which stimulated the experts from different countries to communicate and exchange their ideas and experiences. The presentations and the written contributions documented the present state of technical solutions for design and engineering of unsignalized intersections both regarding scientific research and practical applications. Moreover, numerous unsolved problems were identified.

Thus, the 1988 workshop stimulated new developments in the field of unsignalized intersections in several countries. In the meantime, these investigations have lead to a remarkable progress. For example in Germany a new guideline for unsignalized cross intersections and T-junctions has been finished and is going to be introduced in 1991. New results on roundabout capacity have been worked out as well. Many particularly important developments were made in foreign countries. Especially in the United States, an increasing interest in this subject can be observed. In the annual meetings of the TRB, this item received great attention. Many research institutes in North America have concentrated their activities on that point. A new TRB-circular concerning unsignalized intersections is going to be published. It will contain a new procedure for four-way-stop-control intersections, which seems to be a special feature of North American traffic engineering. However, new results from the US for two-way-stop-control intersections are available as well.

Especially in the field of roundabouts - or "rotaries", which seems to be the American term - some progress has been made. Roundabouts are on the advance worldwide. Many countries like Australia, France, Norway, Sweden, Switzerland and Germany are increasingly using roundabouts as an intersection design which bears comparison with other solutions or which is even more advantageous. This increasing use of roundabouts induces a great variety of new experiences both in the field of intersection capacity and with intersection design and traffic safety.

---

1) W. Brilon (editor): Intersections Without Traffic Signals.
Springer Verlag, 1988, ISBN 3-540-18890-8.

The progress made between 1988 and today supported the idea of holding a second international workshop on unsignalized intersections. This was also confirmed in discussions with experts from different countries. The Ruhr-University Bochum will again host this event. The fact that a second workshop is held on the same item at the same place, however, should not establish this event as another continuous series of traffic engineering congresses.

The general situation in the theory of capacity for unsignalized intersections is still characterized by the discussion between gap-acceptance theory and empirical regression models. Both still have their own specific advantages and drawbacks. Therefore, both approaches are in use in different countries. It can be stated that the more theoretically based gap-acceptance theory is still applied for cross-intersections and T-junctions in many countries. As many of the contributions to the workshop point out, this theory is still capable of further improvement. One reason for the use of this fundamental theory may be that it provides a better understanding of individual operations and of the hierarchy of traffic streams at such intersections. Thus, it enables an extrapolation of solutions for traffic flows which have not been observed in reality. The regression theory, however, is also more and more observed with interest outside Great Britain, which has the longest experience with this general approach. Presumably -with few exceptions- it has occupied nearly the whole field of roundabout capacity calculations. The reason for this could be that this approach is more suitable for handling driver behaviour which differs from pure gap-acceptance patterns like gap-forcing and priority reversal. The idea that critical gaps and move-up times could depend on the traffic load of the intersection could become an approach to overcome these differences.

To detect these kinds of effects, the precise evaluation of critical gaps from traffic observations could again become a field requiring increasing attention in the future.

This point of discussion brings us back to the basic fact that modelling in traffic engineering always contains two aspects: the mathematical construction and solution of the model on one side and the implementation of driver behaviour on the other side. For further improvements of traffic engineering methods, especially the aspects of a more sensitive and realistic modelling of driver behaviour by the models as well as by the parameters being included into the models will become increasingly important. Moreover, some of the differences between the approaches and the experiences from various countries could be due to basic differences in behaviour of road users.

Beyond intersection capacity estimations, however, the improved safety of traffic operations on our roads is the most essential task for traffic engineers. This is especially true for the decisions on control strategies and on types of intersection design. The contributions to the workshop consider this aspect as well. Unfortunately, due to the closely packed timetable, only one contribution which concentrates on safety could be included, although two more interesting papers on this issue had been proposed. However, almost every contribution pays special attention to the aspect of traffic operation safety, which is an argument for more integrated considerations of design, capacity and traffic safety.

For the preparation of the workshop, a number of papers from several countries have been invited to give a current overview of the whole subject considering the respective circumstances of the specific countries. In addition, a call for papers was sent out in September 1990, resulting in an interesting variety of abstracts proposed for the workshop. From these, a series of 15 papers has been selected by the advisory board (Prof. Keller, Munich; Prof. Retzko, Darmstadt and Prof. Brilon, Bochum). The work of this group and especially the excellent discipline of the authors regarding the quite narrow schedule for the preparation of the manuscripts should particularly be commended, since it enables the presentation of the proceedings just in time for the workshop.

The authors and the editor hope that this publication and the workshop itself will contribute to continued efforts within the community of traffic engineering researchers and practicians. Moreover, this event should accelerate the international exchange of new ideas and generate new impulses for further developments on this specific item of unsignalized intersections as one part of the whole field of traffic engineering.

Bochum, April 1991 Prof. Dr.-Ing. W. Brilon

Contents

Capacity and Safety of Roundabouts - German Results

Birgit Stuwe
Lehrstuhl für Verkehrswesen, Ruhr-University, Bochum, Germany

SUMMARY

The objective of the research project described in this paper was to develop a formula for calculating the capacity of roundabouts. These formulae were developed by use of an empirical procedure and regression techniques. Therefore, traffic flow at several roundabouts was observed and recorded by video equipment. Based on the assumption that the entry flow was saturated, traffic flow of entering and cirulating vehicles were analysed and counted in 1-minute intervals. Based on this data samples, a provisorial formula has been developed for the prediction of roundabout capacity. Four different curves have been specified according to dependance upon the number of lanes in the circular and the entry. These curves are used in practice for capacity calculations of roundabouts in Germany.
Furthermore, an investigation of accident data is described in this paper. In particular the accident research confirmed considerable increase in traffic safety at roundabouts compared to signalised or unsignalised intersections. Accident severity is reduced at roundabouts. It was found that the roundabout is a good solution to increase traffic safety.

INTRODUCTION

Places and sites of circular shape have been an instrument of city and town planning since the Middle ages and especially during the Renaissance. Circular places were suited where it was worth to emphasize the significance of a place. It was an idea of Hénard to implement one-way traffic on such greater places and thus the Place de l'Étoile in Paris was the first roundabout.
Also in Germany roundabouts have been constructed especially during the 30ies and 50ies. Later on, many of the existing roundabouts in Germany were rebuilt into junctions with traffic signals. Roundabouts were not paid attention to any more - especially due to a lack of suitable capacity estimations, of bad accident experience and also because of a misinterpretation of the existing rules. Traffic engineers assumed that they would have to treat the entries like waving areas. The length of the waving area determines the capacity of such facilities. So the capacity increases with the length, which means that

the diameter of the roundabout must increase. The resulting conclusion was that roundabouts had to be large. This, however, was a fault. And also some bad accident experience was due to the fact that design standards were not in accordance with traffic rules.

Recently, roundabouts have attracted attention in Germany again. A main reason why the roundabout was rediscovered in Germany was the good experiences regarding traffic safety from other countries. For example in France this type of intersection is increasingly applied mainly because of aspects of traffic safety. In other countries like e.g. Switzerland, Australia or Sweden investigations are being made with regard to the possible applications of roundabouts.

In Germany, roundabouts were introduced rather hesitantly and mainly as a traffic calming measure. But soon its excellent qualities regarding traffic operations were discovered. German traffic engineers recognized that a roundabout can be a good alternative to an intersection with traffic signals.

Thus, the Ruhr-University Bochum performed a research project on behalf of the Federal Minister of Transport to investigate to which extent good experiences from other countries can be confirmed in Germany. Furthermore, the research project should contribute to develop an approach towards capacity calculations.

How does a roundabout work ?

It is a significant rule of roundabouts that the traffic in the roundabout has priority and the traffic in the entries must give way. Otherwise, when vehicles in the entries have priority there is the danger that the roundabout will be jammed by the vehicles in the roundabout. Only for the reason that the traffic in the roundabout has priority, the system roundabout works correctly. Therefore, we can consider the roundabout as a number of T-junctions following one another.

DETERMINATION OF ROUNDABOUT CAPACITY

Method of Analysis

In order to determine the capacity of roundabouts, each entry will be considered seperately. The vehicles in the roundabout with the traffic volume Qk have priority. The approaching vehicles must give way. The capacity of the entry is the maximum possible traffic volume that can enter the main stream (maxQz).

Traffic engineering allows two possibilities to investigate the capacity of intersections without traffic signals. One way leads to the classical theory for calculating the capacity of unsignalized intersections - the gap acceptance theory; the second way - an empirical method which was first developed in England - is to determine the capacity based on observations in real traffic flow by regression techniques. Analyses based on preliminary studies (Brilon, Ohadi, 1988) showed that the gap acceptance theory could not easily applied to roundabout entries. The reasons were firstly that in some cases the move-up times turned out to be greater or as great as the critical gap and secondly that it was nearly impossible to define the correct major streams at multilane roundabouts with multilane entries. Therfore, also for the new German approach to roundabout capacity regression techniques was applied.

Observations and Analyses

Observations were made at 10 different roundabouts in Germany. The diameter of the roundabouts considered was between 28 m and 100 m. The sample included only two single lane roundabouts, seven two lane roundabouts and one roundabout with three circulating lanes. The intersections for this sample were chosen so that at least during a remarkable period of the day there were permanent queues at one or more entries. The traffic flow during these times was recorded by video equipement at altogether 29 entries. The approaching traffic as well as the circulating traffic (the flow on the roundabout crossing the saturated entry) was counted in short time intervalls (here 1-minute-intervalls).
Passenger car units (pcu) were determined for heavy vehicles and two wheelers in the entering and in the circulating flow, with:

- heavy truck : 2,0 pcu
- light truck : 1,5 pcu
- two wheelers : 0,5 pcu

The results of entry and circulating flow Qk and Qz were plotted into diagrams and a regression line was ascertained which represents the entry/circulating relationship.
Since only intervals with saturated entry flows were used, the counted number of entering vehicles is at the same time the maximum number of vehicles which can enter during this interval - and also consequently the capacity during this interval.

Results

In Germany, an exponential regression line was used to describe the relationship between Qk and Qz because of a better agreement with the gap acceptance theory. This function proves to be a little bit better (slightly greater correlation coefficients) than the linear approximation, which was used e.g. in England (Kimber, 1980), France (Louah, 1988) and Switzerland (Simon e.a., 1988).

In Germany, this function was chosen:

$$\max Q_z = A * e^{-B/10.000 * Q_k}$$

with:
- Qz = maximum entry flow [pcu/h]
- Qk = circulating flow [pcu/h]
- A,B = determining parameters

Its mathematical concept is identical with Siegloch's capacity formula (Siegloch, 1973). The parameters A and B were determined from the measurements by regression techniques. Several different types of entries were distinguished. According to the number of lanes in the circular roadway and in the entry the following values were obtained:

Number of lanes				Sample size
circular roadway	entry	A	B	n
3	2	2018	6.68	295
2	2	1553	6.69	3873
2-3	1	1200	7.30	455
1	1	1089	7.42	275

Tab. 1

The sample size is given by the number of 1-minute intervals observed. The corresponding curves are shown in Fig. 1.

Because of the big number of measurement points, the curve of two lane entries to two lane roundabouts can be regarded as reliable to the greatest possible extent. The results of one lane entries to one lane roundabouts as well as the results of two lane entries to three lane roundabouts are of preliminary validity, although especially these equations are of great interest for practice. The reason for this small sample size for single lane roundabouts is the fact that at the time of measurements such small roundabouts with high traffic volume were very rare in Germany.

Nevertheless, the investigated formulae are recommended for practical use in Germany. However, a continued investigation especially on single lane roundabout capacity should be performed to confirm or correct line 4 of table 1 in the near.

The new German formulae have been implemented into the computer program KREISEL which is also able to calculate capacities according to the English, French, Australian and Swiss formulae als well as to gap acceptance techniques.

Comparision of the results with other countries

A comparison of the German capacity formulae for the single lane roundabout with the results from other countries is given in Fig. 2. This figure shows that the capacities investigated in Germany are situated considerably lower than the values predicted by the English formulae. The German results are between 0.7 and 0.8 of the English values. The French and Swiss results correspond with the German curves to a higher degree.

This differences can only be explained by different driver behaviour. It is assumed that drivers in England are more familiar with roundabouts since they have experienced this type of intersection for a long time. Whereas in Germany a roundabout - up to now - is still an exotic solution.

Other factors

Does the capacity of an entry depend on other factors, too ?
Kimber (1980) e.g. identifies a great number of geometrical characteristics which influence roundabout capacity. Moreover, Louah (1988) finds an influence of exiting vehicles on the capacity of the downstream entry.

First results of the analysis mentioned above have shown that neither an influence of the inscribed circle diameter nor an influence of the exiting vehicles could be verified to be significant for the quite extensive sample which was used. This result, of course, was also not due to a concentration of data on special values. In contrary, the variables being investigated varied over a remarkable part of the scale, e.g. roundabout diameter 28 - 110 m, exiting flows: 60 - 1140 pcu/h at single lane roundaboutsand 60 - 2750 pcu/h at two lane roundabouts. Therefore, at present we can state that the only variables determining entry capacities are the circulating flow and the number of lanes under German conditions.

Moreover, some efforts were made to establish results for the influence of pedestrian crossing entries and exits of the roundabouts. There were three roundabout in the sample size carrying enough pedestrians to perform this anaysis. The entrances being studied as well as the circular roadway itself have two lanes. Zebra crossings were available where pedestrians have the right of way over cars and other vehicles. A linear multivariate regression analysis including pedestrian flows as an undependent variable resulted in another capacity formula:

$$\max Q_z = A * e^{-(B*Q_k+C*Qf)/10.000}$$

with:
- Qz = maximum entry flow [pcu/h]
- Qk = circulating flow [pcu/h]
- Qf = pedestrian flow [ped./h]
- A,B,C = determining parameters

Only at one roundabout a significant influence to roundabout capacity by pedestrians could be ascertained. The interpretation of this equation results to the fact that the impeding effects of pedestrians on entry capacity seems to be much lower than it is predicted by the solution from Marlow, Maycock (1982) based on Griffiths (1981) formula. These evaluations, however, have to be continued before a definite solution can be given. Because of the preliminary state of these results and since the results of Marlow are on the safe side the program KREISEL, which has become something like a standard solution for the German speaking countries, still uses the Marlow-Griffiths solution for estimating the influence of pedestrians on exit and entry capacity.

SAFETY OF ROUNDABOUTS

Increased safety is one of the main reasons why roundabouts are introduced in many countries. Roundabouts are supposed to be one of the safest form of intersections. In France, Australia and Switzerland particularly good experiences were made by changing junctions into roundabouts.

Higher safety at roundabout can be expected because of the smaller number of conflict points. A 4-arm junction has 28 conflict points, a roundabout has only 8 conflict points. Due to the one way traffic on circular lanes, the left-turn accidents, which are most of the fatal or serious accidents at cross intersections, are avoided.
Moreover the lower speed in the circular and in the approaching street of a roundabout has been found to be one reason for the high safety of roundabouts.

Accident research

Comparative investigations at intersections and roundabouts were made. Accident data from 14 roundabouts and 14 other junctions were obtained. The junctions selected for the investigations should be comparable with the roundabouts. That means the junctions must be situated in the immediate vicinity of the roundabouts because they should have equal or at least similar conditions for traffic parameters, like traffic composition, driver behaviour and volumes.
The reported data covered several years. All roundabouts considered were operated based on the offside priority rule. The accident data were obtained from the files of police authorities.
The following parameters were evaluated:

- absolute number of accidents e.g:
  total number of accidents
  number of accidents with injured persons
  number of damage only accidents
  distribution of accidents among road users
  distribution of accidents according to the seven types of accidents:
  - single vehicle accident
  - crossing manoeuvre
  - turning manoeuvre
  - accidents with parked vehicles
  - crossing of the road by pedestrians
  - rear end collisions
  - collisions with oncoming traffic,

- relative numbers of accidents, e.g.:
  - accident rate [acc. per 1 million veh.]
  - accident cost rate [DM per 1 million veh.]

The analysis shows that the total number of accidents at roundabouts seems to be higher than at junctions but likewise that the severity of these accidents is lower.
Going into more detail, the investigation clearly shows that two groups of roundabouts must be distinguished: First of all there is the group of large roundabouts with an old design. That means: two lane entries with a curved approach resulting into small entry angles (angle between the tangents of the entry and the circular roadway). These intersections have high numbers of accidents. On the other hand there is the group of modern single lane roundabouts with almost radial entrances and 28 - 35 m inscribed circle diameter. These new intersections had only few accidents with nearly no severe damages.
Table 2 shows the accident rate and the accident cost rate for these two groups of roundabouts in comparison to the junctions.

accident rate and accident cost rate

	Roundabouts		Junctions	
	AR	ACR	AR	ACR
older roundabout/ junction with traffic signals	6,58	24,90	3,35	6,49
newer roundabouts/ junctions without traffic signals	1,24	4,67	1,00	11,96
all roundabouts/ all junctions	4,40	16,66	2,76	19,29

with : AR = accident rate [acc.per 1 million veh.]
ACR = accident cost rate [DM per 1 million veh.]

Tab. 2

Three sites were rebuilt into roundabouts during the time of our observation. Therefore, they offered the possibility to perform a before-and-after comparision.
All roundabouts have diameter between 28 and 35 m. At all of these sites the number of accidents decreased as well as the number of serious accidents. There were no more accidents with serious injuries after the transformation and the number of serious damage only accidents decreased as well. Slight damage only accidents were the most frequent. The number of accidents decreased from 4 accidents per year at junctions to

2,4 accidents per year at roundabouts. The number of personal injury accidents decreases from 3,3 injured per year at junctions to 0,5 injured at roundabouts (cf. Fig. 3 and Fig. 4).

Unfortunately, the sample is too small to allow definite results. Thus, accident research at roundabouts should be continued for German conditions as well.

CONCLUSIONS

Some of the positive experiences with roundabouts made in other countries could also be confirmed in Germany, especially regarding capacity. A group of formulae has been developed for the prediction of roundabout capacity. These formulae are used in practice in Germany.
However, the data sample especially for one lane roundabouts is too small to give definite results. Therefore, further studies are required especially for one-lane roundabouts.
Accident research on the basis of rather narrow samples also gave some promising results. Up to now, the empirical background especially with modern designed single lane roundabouts was too small to ascertain each of the results also on a mathematical statistical basis. However, the interest to create new roundabouts in practice has increased tremendously in the recent past in Germany. According to the author's information more than 30 new roundabouts have been realized in Germany recently and the trend is still going upwards. Therefore, it is necessary to get more insights in traffic operations and safety at roundabouts also on the German background.
Moreover, aspects of energy consumption and emissions of noise and pollutants should be studied in more detail. Nevertheless, also our present knowledge clearly allows us to state that the roundabout - on the whole - is one of the better instruments in the toolbox of traffic engineering.

References

1 ARAGAO, P.de: Les carrefours giratoires, Institut des Transports et de Planification, École Polytechnique Fédérale de Lausanne, Département de Génie Civil, Switzerland, 1988.

2. BRILON, W.: Programm KREISEL, Handbuch, Version 3.0, Lehrstuhl für Verkehrswesen, Ruhr-University Bochum, Germany; 1991.

3. BRILON, W.; OHADI, H.: Empirische Überprüfung der Leistungsfähigkeit von Kreisverkehrsplätzen, Arbeitsblätter AVV Nr. 4, Arbeitsgruppe für Verkehrstechnik und Verkehrssteuerung, Ruhr-University Bochum, Germany, 1988.

4. BRILON, W.; STUWE, B.: Schlußbericht zum Forschungsauftrag Einsatzmöglichkeiten von Kreisverkehrsplätzen und aufgeweiteten Knotenpunkten unter besonderer Berücksichtigung ausländischer Erfahrungen, Teil 1 Kreisverkehrsplätze; Lehrstuhl für Verkehrswesen, Ruhr-University Bochum, Germany; 1990.

5. CETUR: Conception des carrefours à sens giratoire implantes en milieu urbain, France, 1989.

6. CEDERSUND, H.A.: Cirkulationsplatser, Meddelande, Statens väg-och trafikinstitut, Linköping, Sweden, 1983.

7. GAMBARD, J.M.: Safety and design of unsignalised intersections in France, SETRA, France, 1988.

8. GRIFFITHS, J.D.: A mathematical model of a nonsignalized pedestrian crossing, Transportation Science, Vol. 15, No. 3, pp. 223-232, 9/1981.

9. KIMBER, R.M.: The traffic capacity of roundabouts. TRRL LR 942. United Kingdom, 1980.

10. LOUAH, G.: Recent French studies on capacity and waiting times at rural unsignalised intersections. SETRA, France, 1988.

11. MARLOW, M; MAYCOCK, G.: The effect of Zebra crossing on junction entry capacities, TRRL SR 724, United Kingdom, 1982.

12. SIEGLOCH, W.: Die Leistungsermittlung an Knotenpunkten ohne Lichtsignalsteuerung, Straßenbau und Straßenverkehrstechnik, Heft 154, Germany, 1973.

13. SIMON, M.; RUTZ, H.: Sicherheit von Verkehrskreiseln innerorts. Schlußbericht, Emch + Berger AG, Aarau, Switzerland, 1988.

14. SIMON,M.; HINTERMEISTER, O.; Lehmann, U.: Empirische Ermittlung der Leistungsfähigkeit von innerörtlichen Kreisverkehrsplätzen, Emch + Berger AG, Aarau, Switzerland, 1988.

15. TROUTBECK, R.J.: Traffic interactions at roundabouts, Proceedings 15th ARRB Conference, Part 5, Australia, 1990

16. TUDGE, R.T.: Accidents at roundabouts in New South Wales, Proceedings 15th ARRB Conference, Part 5, Australia, 1990

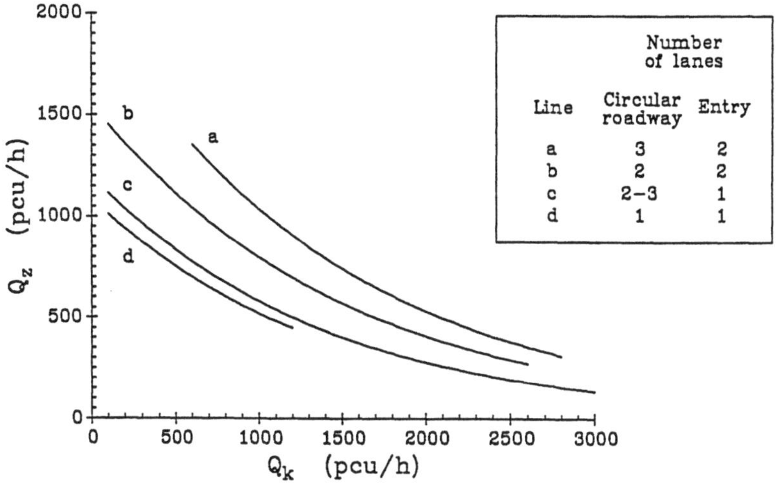

Fig.1. Regression curves for calculating the capacity of roundabouts.

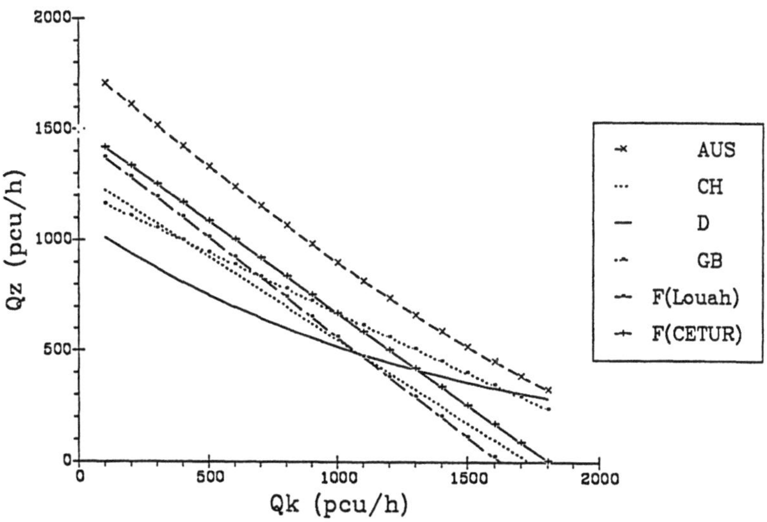

Fig. 2. Comparison of capacity formulae for single lane roundabouts of several countries

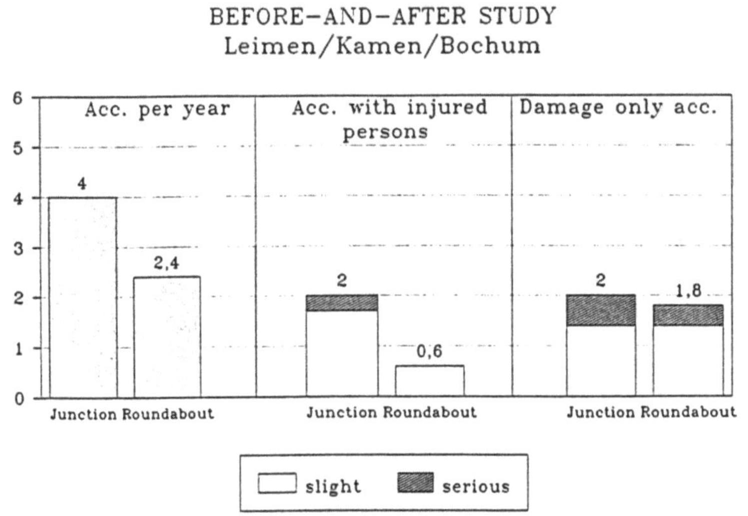

Fig. 3. Absolute number of accidents per year

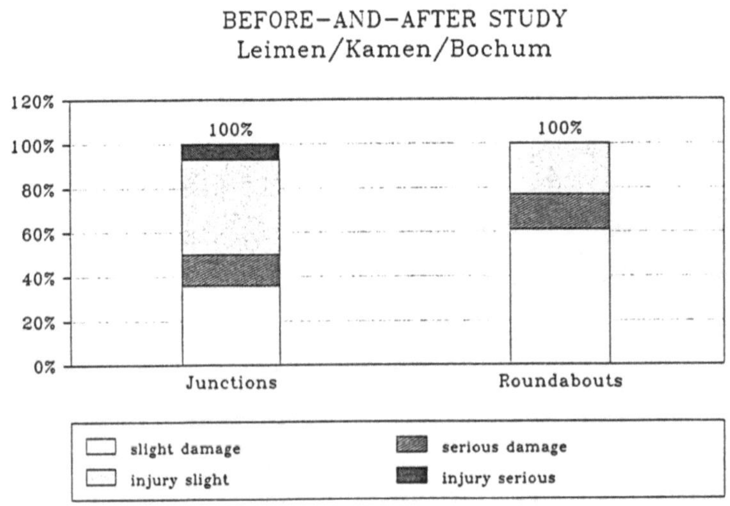

Fig. 4. Portion of accidents with injured persons or damage only to the total number of accidents per year

The Design of Major/Minor Priority Junctions in Poland

Prof. M. TRACZ

Cracow Technical University
Civil Engineering Department
Institute of Roads, Railways and Bridges
Poland

Abstract

The paper describes a method recommended by the National Road Administration to calculate the capacity of minor road approaches and to evaluate traffic conditions. Level of service concepts based on average delay and the reserve capacity have been assumed. The method is based on simulation and empirical studies and on available guidelines. The SBSS computer program is based on the described method.

Introduction

Major/minor priority junctions are by far the most frequently encountered types of junctions in Polish road network. There are 3- arm "T" and "Y" junctions and 4- arm-crossroads, while staggered junctions and roundabouts are not numerous. There are only few small or large roundabouts with diameters in the range of 20 - 120 m. The attractiveness of roundabouts has decreased since the introduction of new traffic rules in 1984, giving priority to vehicles entering a roundabant. In practice, however, the same rule as before still exists, due to adding "YIELD" signs at all approaches. New design roundabouts have not been accepted yet. Therefore, an emphasis is laid on methods of traffic performance evaluation of major/minor junctions. Capacity of major / minor junctions were calculated in Poland for several years using the method derived at Cracow Technical University (1978). This method has been modified based on simulation and a limited number of validation studies [1] and the revised version was published in 1989 [2]. The basic structure of the new method is similar to that of an old German method published in 1972 and adopted later in HCM-85 [3]. The new method contains procedures for the capacity analysis for YIELD- and two- way STOP-controlled junctions, level of service estimation and also provides special procedure for junctions with storage space in the central reserve. In the same time the guide for geometric design of junctions was published. It was assumed in the description of the method that the reader is familiar with HCM-85, chapter 10 terminology.

The structure of the method

The method is based on the gap-acceptance theory and the simulation model. Thus the capacity graphs of minor road movements were derived using simulation, whereas the model itself was calibrated on empirical way. The structure of the method is shown in Fig.1 and it includes:

1. Input module (not shown in Fig.1) in which existing geometric and traffic conditions are defined and traffic volumes adjusted to calculation.
2. Conflicting traffic module: conflicting streams through which each minor stream must cross are identifid and the total primary volume is calculated.
3. Critical gap module: critical gap size for each minor stream is determined and adjusted.
4. Capacity module: potential capacities for minor streams and lanes are determined, and adjusted to prevailing traffic and roadway conditions. Then the capacity of shared lanes and approaches is calculated.
5. Level of service module: for existing traffic volumes converted to pcu/h, the level of service is determined. Some of these steps are discussed in the paper.

Conflicting traffic and critical gap size

In the method, calculation of the potential capacity of a minor stream is done by considering each minor stream individually in relation to conflicting traffic Q_n that it has to cross and to the critical gap t_g for the minor stream drivers.

The conflicting volumne for movement "i" is the total volume which conflicts with this movement in veh/h. Each of four types of minor movements faces a different set of priority movements. In the method the same sets of conflicting traffic movements were assumed as in the HCM-85 [3] and in the new German quidelines [5]. One-half of the right turn movement from the major road is also included in the total primary volumes, because of an inhibiting effect on all minor road movements. This effect has been confirmed in the author's observations of gap-acceptance by minor road drivers and was included in the simulation model.

The critical gap is used to extract the potential capacity from the family of curves.

Adjustments and modifications to the basic sizes of critical gaps are provided to incorporate local conditions:

- for RT from minor road - curb radius, and acceleration lane as in [3].
- TH - cross one-way major road - t_g = -0.5 s,
- lower major road speed (40 km/h) - t_g = -0.5 s,

Total decrease in critical gap cannot be greater than 1.0 s.

The size of critical gap for the considered minor stream reflects a number of factors affecting capacity:

- type of minor road control (STOP or YIELD sign),
- average major road speed,
- turning radius and visibility conditions.

Values of critical gap are selected from Table 1. The values are based on Polish measurements and results of some other European studies. They vere modified in process of the model development [4] and its sensitivity analysis, including comparison with PICADY and HCM-85 [1].

Potential capacity

The potential capacity C_{or} is defined as the ideal capacity for a specific subject movement, assuming that major movements do not impede the subject movement and a separate lane is provided for the exclusive use of each minor road movement under consideration.

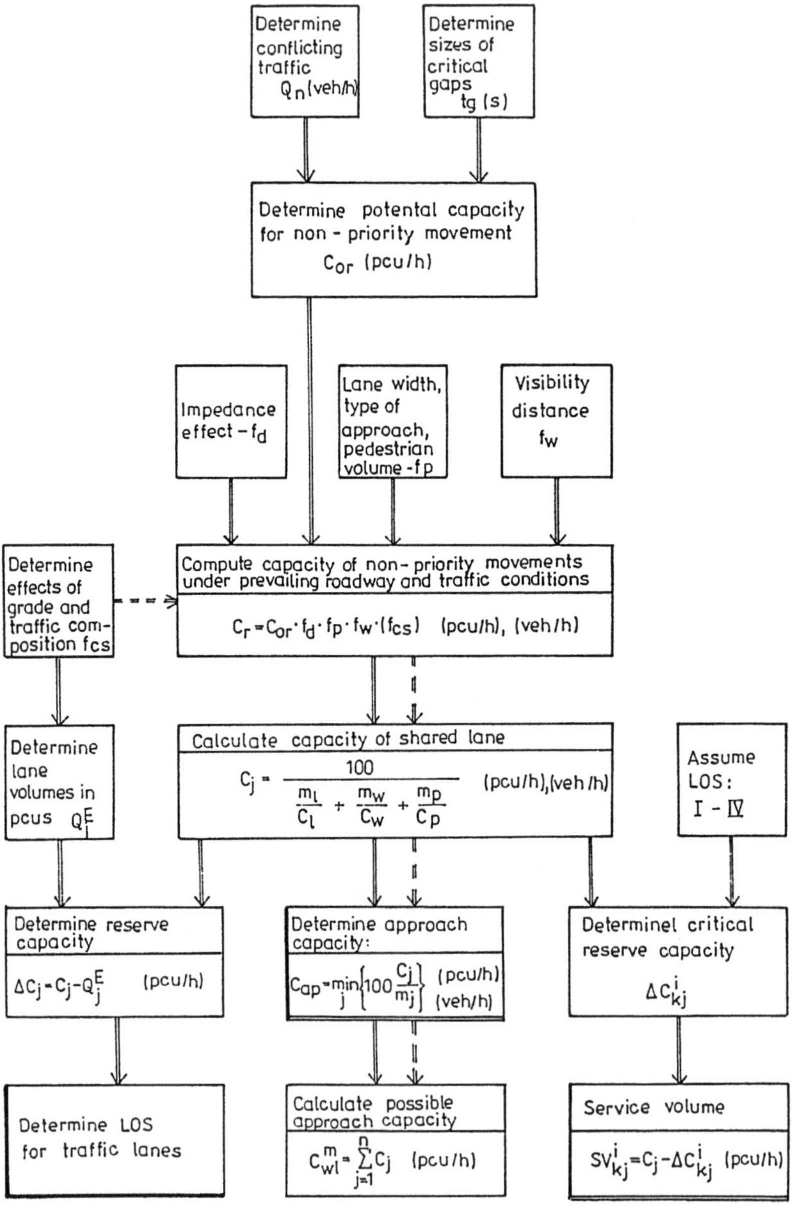

Fig. 1.: Structure of capacity calculation procedures

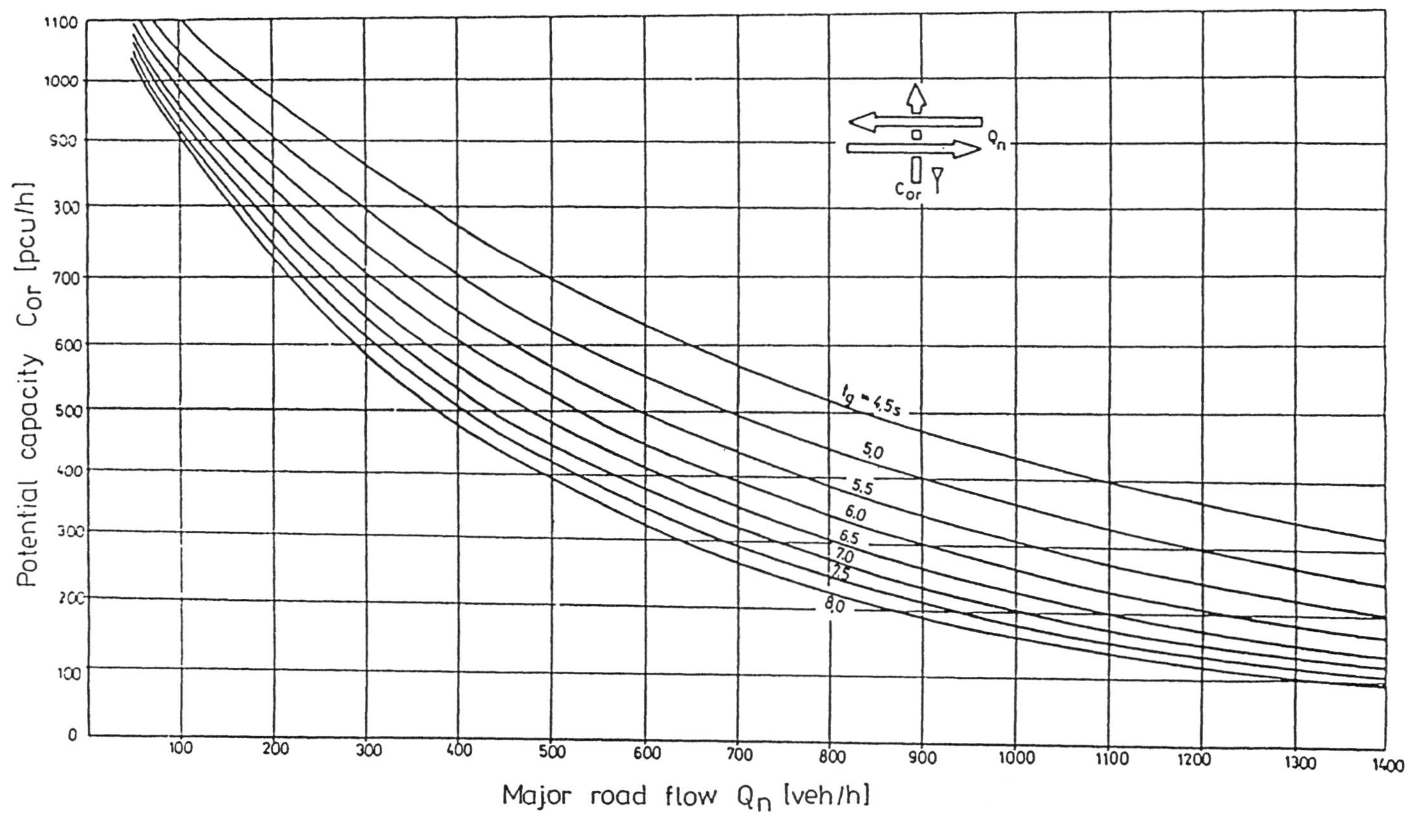

Fig. 2.: Charts of potential capacity for minor road movements

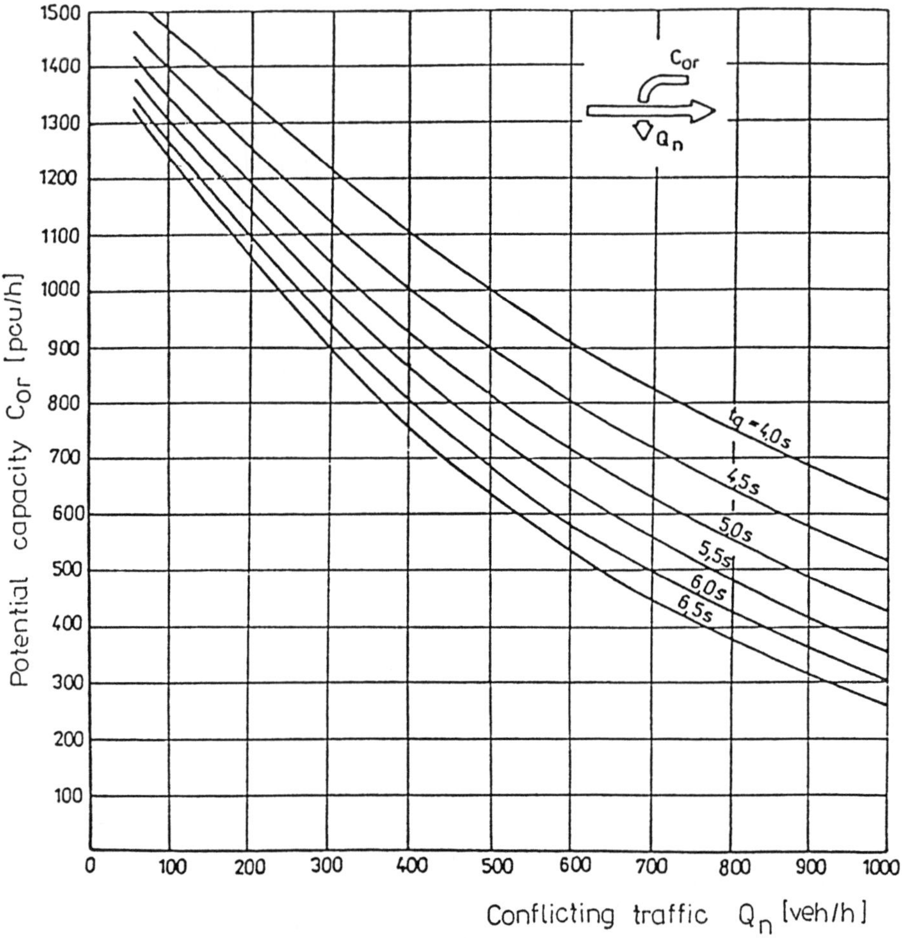

Fig. 3.: Charts of potential capacity for left turn movement from major road

Table 1. Basic critical gaps for passenger cars t_g(s).

Traffic streams and type of control	Average major road speed (km/h)		
	50	70	90
LT from major road	5.0	5.4	5.8
RT from minor road			
- YIELD (A-7)	4.8	5.5	6.0
- STOP (B-20)	5.5	6.0	6.5
LT from minor road			
- YIELD	6.0	6.5	7.0
- STOP	6.5	7.2	7.7
TH from minor road			
- YIELD	5.5	6.0	6.5
- STOP	6.0	6.5	7.2

The potential capacity C_{or} is read from the graphs shown in Fig. 2. - for minor road movements and in Fig. 3 for left turn from major road, volume Q_n is expressed in veh/h, whereas the potential capacity is expressed in pcu/h.

These graphs were desired from simulation, proceeded by extensive studies [4] of model factors affecting capacity, ic. non-stationarity and platooning of traffic flow demands, parameters of critical gap distribution. This studies were undertaken due to a certain underestimation of the potential capacity in the range of high major roads flows in the first version of the method [1]. The model still assumes that major road traffic is not affected by minor road flows.

The value of C_{or} is then adjusted to reflect prevailing roadway and traffic conditions:

$$C_r = C_{or} \cdot f_d \cdot f_p \cdot f_w \quad \text{pcu/h} \tag{1}$$

where:

C_r- capacity for a subject minor movement in under prevailing roadway and traffic conditions,

f_d- adjustment factor for the impedance effect,

f_p- adjustment factor for lane width, type of approach and pedestrian volume,

f_w- adjustment factor for visibility distance.

Impedance effect

The potential capacity of minor road movement depends are the volume of the conflicting traffic which contains high and lower priority movements. Drivers use gaps in a prioritized manner. Thus drivers turning left from the minor road may use gaps in the conflicting movements and are not impeded by these movements. When traffic becomes congested, major road left turning vehicles can impede lower priority movements, i.e. left turns and through traffic from the minor road, from utilizing gaps in the conflicting traffic, and thus reducing potential capacity of these movements. Left turns from the major road impede both through movements and left turns from the minor road, while through and right turn movements from the minor road impede opposed left turns.

In the method, impedance expressed by the impedance factor f_d reflects the difference between a real traffic situation at the junction (with mutual relationships between high priority movements) and situation assumed as a basis for determination of potential capacity from Fig 2/3. Figure 4 illustrates now the conflicting traffic Q_n is equivalent to two major road priority streams with traffic volume equal to total volume which conflicts with the considered movement (here LT from the minor road).

An impact of impedance depends on the ratio of potential capacity used by the existing demand of the impeding movement: $\rho = Q_{rd}/C_{rd}$. This effect is represented in the calculation method by the impedance factor f_d which is defined as a ratio of capacity of minor road stream calculated by taking into account the impeding movement (C_{cw} in Fig.4) and capacity of the same movement derived for the equivalent through volumes ($C^o{}_{cw}$ in Fig.4).

The relationship between the impedance factor f_d and ratio of capacity ρ used by the existing demand for different impeding streams is shown in Fig. 5. In general, the impact of impedance is adressed by multiplying the potential capacity by a set of impedance factors f_d:

$$f_d = f_{d1} \cdot f_{d2} \cdot f_{d3} \cdot f_{d4} \qquad (2)$$

for each impeding movement i = 1,2,3,4 (Fig 4 and 5).

Curves in Fig 5 are related to different impeding streams and traffic lane configurations. It can be seen that lower impedance (decrease up to 20%) can be expected for layouts where separate lanes for left turns from the major road are provided. At 4-arm junctions the mutual impedance forces the analyst to apply an iteration procedure.

Shared lane and approach capacity

After adjusting for impedance factors, capacity is also adjusted for lane width, type of approach and pedestrian volume (factor fp) and for visibility conditions, depending on junction environment and vertical alignment of major road (factor f_w). Some results of British capacity studies [7] have been adopted here. When two or three movements share a single lane on the minor approach, its combined capacity can be derived using the following equation:

$$C_j = \frac{100}{\frac{m_L}{C_L} + \frac{m_T}{C_T} + \frac{m_R}{C_R}} \quad \text{pcu/h} \qquad (3)$$

where:

C_L, C_T, C_R - capacities for LT, TH and RT streams in a shared lane are calculated under the assumption that each minor road movement has an exclusive use of lane,

m_L, m_T, m_R - percentage of left turn, through and right turn movements in shared lane.

If the shared lane includes only two movements, values of m_i and C_i for the third movement are deleted in the equation. Equation (3) was verified in the simulation way. If the minor approach has two lanes with capacities C_j carrying m_j percent of approach volume, its real capacity is equal:

$$C_{or} = \min\{100\, C_j/m_j\} \quad \text{pcu/h} \qquad (4)$$

Possible capacity of approach is equal the total of lane capacities C_j calculated from (3).

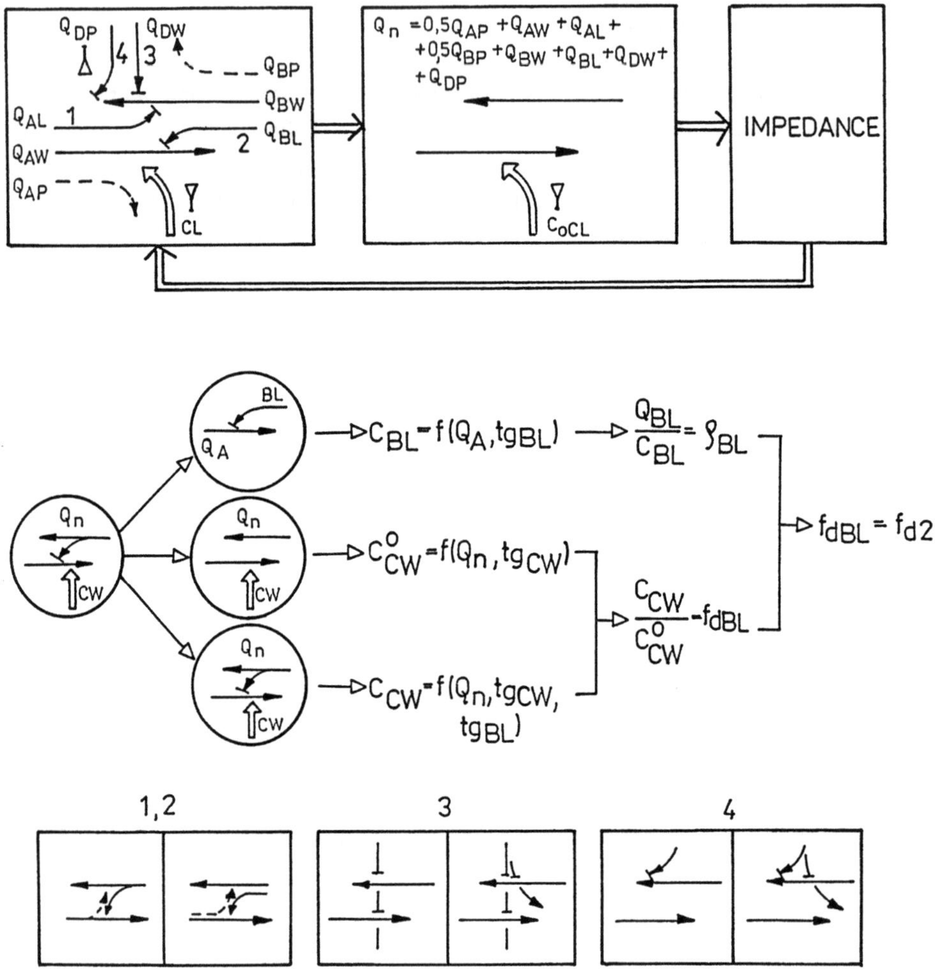

Fig. 4: Impedance of impedance effect analysis

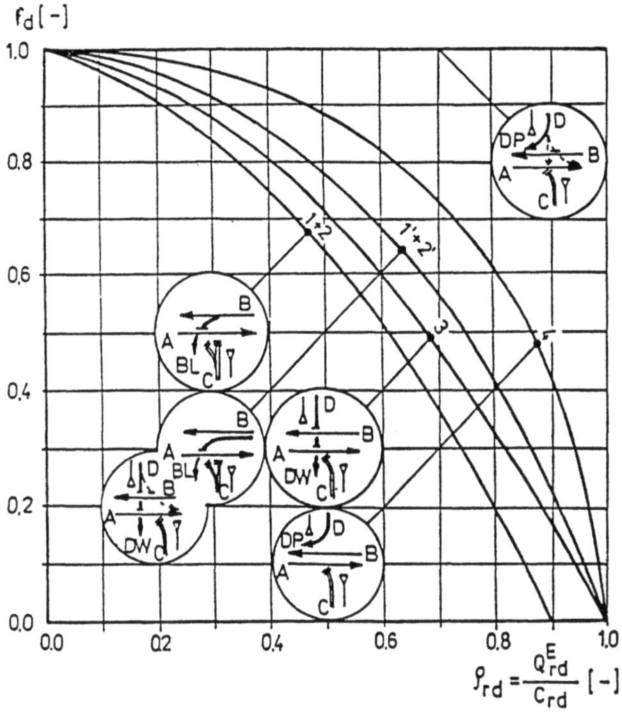

Fig.5. Impedance factor graphs.

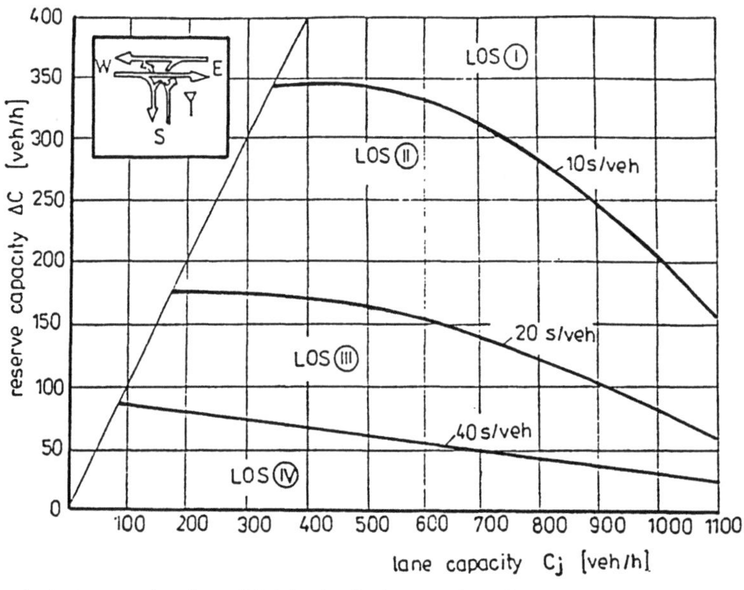

Fig.6. Determination of LOS of priority type junction based on the reserve capacity and delay.

Level of service criteria

The Level of service module brings together capacity and demand on major road (left turns) and minor road approaches, producing level of service estimates. Non-priority traffic streams are converted into passenger car units and compared with the calculated capacity. Then the reserve capacity is calculated as the difference between the traffic volume Q_j and capacity of traffic lane C_j:

$$\Delta Cj = Cj - Qj \qquad \text{pcu/h} \qquad (5)$$

This is the measure which forms the basis for the level of service classification.

On the basis of design practice a fewer number of levels than in HCM-85 has been assumed, each representing a wider range of traffic condition variation at a lower degree of traffic movement determination at junctions. At engineering design stage the detailed characteristics of flow demand non-stationarity are usually unknown and therefore these effects can not be taken into account. It has an effect on uncertainty of results of traffic conditions evaluation and very detailed qualitative classification seems in this case unfavourable. Four levels of service for priority type junctions, have been initially assumed. These levels are based on the following general ranges of average vehicular delay (Fig.5)

LOS I	d =	0 - 10 s/veh,
LOS II		10 - 20 s/veh,
LOS III		20 - 40 s/veh,
LOS IV		> 40 s/veh.

These values relevant only for isolated priority junctions are connceted with the reserve capacity and its critical values. Depending on the value of ΔC the average delay of the minor stream vehicles is kept within defined ranges. To determinate LOS on considered lane, the reserve capacity $\Delta C_j = C_j - Q_j$ is calculated and then LOS can be determined from Fig.6. These criteria should be used with some caution.

Quantitative evaluation of traffic condition based on the threshold values of the selected measure (average delay) should reflect a qualitative rating of traffic conditions by drivers. Only a few surveys conducted to establish a rational relationship between qualitative user's rating of operating condition and objective qualitative measure can be found in literature. A computer simulation to animate traffic at junctions described elsewhere [8] has been developed in CTV Krakow to conduct such survey. Simulation involved 16 (8+8) various traffic situations for both priority type and signalized junctions. The tested driver was requested to observe a movement of a selected vehicle from its occurence on the screen and then:

- to estimate the delay experienced by the observed vehicle,
- to valuate driving conditions on the opinion scale including the following qualitative description: very good, good, fair, poor and very poor, and indicate their opinion regarding vehicular delay.

Results has shown a meaningful similarity of the delay / qualitative ranking regression curves for both types of junctions. Very poor rating was obtained at average delay of 80 - 90 s/veh. On the basis of this initial results new proposal of levels of service criteria are recently under consideration. Two alternatives of threshold values of vehicular delay are suggested. One of them relevant for four levels is as follows:

LOS	average vehicular delay s/veh
I	0 - 15
II	16 - 40
III	41 - 65
IV	> 65

Another classification maintains six levels of service. The final decision will be based on wider user's survey, and experts opinion with regard to international aspects of the method.

Computer program

SBSS is a computer implementation of the method described here. It is written in modular form, in PASCAL, for micro-computers. It reguires about 180 Kbytes of memory available. The program models capacities and evaluates traffic conditions at normal and channelized three- and four-arm junctions with- and without central reserve.

Calculations are conducted for 1 hour period, which can be divided into a number of time segments 5-15 minutes long which are treated consecutively. Different inputs can be provided for each time segment. It is also possible to calculate the possible capacity of minor road approaches by applying the iteration procedure. Composition of traffic streams in percent and major/minor volumes proportions are assumed as an input and iteration calculations are conducted to reach capacity at one of minor road approaches. The program operates in the typical manner: the user selects one possible option given on the screen, answers questions and puts in the required data. If necessary, the user can change values of the adjustment factors calculated by the program, in non-typical situations.

Final remarks

The presented method is based on simulation model which occured very useful in investigations of traffic streams interrelations and measures of effectiveness estimation. The accuracy of the new method have not yet been judged, although some preliminary validations appear to be promising.

Further work and decisions remains to be done on traffic conditions classification, including both number of LOS and vehicular delay threshold values.

References

1. Tracz,M.: Research of traffic performance of major/minor priority intersections, in Intersections without Traffic Signals, W. Brilon - editor, Springer Verlag, Berlin 1988: 62 -79.

2. Tracz, M.; Chodur, J. et all: Capacity of Unsignalized Intersections, Polish Capacity Manual, Cracow 1989 (in Polish).

3. Highway Capacity Manual - Chapter 10, Transportation Research Board, Special Report 209, Washington 1985.

4. Chodur, J.; Gaca, S.: Simulation studies of the effects of some geometrical and traffic factors on the capacity of priority intersections, in Intersections without Traffic Signals, W. Brilon - editor, Springer Verlag, Berlin 1988: 197 -213.

5. Brilon, W.: Recent Developments in Calculation methods for Unsignalized Intersections in West Germany, in Intersections without Traffic Signals.

6. Brilon, W.; Grossman, M.: Empfehlung für ein neues Berechnungsverfahren für Knotenpunkte ohne Lichtsignalanlagen, Ruhr Universität Bochum, 1990.

7. Kimber, R.M.; Coombe, R.D.: The traffic capacity of major/minor priority junctions, TRRL Report SR 582, 1980.

8. Tracz, M.; Chodur, J.; Gondek, S.: The use of reserve capacity and delay as the complemantary measures of junction effectiveness. Proceedings of the 11th International Symposium on Transportation and Traffic Theory, Yokohama 1990.

Some Aspects of Capacity Analysis of Unsignalized Intersections

Dr.-Ing.habil. Werner Schnabel
Hochschule für Verkehrswesen "Friedrich List"
O - 8010 Dresden

For the determination of the capacity of unsignalized intersections or service volumes, a calculation model is required which corresponds to the traffic operations observed. In this connection, it is especially important to consider the level of priority of the different streams. This method by HARDERS /1/ and SIEGLOCH /2/ takes these aspects into account.

Due to its low calculation requirements and its clearness, the SIEGLOCH method is generally used for the calculation of unsignalized intersections.

In the study of unsignalized intersections, there are two major aims:

1) Determination of their total capacity
2) Studying the possibility of traffic operations in the minor road approaches according to quality requirements.

1. Capacity Analysis of Unsignalized Intersections

For capacity analysis, it is necessary to fix the percentage of the single traffic streams. For this, two calculations are sufficient, i.e. a first calculation based on a relatively high traffic volume which exceeds the expected intersection capacity and a second one with a low traffic volume (which is below the limit of capacity). The intersection capacity can then be determined by interpolation. The calculation is finished when the traffic volume assumed for the traffic streams with the highest order (lowest priority) agrees with the calculated capacity.

In Appendix 1 the principal steps of calculation are given for a very simple intersection (crossing of two one-way roads).

Capacity of Non-Priority Approach Lanes for Different Traffic Flow Conditions of Major Streams

Although the calculation model by HARDERS/SIEGLOCH is relatively mature, there are still a number of open problems. The Poisson flows assumed here can be treated mathematically in a relatively simple way, and therefore the practical calculation is based on the formulae derived for this purpose. It is generally known that in the urban network the traffic stream is no longer randomized. After traffic signal installations, there is a platoon of vehicles. The traffic bunching can be described as "over"-randomized with the following formula, the pertinent cumulative distribution function of gaps is described in fig. 1a.

$$P(ZL \geq t) = fe^{-2fqt} + (1-f)e^{-2(1-f)qt}$$

where

$P(ZL \geq t)$ - probability of a gap t to occur

q - traffic flow in vehicles per second

f - partial value in the composite exponential function

$$f = \frac{1}{2}\left(1- \sqrt{\frac{j-1}{j+1}}\right) \quad \text{mit } j = s^2/\bar{t}^{\,2}$$

$\bar{t}$ - average gap in traffic stream

s^2 - scattering of gaps

The density of probability is

$$f(t) = 2f^2qe^{-2fqt} + 2(1-f)^2qe^{-2(1-f)q \cdot t}$$

Thus the capacity of non-priority approach lanes C based on

$$C = M \int_{t_o}^{\infty} g(t) \cdot f(t)\, dt$$

$g(t)$ - arrival function of non-priority vehicles $g(t) = (t-t_o)/t$

t_o - zero gap

t_f - move-up gap

is

$$C = \frac{1800}{t_f} \left(e^{-2ft_oq} + e^{-2(1-f)t_oq}\right)$$

in the case of "over"-randomization.

With $M = 3600 \cdot q$ in veh/h we obtain the following formula

$$C = \frac{1800}{t_f} \left(e^{-ft_oM/1800} + e^{-(1-f)t_oM/1800}\right) \qquad (1)$$

The well-known capacity formula for random traffic streams is obtained if in (1) f is equal to 0.5 or j is set equal to 1, i.e.

$$C = \frac{3600}{t_f} \; e^{-t_oM/3600} \qquad (2)$$

In the case of stationary traffic bunching in longer road segments, the gaps in the traffic flow are relatively steady. It is possible to express the cumulative distribution function of gaps for "sub"-random streams (fig. 1a) according to ERLANG /4/ with the following formula

$$P\ (ZL \geq t) = e^{-kqt} \sum_{n=0}^{n=k-1} \frac{(kqt)^n}{n!}$$

where k is determined from

$$k = \bar{t}^{\,2} / s^2$$

By means of the probability density

$$f(t) = kq \frac{(kqt)^{k-1}}{(k-1)!} \; e^{-kqt}$$

it is possible to derive the following general capacity formula for "sub"-random traffic:

$$C = \frac{M(-1)^k}{qt_f k!} \; e^{-kqt_o} \left(-(-kqt_o)^{k-1} + 2(k-1)(-kqt_o)^{k-2} - \right.$$

$$\left. -3(k-1)(k-2)(-kqt_o)^{k-3} + - \ldots\right)$$

$$C = \frac{3600(-1)^k}{t_f \cdot k!} \; e^{-kt_oM/3600} \left(-(-kt_oM/3600)^{k-1} + \right.$$

$$\left. + 2(k-1)(-kt_oM/3600)^{k-2} - + \ldots\right) \qquad (3)$$

The equations (1),(2) and (3) are analyzed in Appendix 2. Obviously, these capacity formulae permit a better adaptation to real traffic conditions. But the traffic engineer should be given some information about traffic flow conditions depending on the influence of adjoining intersections.

A full consideration of traffic conditions in the individual approaches would result in an additional complexity of calculation statements because of a superposition of different gap distributions.

Influence of Single Lane Traffic on the Capacity of Non-Priority Approach Lanes

The design of unsignalized intersections is generally on a low standard. In many cases, there is only one lane for several major streams, but this fact is not taken into consideration for capacity analysis.

The cumulative distribution function of gaps for single-lane traffic is derived by a shifted exponential function (fig. 1b):

$$P(ZL \geq t) = \begin{cases} 1 & \text{for } t \leq t_c \\ e^{-(t-t_c)/(\bar{t}-t_c)} & \text{for } t > t_c \end{cases}$$

t_c - minimum gap

$\bar{t}$ - average gap in the traffic stream.

Based on the adequate density function

$$f(t) = \frac{1}{\bar{t} - t_c} e^{-(t-t_c)/(\bar{t}-t_c)} \qquad t \geq t_c$$

the capacity of a minor stream can again be derived if there is only one shared lane for the major streams:

$$c_1 = (\bar{t} - t_c)\frac{M}{t_f} e^{-(t_o-t_c)/(\bar{t}-t_c)}$$

With t = 3600/M, the basic formula is obtained in the following form

$$c_1 = \frac{3600-Mt_c}{t_f} e^{-M(t_o-t_c)/3600-Mt_c} \qquad (4)$$

Two or more lanes for major streams result in the well-known capacity formula by SIEGLOCH

$$c = \frac{3600}{t_f} e^{-Mt_o/3600} \qquad (2)$$

A comparison between the capacities of non-priority approach lanes depending on the number and use of lanes in the major approaches shows that great differences between the minor streams exist (Table 1). With increasing major stream, the deviation increases as well. Due to this fact, the question is raised whether the calculation of unsignalized intersections should be kept up in the way just described (Appendix 2).

2. Quality Aspects of Traffic Operations

Traffic layouts are no longer economical if the traffic volume is near the capacity. That is why the definition of the service volume is necessary. The computational formula for the service volume derived by HARDERS is very useful because of its simplicity. HARDERS has defined the two levels of quality by regarding traffic safety aspects, i.e. he has seen a connection between the delay and the driver behaviour of minor road users. In any case, this is a more reasonable point of view. But there is an important difference between the acceptable delay for unsignalized intersections according to HARDERS and the analogous value for signalized intersections (about 45 sec.). In practice, these given delays at unsignalized intersections are exceeded especially for the left-turn movements from the minor stream.

That is why a differentiated fixation of service volume is necessary especially with regard to aspects of economy of intersections. The delays are given in steps of 15 sec for the individual levels of service. For the determination of the service volume of that levels of service, it is necessary to substract the values given in table 2 from the capacity in the following way:

$$M_{zul.LOSi} = C - K_{LOSi}$$

where

$M_{zul.LOSi}$ - service volume for a fixed level of service

C - capacity of the lane under consideration

K_{LOSi} - substracted traffic volume (reserve capacity) according to the permissible delay for the level of service in question.

In the author's opinion, the criteria of quality for minor streams indicated in /4/ (Tab. 3) do not correspond with practical requirements, the reserve capacities being too high.

The level of delay at unsignalized intersections is generally different for the individual lanes. It increases with a decrease of priority of the streams concerned.

3. Results of a Practical Intersection Study

When studying the most appropriate way of operation at an unsignalized intersection (appendix 4), the delays of unsignalized traffic operation for the individual traffic streams have been determined by means of a license plate. In this investigation, differences appear between the delays which are computed theoretically and the measurement results. The reason for these differences is attributed both to the insufficiency during the capacity computations and the critical gaps and the move-up gaps on which the computations are based.

A simple measurement method for investigation of critical and move-up gaps seems to be useful, for example analogous with the determination of the saturation flow for signal-controlled approaches by time demand measurements. In this way, it would be possible to take into consideration the typical features of various intersections.

The measurement results show once more that it is necessary to make a differentiating valuation of the delays for the individual traffic streams. Although traffic operations without the use of signal controls is more economical than in the case of its use, the average delay for the heaviest left turn movement from the minor road is very high.

4. Summary

Although the computation model for unsignalized intersections has reached a high level, improvements are necessary. This refers to the consideration of approach conditions and traffic operations of major streams on one or several lanes.

Delay studies have shown that a splitting into different levels of service is reasonable. In this connection the limits for maximum permissible delays depending on the priority of the traffic stream considered have to be defined. A proposal is submitted for the determination of permissible traffic volumes of minor streams.

A simple method for the measurement of critical and move-up gaps should be developed.

References

/1/ Harders, J.: Die Leistungsfähigkeit nicht lichtsignalgeregelter städtischer Straßenknoten. Dissertation, Technische Hochschule Hannover 1968.

/2/ Siegloch, W.: Die Leistungsermittlung an Knotenpunkten ohne Lichtsignalsteuerung. Forschungsberichte Straßenbau und Straßenverkehrstechnik, Heft 154, Bonn 1973.

/3/ Potthoff, G.: Die Bedienungstheorie im Verkehrswesen. transpress VEB Berlag für Verkehrswesen, Berlin 1965.

/4/ Highway Capacity Manual. Transportation Research Board, Special Report 209, Washington 1985.

Traffic volume of major streams in veh/h		200	400	600	800	1000
Capacity of minor stream in veh/h	a) according to eq. (2)	1076	869	701	566	457
	b) according to eq. (4)	1057	802	573	574	213
Deviation of a) and b) in %		2	8	22	51	114
Calculation-based gaps: t_o = 3,85 sec., t_f = 2,7 sec. Case a) Major streams in multiple lanes Case b) Major streams in a single lane						

Table 1

Comparison of minor stream capacity in the case of different traffic operations of major streams

Level of service	A	B	C	D	E	F
Average delay of vehicle in sec.	15	30	45	60	60	very long
Substracted traffic volume K in veh/h	240	120	80	60	40	
Hints for application A: apply B: apply: operations still safe C: apply: still permissible D: exception E: do not apply F: do not apply						

Table 2

Determination of levels of service to indicate permissible traffic volumes of minor streams

Reserve capacity	Level of Service	Expected delay to minor road traffic
400	A	Little or no delay
300 - 399	B	Short traffic delays
200 - 299	C	Average traffic delays
100 - 199	D	Long traffic delays
0 - 99	E	Very long traffic delays
a	F	a

a) when demand volume exceeds the capacity of the lane, extreme delays will be encountered with queueing which may cause severe congestion affecting other trafficmovements in the intersection. This condition usually warrants improvement to the intersection.

Table 3

Level of service criteria for unsignalized intersections /4/

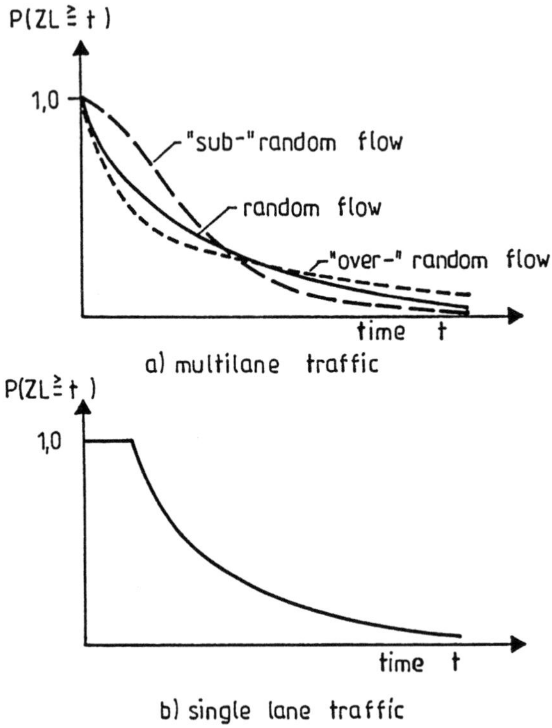

Fig. 1
Gap sum lines

Situation

Crossing of two one-way roads, left turns from minor road not permitted

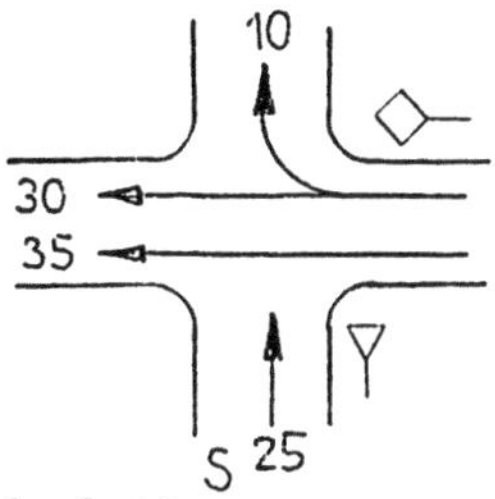

Traffic volume in %
Necessary gaps for crossing
Zero gap $t_0 = 4,4$ s
Move-up gap $t_f = 3,2$ s

Calculation

1. Assumed traffic volume for the intersection with M = 1500 veh/h
Stream of 2nd level, existing traffic volume N = 375 veh/h
Traffic volume of major streams M_H = 1125 veh/h

$$C = \frac{3600}{3,2} e^{-1125.4,4/3600} = 284 \text{ veh/h}$$

N being grater than C, the traffic volume of the intersection must be reduced. Further calculation is shown in the following table.

No.	Assumed traffic volume in veh/h Intersection	Major Streams	Minor Streams S_g	Capacity S_g	Remarks
1	1500 (M_1)	1125	375 (N_1)	284 (C_1)	N > C
2	1200 (M_2)	900	300 (N_2)	374 (C_2)	N < C
3	1335	1001	334	331	N ≈ C

The intersection capacity is between the two traffic volumes assumed M_1 and M_2. By interpolation one abtains

$$C = M_1 + \frac{(M_2 - M_1)(C_1 - N_1)}{(C_1 - N_1)-(C_2 - N_2)} = 1500 + \frac{-300.(-91)}{-91 -74} = 1335 \text{ veh/h}$$

The interpolation result is confirmed by the calculation. The intersection capacity amounts to 1335 veh/h.

Appendix 1

Principle of capacity calculations for unsignalized intersections

Flow	Traffic flow of major stream in veh/h 200	400	600	800	1000	1200	1400
"over-"random							
j=4	1091	917	790	696	624	568	524
j=3	1088	909	775	674	594	532	487
j=2	1084	896	751	637	548	476	423
random							
k=j=1	1077	869	701	566	457	369	298
"sub-"random							
k=2	1055	809	607	447	325	234	167
k=3	1050	787	564	390	262	171	110
k=4	1049	777	540	355	223	134	78

assumed gap values: $t_0 = 3{,}85$s zero gap; $t_f = 2{,}7$s move-up gap

Capacity of minor streams in veh/h

The calculations are based on the following capacity formulas:

a) "Over-" random flow

$$C = \frac{1800}{t_f}\left[e^{-f \cdot t_0 \cdot M/3600} + e^{-(1-f) t_0 \cdot M/3600}\right]$$

j =	4	3	2
f =	0,1127	0,1464	0,2113

b) Random flow

$$C = \frac{3600}{t_f} e^{-t_0 M/3600}$$

c) "Sub-" random flow

$$C = \frac{1800}{t_f} e^{-t_0 M/1800} (2 + t_0 M/1800) \qquad \text{if } k=2$$

$$C = \frac{600}{t_f} e^{-t_0 M/1200} [(t_0 M/1200)^2 + t_0 M/300 + 6] \qquad \text{if } k=3$$

$$C = \frac{150}{t_f} e^{-t_0 M/900} [(t_0 M/900)^3 + 6(t_0 M/900)^2 + t_0 M/50 + 24] \qquad \text{if } k=4$$

Appendix 2

Minor stream capacity in the case of different flow conditions in the major stream

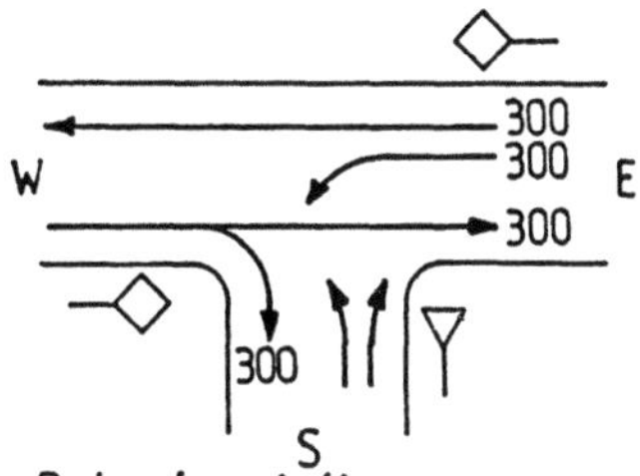

Data in veh/h

Approach W : single lane
Approach E, S: two-lane

Capacity in veh/h	Single lane traffic conditions considered	
	yes	non
E_l	543	701
S_r	742	824
S_l	214	215

Appendix 3
Capacity of minor streams at a T-junction with and without consideration of single lane traffic conditions

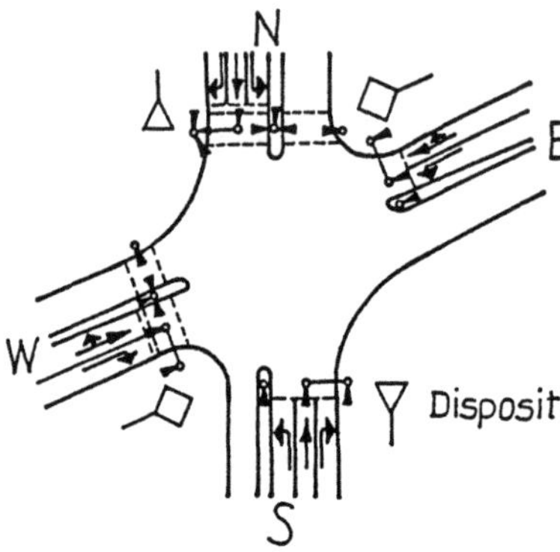

Disposition plan

Intersection in a suburban area, buildings only on one side, minor influence of neighbouring intersections, permissible speed on major road 60 km/h

Global Measurement Results

	Traffic Signal Installation off: unsign. intersection	on: FZS [1]	on: VAS [2]
Traffic volume of intersection in veh/h	1248	1095	1273
Total delay in h during one hour	6	9	8
Number of stops	711	733	853
Fuel consumption in l/h [3]	30	41	43

1) FZS = fixed time control
2) VAS = traffic-actuated control
3) calculated by means of measurement results and corresponding fuel consumption values

Comparison between measured and calculated delays

	Stream having to wait	Measurement Results: Volume N in veh/h	Measurement Results: Average delay in s	Calculations: Average delay [1] in s	Remarks
Major road	W_l	22	7,0	4,3	
	E_l	39	1,9	4,4	
Minor road	N_r	17	8,0	5,2	
	N_g	112	28,8	15,3	large devitations between delays
	N_l	22	31,0	13,6	
	S_r	52	2,2	4,2	
	S_g	66	16,3	10,9	
	S_l	196	66,3	75,0	

The average delay has been determined by approximation according to the formula

$$t_W = \frac{3600}{N} \cdot \frac{1-P_0}{P_0}$$

, where P_0 is the probability that no stream vehicle is queueing

Appendix 4
Measurement results of an unsignalized intersection

Roundabouts in Switzerland

Recent Experiences, Capacity, Swiss Roundabout Guide

Michel J. Simon

Head of Planning and Transportation, Emch + Berger Zürich AG, Civil Engineers, Switzerland

Summary

A brief introduction describes the development of modern, small roundabouts in Switzerland. Recent case studies show the effects on safety, Public Transportation an environmental impact. Two different methods for the estimation of capacity have been elaborated and are compared to international formulas. Finally a concise overview on the recently published Swiss Roundabout Guide is given.

1. Introduction

Switzerland is a small country in Western Europe with a ground surface of 41'000 km^2 and a total length of roads of some 69'000 km. The total population comes to 6.5 million inhabitants. The density of population is an overall of 157 inhabitants per km^2 but up to 5'500 in the large cities. The motorisation too is very high, with an overall of 430 cars per 1000 inhabitants in 1989. Although a small country, Switzerland suffers of this high density of cars and the traffic congestion related to it.

The Swiss people is renowned to be skillful, accurate and technically gifted. These attributes have led to a high level of transportation techniques aiming at a perfect management of road traffic until the mid 70-ies. In this philosophy all efforts were made to modernize and sophisticate traffic signals with a technique of software packages responding to every traffic demand. This technistic attitude also is the main reason why until the late 70-ies you could hardly see any roundabout in Switzerland. In fact, the basic rule of auto-responsibility and self-regulation in roundabouts was not compatible with the attempt of perfect traffic regulation.

However - when the criticism to our technistic societies and the awareness of urban quality and environmental problems grew, the need for more compatible traffic solutions also grew. The first roundabouts appeared in the occidental, french speaking part of Switzerland in the late 70-ies, when a few isolated intersections were modified into roundabouts in order to improve safety.

In the early 80-ies there was a larger impact on roundabout ideas coming from France, where more and more intersections of this type where installed and where the famous "right of way" - rule was even abandoned by law and replaced by a "give way"- rule on entries of roundabouts in 1983. This evolution was followed by Swiss traffic engineers but it was not before 1985 that the first "modern" roundabouts appeared in Switzerland.

2. Recent Experiences

2.1 Incentives for roundabouts

In the last 5 years more and more roundabouts have been introduced in Switzerland. By now, their number includes about 100 in operation and several hundreds of planned or designed intersections. So far there have been designed only small roundabouts with inscribed diameter of 25 to 40 meters. Most of these intersections have been transformed from an unsignalized or signalized intersection into a roundabout. Almost all these intersections are located in urban or suburban aereas, a few in residential aereas and almost none in the countryside.

A lot of scepticism remaining within the authorities (county or community), many of these roundabouts have been built at little cost as so-called experimental or temporary roundabouts.

The main objectives pursued when introducing roundabouts at urban intersections are the following:

- Improve the urban quality of the public space at intersections.
- Achieve fluent traffic conditions and high capacity under changing traffic demand at peak hours and in beetween.
- Improve safety for all users by reducing the speed of motorized traffic.
- Reduce environmental impact (noise, pollution) thanks to fluent traffic flows at limited speed and reduced delays.
- Extend the flexibility of movements by allowing U-Turns.

Although these general objectives are largely admitted by now, there are some aspects which need to be consolidated. In the following chapter the aspect of capacity will be emphasized according to its overall importance for traffic engineers. Before however, some other points of interest will be stressed: safety, management of public transportation and environmental impact.

2.2 Safety

Urban traffic in Switzerland usually includes a considerable number of cycles (often 15 -25 % of total traffic) and important pedestrian flows crossing the branches of an intersection. This is why already in 1987 a research on behalf of the Swiss Foundation for Traffic Safety was conducted in order to analyze the behaviour and safety of cyclists and pedestrians in small roundabouts (1). This research included a detailed analysis of the first temporary roundabout of 1985 and behaviour-studies on three other intersections. The main results are listed hereafter:

- Roundabouts generally lead to a reduction of the casualty-rate and in accident-spots also to a reduction of the accident-rate.

- To achieve safety for cyclists a good visibility on the approaches is requested. Vertical obstacles should be avoided.
- Permanently built roundabouts are preferable to temporary ones arranged at low cost.
- No special problems for cyclists have been observed on double-lane-approaches.
- At zebra-crossings priority is often refused to pedestrians crossing exits.
- All branches of roundabouts should be equipped with traffic islands as soon as the directional flow exceeds 300 vehicles / hour. These islands enable pedestrians to cross the road in two times.

In 1989 another research led by the Zuerich County Police included a before / after analysis of three roundabouts (2). The figures are given in Fig. 1 below.

CRITERION	BEFORE transformation	AFTER transformation
Period of investigation (months)	153	31
Number of accidents	100	14
accidents / month	0.65	0.45
casualties	20	2
casualty-rate	0.13	0.06
cyclists involved / month	0.08	0.06
pedestrians involved / month	0.01	0.00

Fig. 1. Main Results of before / after analysis on three roundabouts.

Other studies, e.g. the first ever conducted in Switzerland on the older roundabouts of the County of Vaud (3), give similar results. As a general conclusion the following can be said:

- Small roundabouts have a lower casualty-rate than other types of intersection.
- At accident-spots small roundabouts generally lead to a clear reduction of the accident-rate.
- Small roundabouts are safe solutions for cyclists and pedestrians. However to achieve safety, a comprehensive design and a careful construction are requested. Temporary arrangements need a close follow-up.
- Cyclists should be integrated in the roundabout lanes: no cycle lanes, no separate cycle routes. Motorized traffic must be forced to reduced speed by layout. Therefore diameters of not more than 30 meters are recommended.
- To increase pedestrian safety, traffic islands should be provided at every approach.

2.3. Public Transporation

In Switzerland Public Transportation in urban aereas is of great importance and traffic management gives a high priority to accelerating public buses by special regulations.Special bus lanes are therefore often arranged at approaches of intersections. At traffic signals special software routines provide priority to vehicles of Public Transportation. In roundabouts management of Public buses is somewhat contradictory to the self-regulating traffic conditions. This is why some experiences with special bus lanes on apprаches to roundabouts were followed with great interest. The results are convincing: the buses easily overtake queueing vehicles and insert into the circular lane without major conflicts with the entering traffic on the other lane. As a benefit of the short average delay the length of bus lanes can be reduced compared to lanes at signallized intersections.

2.4 Environmental impact

In these days public opinion is focused on the environmental impact of transportation. Regulations in Switzerland are the most severe in Europe in matters of limiting noise and air pollution. Therefore a particular interest is given to the aspects related to environmental impact. However research in this field is complex and before / after-analyses request a considerable amount of data collection. Up to now the pracitical knowledge is very limited. Theoretical work however (4) and global observations of traffic flow in roundabouts reveal that the dynamic behaviour of motorvehicles is fundamentally different in roundabouts.

Some empirical work has been achieved at the Higher Institute of Technology in Bienne in the field of vehicle conduct (5). All pollution is related to speed level, acceleration and deceleration. The studies included comparative analysis of these variables in the situation of a congested intersection and the transformation to a roundabout at the same site. Fig. 2 shows both diagrams at the same time of day and the same traffic demand. It can be seen that the typical stop-and-go-mode before transformation is replaced by a fluent move in the roundabout situation. In the same time a general reduction of fuel consumption of up to 16 % for the total traffic flow has been estimated. This indicates that a reduction of the poisonous gases, CO and CO_2 and to a minor extent HC and NOx can be assumed. Furthermore the reduced amount of decelerating and accelerating would lead to lower noise values. Although further research is needed these first results confirm the theoretical thesis of lower environmental impact in roundabouts.

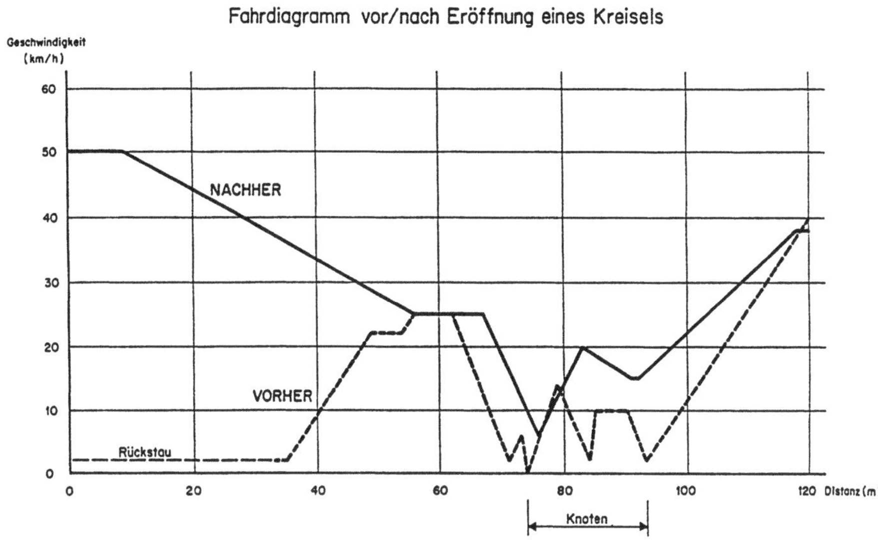

Fig. 2. Typical speed diagramm before / after transformation into roundabout.

3. Capacity

3.1. General

Since the planning of the first Swiss roundabout began there has been always a special interest in the capacity of this type of intersection. Questions and arguments always pointed out that individual behaviour determinates capacity and that therefore the English empirical formulas for capacity estimation could not be suitable for Switzerland. Therefore in 1989 the VSS Association of Swiss Road and Traffic Engineers decided to launch an empirical case study on capacity of roundabouts. This research work, done by Emch + Berger AG included the analysis of 16 entries of 7 different roundabouts spread over the three main linguistic parts of Switzerland (6).

3.2. The method of VSS / Emch + Berger

For the estimation of capacity an approach was chosen that uses regression calculations. This method is typical for the English capacity formulas of Kimber, the French methods of CETUR and SETRA and last not least the empirical research work done by Prof. Brilon in Germany. In fact an extensive cooperation with Prof. Brilon has taken place during this stage of the study.

The evaluation of regression coefficients for different types of equations has led to the conclusion that the most appropriate model is a linear realationship between Qe, the entry flow from a continuous queue and Qc, the circulating flow. No evidence was given for any relationship with the exiting flow Qs. The capacity of a single-lane entry Ce 1 is determined by one of the two followings formulas according to the field of application descripted below:

1) General Formula CH 1

$$Ce\ 1 = 1'300 - 0.75 * Qc \quad (pcu / h) \quad [CH\ 1]$$

Range of application:

- urban roundabouts (diameter 25 - 40 m, one circular lane)
- no special geometric or traffic conditions

2) Special Formula CH 2

$$Ce\ 1 = 1'450 - 0.95 * Qc \quad (pcu / h) \quad [CH\ 2]$$

Range of application:

- urban roundabouts (diameter 25 - 40 m, one circular lane) and entries with
  - one entry lane and one bus lane
  - one entry lane which is enlarged
  - one entry lane with an entry flow of more than 1´000 pcu / h

Furthermore an estimation was given for double-lane-entries, determining their capacity Ce 2 as follows:

$$Ce\ 2 = 1.4 * Ce\ 1$$

To calculate the passenger car units the following coefficients were found to suit best :

1 Cycle	=	0.5 pcu
1 track, trailor	=	2.0 pcu

To determine delays and queues it is recommended to use the methods of Kimber / Hollis or CETUR.

3.3. The method of Bovy

At the same time a second research program was conducted at the Federal Institute of Technology in Lausanne by Prof. Bovy, Tan and others (7). Based on the experience of the French CETUR they tried to take into account the exiting traffic flow Qs which visibly seems to influence the entry flow. A simulation model was elaborated in order to consider the specific conditions of the roundabout and to take into account the interaction of the different traffic flows (entering, exiting, circulating). The model was calibrated by the empirical data collected in the VSS-research descripted above. As a result the capacity is given by a linear relationship between the entry flow and the general disturbing traffic flow Qg.

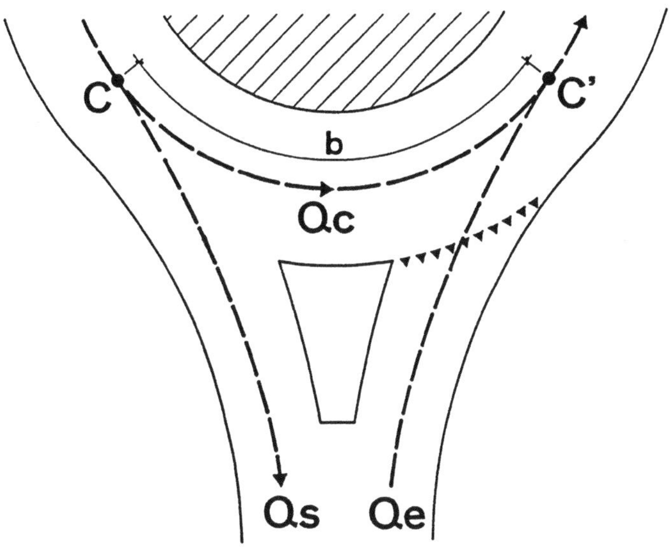

Fig. 3. Definition of flows and conflict points

The formula is:

$$Ce = 1'500 - 8/9 * Qg \text{ (pcu / h)}$$

with

$$Qg = \beta * Qc + \alpha * Qs \text{ (pcu / h)}$$

where Qc: circulating flow

Qs: exiting flow

The meaning and determination of the coefficients α and β are the following:

α takes into account the disturbance of the entry due to the exiting flow. It has been determined by the simulation model to be a function of the distance between the conflict points of exit and entry (see Fig. 3). The values are to be taken from the diagram below:

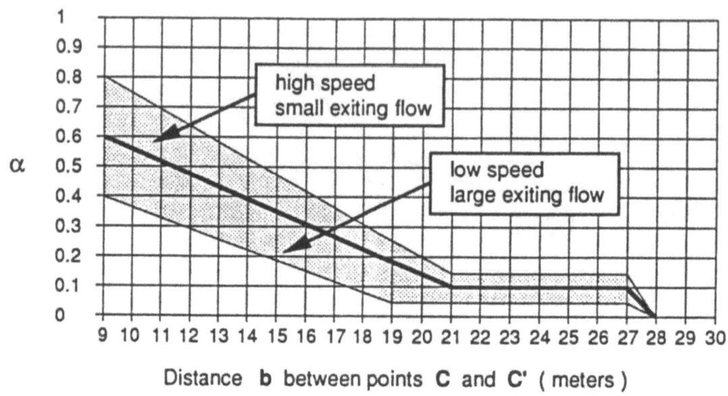

Fig. 4. Determination of α

β takes into account the number of circular lanes as follows:

one circular lane:	$\beta = 0.9 - 1.0$
two circular lanes:	$\beta = 0.6 - 0.8$
three circular lanes:	$\beta = 0.5 - 0.6$

In order to determine the capacity in case of several entry lanes a coefficient of used capacity (TCU) is determined at the entry point (e) and at the point of conflict on the circular lane (c):

$$TCUe = \frac{\gamma * Qe}{Ce} * 100\ [\%]$$

$$TCUc = \frac{\gamma * Qe + 8/9 * Qg}{1'500} * 100\ [\%]$$

γ takes into account the number of entry lanes and its value is:

one entry lane:	$\gamma = 1.0$
two entry lanes:	$\gamma = 0.6 - 0.7$
three entry lanes:	$\gamma = 0.5$

3.4 A comparision of methods

Extensive comparative analyses have been undertaken by the two teams elaborating the Swiss capacity methods. In a general appreciation it can be said that the two methods give very similar results, the one of VSS / Emch + Berger being very easy to use under standard conditions and the one of Bovy allowing to evaluate even very complex designs with several entry and circular lanes. A comparision with methods of other countries is shown in Fig. 5. Under the specific conditions the performance of roundabouts in Switzerland is very similar to France (CETUR). Compared to Great Britain (Kimber) and Germany (Brilon) the capacity in the range of heavy entry flow is considerably higher in Switzerland.

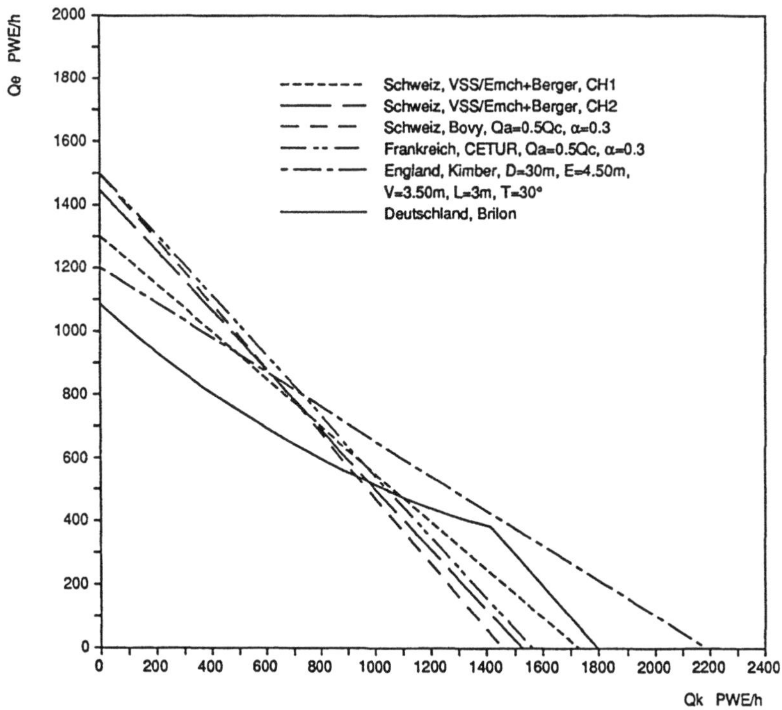

Fig. 5. Comparision of capacity formulas for single-lane entries of small roundabouts

4. Swiss Roundabout Guide

4.1. General

In order to cope with the general emphasis on roundabouts in Switzerland the VSS and the Federal Institutes of Technology decided to elaborate a suitable design guideline for roundabouts. As

to collect all information related to roundabouts in a larger book to be called a "guide". This guide, published in march 1991, is a state of the art report at the end of 1990. It is understood that further experiences should complete the knowledge and an amended edition be published in a few years.
According to the general practice in Switzerland the guide is focused on so called "small roundabouts" with inscribed diameters between 22 and 35 meters.

4.2. The procedure

The Swiss Roundabout Guide (7) intents to stress the roundabout design as a coordinated action between the traffic engineer and the town planner. Thus, the visual and spacial aspects of a roundabout become equally important with traffic and road design aspects. The recommended procedure aims at providing solutions that could satisfy both the road user and the user of public domain. Fig. 6 shows the procedure of design and construction of a roundabout as a concerted move and a step-by-step action.

The design phases need an intensive communication and cooperation between the traffic engineer and the town planner. Only the first step, the so-called geometric feasibility study is a more technical and limited work. It consists mainly of a geometric check to see wether or not the space is sufficient for a roundabout. In addition this step leads to a listing of the main concerns for the design phase.

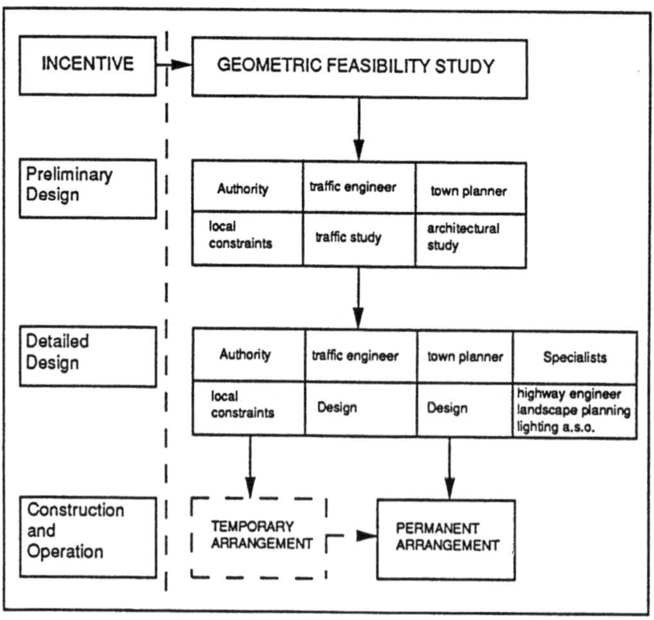

Fig. 6. Roundabout design procedure

4.3. The contents of the Guide

The Guide consists of three parts:

- General bases
- Roundabout design
- Technical Appendices

The first part gives an overview on definitions and fields of application, explains the importance of the spacial and urban quality of the roundabouts and finally enumerates the typical elements of roundabouts.

In the second part the procedure is described from the geometric feasibility study to the operation. Special techniques are illustrated and show how to take into account the aspects of town planning and architecture. For the operation of a roundabout two modes are recommended:

- temporary operation
  in order to check the suitability and the acceptance by users
- permanent operation
  including a permanent construction, often deduced of a temporary operation.

However the state of temporary operation must be well prepared in order not to create safety risks.

The technical appendices of the third part include computer graphics, capacity methods and software programs, bibliography and some special items related to mini-roundabouts.

5. Conclusions

The experiences gained in Switzerland over the last five years are rather promising. However, in roundabouts more than in any other type of intersection, the success of the traffic regulation depends on the behaviour of any single user. This is why a general appreciation of the effects of roundabouts on the traffic conditions in villages and towns cannot yet be made. Especially in the domains of safety, environmental impact and compatibility with Public Transportation further research is needed to give a definitive answer. However it can be said already now that the incentive for a roundabout often issues out of public opinion, thereby testifying an enhanced acceptance of this type of intersection. A recent experience in transport planning of the small town of Bulle (Fribourg) (some 9'000 inhabitants) shows that in this type of communities the roundabout can be much more than a single-spot measure. With 8 roundabouts put to operation in the last four years and a total of 15 more planned there will be developped a real network with roundabout nods. The effect by now is very convincing. It seems that for this type of town the repeated use of roundabouts in the road network could lead to the fluent traffic regulation at low speed thereby achieving safety and environmental standards at high capacity, a goal hard to reach with signallized or unsignallized intersections.

References

1) M. Simon, H. Rutz: Sicherheit von Verkehrskreiseln innerorts, Fonds für Verkehrssicherheit / Emch + Berger AG Aarau, Bern 1987

2) R. Jäger, M. Weissert: Verkehrskreisel, sichere und umweltgerechte Lösung im Strassenverkehr, VTA Nr. 14, Kantonspolizei Zürich, Juni 1989

3) Etat de Vaud, Service des Routes: Exploitation des carrefours giratoires - Expériences du Canton de Vaud, Exposé, journée d'étude SIA, Bienne 1986

4) W. Hüsler et. al: Verminderung der Umweltbelastung durch verkehrsorganisatorische und verkehrstechnische Massnahmen, EVED/ASB, VSS Zürich, August 1987

5) St. Luethi: Untersuchung, wie sich die Inbetriebnahme eines Kreisels in Nidau auf den Verkehrsfluss auswirkt, HTL Biel, September 1987 (unveröffentlicht)

6) M. Simon et al.: Empirische Ermittlung der Leistungsfähigkeit von innerörtlichen Kreisverkehrsplätzen, EVED/ASB, VSS Zürich, November 1990

7) Ph. H. Bovy, Prof. et al: Guide Suisse des Giratoires, VSS/OFR + FSR, Lausanne, Mars 1991

Prediction of Junction Delays and Accidents in Great Britain

J. R. PEIRCE

Transport and Road Research Laboratory
England.

Summary
A brief outline of the development of ARCADY and PICADY, the British junction design programs for roundabouts and priority junctions, is given. Empirical relationships linking injury accidents at junctions to the geometric layout have been developed and are discussed. The case of rural 'T'-junctions is used to present the methodology used in the analysis. Road accident costs in Great Britain exceeded £6,000m in 1989. Over half of road accidents occur at or near junctions. TRRL has estimated that the use of these safety relationships could reduce junction accidents by up to 30% at junctions with particularly bad records.

Introduction
Research carried out in the 1970s by TRRL (1,2,3,4,5), showed that under British road conditions, the capacities and delays at priority junctions were related to junction geometry. TRRL was able to identify the relationships between capacity and individual components of junction geometry. The relationships were of an empirical nature, evaluated from many on street measurements. The outputs from this work are two computer programs (6,7,8,9) for the design of particular junction types:-

ARCADY - for the design of roundabouts
PICADY - for priority junctions - cross roads, 'T'-junctions and staggered junctions.

There is also a third, similar program, OSCADY (10), for signalised junctions, but this is outside the scope of this conference.

The capacity relationships developed are linear, of the form shown in Fig 1 and equation 1. The values of the intercept (F) and the slope (f) depend only upon the junction geometry.

$$C = F - f.Q \qquad (1)$$

where C = capacity of entry
Q = controlling flow
F = intercept on vertical axis
f = slope

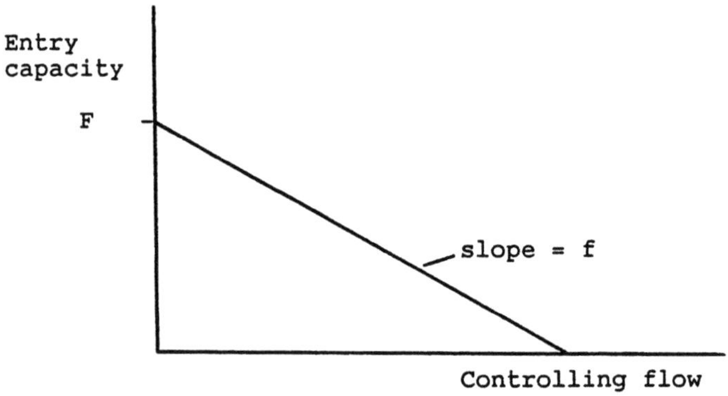

The relationships are robust, and the programs are now in widespread use in Great Britain, both by local authorities and by consultants. Outputs from these programs are required at Public Enquiries into junction developments. Over 350 copies of each program are in use in Britain.

Because of the empirical nature of the relationships, based exclusively on measurements within Britain, the resulting programs are only proved to be fully applicable within Britain. One would suspect that similar types of relationships could be established elsewhere, if similar driver behaviour and rules of the road existed, but not otherwise.

Safety and accident prediction

A large proportion of accidents occur at or near junctions. It is important that junction design programs should be able to evaluate the safety performance of junctions, as well as the capacities and delays. Only with both of these aspects correctly evaluated, can a proper balance between traffic efficiency and safety be reached.

Research at TRRL examined links between the geometry of junctions and injury accidents. Again, the method has been empirical. Data has been collected and analysed for many junctions. Five years of injury accident data are needed for each junction. So far, studies have been completed at roundabouts (not including mini-roundabouts) (11) and at rural 'T'-junctions (12). At the time of writing, only ARCADY, the roundabout design program, is available with the accident prediction relationships incorporated; by June 1991, PICADY, for the design of priority junctions, should be available with the rural 'T'-junction results incorporated.

Only limited types of junction are covered because of the high cost and considerable time taken to study a given type of junction. For example, the roundabout studies involved visiting all roundabouts featuring in the national accident records or listed by local authorities. Eighty-four were selected for detailed study. All geometric characteristics which could possibly have an input to accident records had to be carefully recorded, and the accident records obtained and carefully scrutinised and categorised for the period 1974 - 1981.

Similar work for the rural 'T'-junctions was carried out in the mid-eighties. It is estimated that 'T'-junctions account for 74% of some 15,000 junctions on rural single carriageway Trunk and Principal roads in England. Currently, studies of urban cross-roads and urban 'T'-junctions are under way. A study of injury accidents at mini-roundabouts is about to start. As these studies reach completion, so the results will be incorporated in new versions of the relevant junction design programs.

Methodology of accident analysis

The objective of the analysis is to relate the frequency of accidents to a range of 'explanatory variables' - in this case, traffic flow and speed, site and geometric variables. A form of multiple regression analysis was used. The 'generalised linear modelling 'technique, available in such programs as 'GENSTAT' or 'GLIM', has been used.

The examples used in this section are drawn from the study of accidents at rural 'T'-junctions. This study covered 970 personal injury accidents at 300 junctions, over a five year period. Of these accidents, 674 were within 20m of the junction.

TABLE 1 0 - 20M ACCIDENT GROUPS

1	Rear end shunts on major road, left to right
2	Rear end shunts on major road, right to left
3	Rear end shunts on minor road inbound
4	Right turn from major with major right to left
5	Right turn from major with major left to right
6	Right turn from minor with major right to left
7	Right turn from minor with major left to right
8	Left turn from minor
9	Single vehicle on major, left to right
10	Single vehicle on major, right to left
11	Single vehicle on minor
12	Head-on and overtaking on major
13	Others (ie pedestrian, multiple, shunts, outbound on minor, left turns from major, right with right, overtaking and head-on on minor

All the accidents must be categorised by type. Table 1 shows the categories used for accidents within 20m of the junctions. The descriptions assume the British left-hand rule of the road, and that the 'T' is oriented with the 'stalk' or minor approach downward.

The first stage of the analysis is to develop the best relationships between accidents in each category and suitable flow functions. For each accident category the relationships take the form shown in equation 2.

$$A = kQ^{\alpha} \qquad 2$$

where A = accident frequency (accidents/year)
Q = the flow function, an algebraic combination of the flow movements through the junction, expressed as thousands of vehicles / 24 hours.
k, α are parameters to be estimated.

Where the product of two flows was tested, an alternative form allowing the exponents of the two flows to take separate values was also tested:-

$$A = kQ_1^{\alpha}Q_2^{\beta} \qquad 3$$

For a variable to be included in the predictive equations, its effect must be significant at at least the 5% level. Most of the variables included were significant at better than the 1% level.

Once the accident - flow relationships have been settled, the models can be expanded to include geometric variables, using equations of the form:-

$$A = kQ^{\alpha}\exp(b_1G_1 + b_2G_2 + b_3G_3 + \ldots) \qquad 4$$

where G_1, G_2, G_3 ... are the continuous geometric variables
and k, α, b_1, b_2, b_3 ... are parameters to be estimated.

Where necessary, Q^{α} can be replaced by $Q_1^{\alpha}Q_2^{\beta}$

Equations of this type are a way of presenting the effects of variables that can take values over a continuous range - for example, road width. There are also some parameters for which just presence or absence is the important factor. Ghost islands on the main road are an example; these were found to have a major influence on certain types of accident. The best way of representing this type of factor was found to be to allow the constant 'k' to have alternative values to represent present/not present in the appropriate accident type equations.

Summing the equations for the 13 categories of accidents gives the overall accident prediction relationship for priority 'T'-junctions, linked to vehicle throughput and junction geometry.

The range of geometric variables considered is shown in Table 2; those which are included (for at least one arm of the junction) in the final predictive equations are marked with an asterisk. In all, there are 41 geometric variables when allowance is made for the arm-by-arm nature of many of these parameters. In addition, 16 speed related and 23 flow proportion variables were used in the analysis; of these, 12 appear in the final equations. Three 'factors' present or not present at the junctions were also tested. Only one (ghost islands) was retained in the final equations.

TABLE 2 GEOMETRIC VARIABLES TESTED FOR 0 TO 20M ACCIDENTS

1.	Stopping site distance	
2.	Gradient	*
3.	Arm width at 30m	*
4.	Approach width at 30m	*
5.	Visibility from the minor arm	
6.	Length of diverging lane	
7.	Length of merging lane	
8.	Effective number of lanes at minor entry	
9.	Number of lanes on minor entry	*
10.	Set back of island on minor road	
11.	Angle between major and minor roads	
12.	Overall junction length	*
13.	Major road width at junction	*
14.	Length of right turn lane in ghost island	
15.	Width of right turn lane in ghost island	
16.	1/(Stopping site distance)	
17.	1/(Visibility on major arms)	*
18.	Exit width	

* = Variable represented in final equations on at least one approach

The range of variables considered can be seen to cover all the main features of a priority 'T'-junction.

Similar analysis was performed for accidents between 20m and 100m from the junctions. Accident rates were found to be some 3 to 4 times the average link accident rate for similar roads. Including these accidents increased the number of accidents attributable to junctions by 40%.

Major results of the 'T'-junction studies

The analysis showed that the strongest predictor of accident frequency was the flow level. This was particularly evident when individual accident categories were related to the appropriate flow components. As was found in the similar roundabout study, motor-cycles were over-represented in the accident statistics. Junction characteristics found to be of particular importance were major road width, gradient and the presence/absence of ghost islands.

Costs of accidents

Injury accidents in Britain cost the nation over £6,000m annually at 1989 prices (13). In 1989, there were 260,000 injury accidents on British roads, of which 158,000 were at junctions. Assuming that half of the cost of injury accidents is attributable to accidents at junctions, then each 1% reduction in injury accidents at junctions would give an annual benefit of £30m, several times the cost of the junction safety research programme.

TRRL has estimated that the use of these safety relationships in the junction design process could reduce junction accidents by up to 30% at junctions with particularly bad records.

Conclusions

The junction design programs ARCADY and PICADY have been well regarded and very widely used in Britain for some years. They have so far been mainly used for the prediction of capacities, queue lengths, and delays. Until recently, only ARCADY, for roundabouts, has incorporated an accident prediction routine, but PICADY will soon have this capability. Both programs are based on empirical relationships derived from traffic observations on public roads. As accident studies at more junction types are completed, so the relationships will be incorporated into the programs.

Accident studies currently under way or about to start cover urban cross-roads, urban 'T'-junctions, and mini-roundabouts.

The use of design programs covering both accident prediction and capacity/delay calculations permits proper account to be taken of safety, as well as traffic performance.

Britain has road accident casualty reduction targets of 30% by the year 2000. The use of the safety relationships in the design programs is seen as one step towards that

Acknowledgement

The work described in this paper forms part of the programme of the Transport and Road Research Laboratory and the paper is published by permission of the Director

References

1. Kimber, R.M.: The capacity of some major/minor priority junctions. TRRL Report LR 735 (1976)

2. Kimber, R.M.; Hollis, Erica M.: Traffic queues and delays at road junctions. TRRL Report LR 909 (1979)

3. Kimber, R.M.: The traffic capacity of roundabouts. TRRL Report LR942 (1980)

4. Kimber, R.M.; Coombe, R.D.: The traffic capacity of major/minor priority junctions. TRRL Report SR 582 (1980)

5. Semmens, Marie C.: The capacity of some grade-separated roundabout entries. TRRL Report SR 721 (1982)

6. Hollis, Erica M.; Semmens, Marie C.; Denniss, Sharon L.: ARCADY: A computer program to model capacities, queues and delays at roundabouts. TRRL Report LR 940 (1980)

7. Semmens, Marie C.: PICADY:a computer program to model capacities, queues and delays at major/minor junctions. TRRL

Report LR 941 (1980)

8. Semmens, Marie C.: ARCADY/2: An enhanced program to model capacities, queues and delays at roundabouts. TRRL Report RR 35 (1985)

9. Semmens, Marie C.: PICADY/2: An enhanced program to model capacities, queues and delays at major/minor priority junctions. TRRL Report RR 36 (1985)

10. Burrow, I.J.: OSCADY: A computer program to model capacities, queues and delays at isolated traffic signal junctions. TRRL Report RR105 (1987)

11. Maycock, G.; Hall, R.D.: Accidents at 4-arm roundabouts. TRRL Report LR 1120 (1984)

12. Pickering, D.; Hall, R.D.; Grimmer, M.: Accidents at rural T-junctions. TRRL Report RR 65 (1986)

13. Road accidents in Great Britain 1989 - the casualty report. Department of Transport (1990)

The New German Guideline for Capacity of Unsignalized Intersections

Werner Brilon, Michael Großmann

Lehrstuhl für Verkehrswesen, Ruhr-Universität, Bochum, West Germany

Abstract

This paper describes the background of a new German guideline for the estimation of the 'capacity at unsignalized intersections. It contains information on fundamental aspects of the theory as well as a description of the proposed procedure which results from several research projects carried out on behalf of the West German Federal Minister of Transport. The new guideline will be a revision of the German procedure introduced in 1972, which has been transferred as chapter 10 of the Highway Capacity Manual (HCM) in 1985. Some of the unsolved problems being identified have now been solved based on simulation studies. The new guideline has also been developed in close contact with the HCM-subcommittee for unsignalized intersections. According to this more emphasis has been put on the estimation of delays. The new guideline again has the concept of a paper-and-pencil-method. However, for getting more realistic results the application of the KNOSIMO-simulation tool is recommended for practical use. Both methods have successfully been tested with a series of realistic cases obtained from practice.

1. Introduction

Unsignalized intersections of the two-way STOP- or YIELD-controlled type represent one of the most frequent types of junctions at rural roads or urban streets. This type of junction has limited capacities regarding to traffic loads above which, e.g., the installation of traffic signals becomes necessary to avoid long queues and delays. Therefore, studies of traffic performance at priority type intersections are of particular interest to traffic engineers. By the use of various theoretical models researchers in the past have attempted to get informations about capacities of traffic streams and criteria for the estimation of the quality of traffic flow (e.g. delays and queue lengths). Traditional calculation methods for practical application are based on the so-called gap acceptance theory.

In German research a first important approach into this direction had been developed by Grabe [1] in 1954 with the capacity calculation of non-priority streams. By improving this theoretical model, Harders [2] in 1968 created the starting-point for the development of a calculation method for practical application. Also in 1968, Jessen [3] revised the method in such a way that the resulting version was regarded as a guideline for calculating the capacity of unsignalized intersections. After its publication in 1972 as part of the instructions for traffic signals at rural roads [4] this method came into application in German practice.

However, the results calculated by the method for practical purposes seemed to underestimate the real capacities. This led to the assumption that the theoretical basis was partially incorrect. After that, the method's practical application was stopped in Germany in most cases. Siegloch [5] developed some theoretically based variations and corrections. Among other authors, Brilon [6] also gave critical comments upon the method.

After the transfer of the almost unchanged German 1972 guideline [4] into the Highway Capacity Manual (HCM) in 1985 [7] the problem of capacity at unsignalized intersections received new attention in Germany. Considering some recent findings, Brilon, Harders, Jessen and Siegloch created a first draft for a new guideline in 1987 [8]. Furthermore, in 1988, an international interchange of ideas and experiences regarding this research field took place in Bochum, West Germany. The proceedings of this workshop [9] give a comprehensive survey of the state of the art up to that time. Although, there still remained some uncleared problems and questions which can be summarized as (cf. Brilon [10]):

1. differences between realistic traffic flow and the simple Poisson case
2. concept and size of critical gaps and move-up times
3. correct determination of impedance factors
4. double introduction of rank-2 and rank-3 traffic streams
5. delays, reserve capacities and level of service

Then, after 1988, a new guideline has been developed for practical application in the scope of a recent research project (cf. Brilon, Großmann [11]). The results of these investigations have been introduced into the German commitee's discussions in 1991. They are summarized in this paper together with some comments on the five points mentioned above.

2. Capacity Calculation

The former versions of guidelines [4] and [7] have their background in probability and queueing theory. They are based on a simple Poisson model for each of the traffic streams at an intersection. In addition they use constant values for critical gaps and move-up times. Moreover, these procedures are only valid for streams with non-time dependent traffic volumes.

Siegloch [5] has studied the influence of non-Poisson properties of traffic streams on the main road. He found that more realistic headway distributions could change the capacity of the intersection considerably. On the one hand, slight disturbances of the headway distribution decrease the capacity. However, normal or heavy disturbances (i.e. plattooned traffic) can increase capacity considerably, e.g. up to 1.2 of the Poisson case (cf. figure 4 in Brilon [10]). These effects have been confirmed by several authors (e.g. Zhang [12], Chodur, Gaca and Tracz [13], cf. also Wegmann's contribution to this book), but Siegloch's [5] figures and tables still seem to be the most extensive documentation of this phenomenon.

Furthermore, constant average values for critical gaps and move-up times are used to describe non-priority drivers' behaviour at intersections. As is known however, these parameters show remarkable variations among drivers and even for one driver over time which cause another influence on capacity. This effect has been studied by Ashworth [14], to mention just one author (cf. also Wegmann's contribution to this book). Theoretically he found a relationship between the reduction of minor road capacity and the variance of the distribution of critical gaps for different levels of major road volume. The reduction in capacity can vary up to 40 % depending on the other parameters.

In a recent research project Brilon and Großmann [11] also found both effects mentioned. They have shown that these contrary effects on capacity compensate each other under realistic traffic conditions. This is the reason why the solution for the simple Poisson case still gives realistic results for minor street capacity. Therefore, the Poisson solution will still be the basis of the future German guideline. The intention is to use Siegloch's capacity formula to describe the maximum number of minor stream vehicles which can pass through a priority traffic stream of a given volume:

$$G = \frac{3600}{t_f} \cdot e^{-p \cdot t_0} \tag{1}$$

where

G = capacity of the minor stream (passenger car units per hour = pcu/h)

p = $q_p/3600$

q_p = volume of the priority stream (vehicles per hour = veh/h)

t_0 = $t_g - t_f/2$

t_g = critical gap (= minimum headway between vehicles in the main street traffic streams which enables at least one vehicle in a minor traffic stream to pass through the intersection)

t_f = move-up time (= minimum headway between vehicles in a minor traffic stream entering into the same main stream gap)

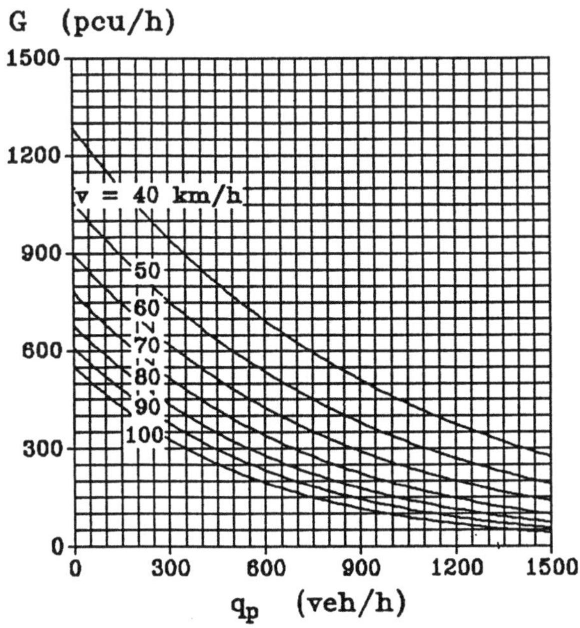

Figure 1. Capacity calculated from eq. 1 in relation to priority volume q_p and average main road velocity v_m for through traffic on the minor road.

The results of this formula are quite similar to Harders's [2] original formula

$$G = q_p \cdot \frac{e^{-p \cdot (t_g - t_f)}}{e^{p \cdot t_f} - 1} \qquad (2)$$

which was the basis of the HCM-chapter 10 procedure. The differences between the two formulae (eq. 1 and 2) are within a margin of 10 pcu/h. Results of eq. 1 are illustrated by figure 1. In the final version of the new guideline, a series of four similar figures will be included for left turning traffic from the major road, right turning traffic from the minor road, through movements and left turning traffic from the minor road.

It should be pointed out that the German guideline only applies to single lanes for each movement. According to other German guidelines (e.g. [15] and [16]), unsignalized intersections in multi-lane streets are to be avoided due to traffic safety.

During recent years, traffic performance at intersections with time-dependent traffic volumes (peak hour effect) has been studied by several authors. Results from Stamm [17] and Tonke [18] are also discussed in Brilon [10]. Zhang [12] and Gaca, Chodur and Tracz [13] should also be mentioned as more recent results. In qualitative terms, these results can be summarized as follows:

- The total capacity would increase considering time-dependent priority streams.
- If, however, the peaks of traffic volumes in priority streams and minor streets coincide, capacity decreases and delays increase

(for more details cf. Brilon [10]).

Therefore, traffic performance depends on the offset between traffic peaks on the priority street and the minor street to a high degree. It is acknowledged that details of these relations cannot be expressed by a simple paper-and-pencil procedure.

In a recent publication, Kimber [19] reiterates his former argument that the concept of mathematical models with gap acceptance theory leads to unrealistic estimations of capacity. He has developed an empirical approach to estimate capacities by using regression techniques. These ideas, of course, should be considered and evaluated carefully and comparisons should be begun as soon as possible.

However, it will take further study over a number of years to develop this empirically supported concept as an alternative to the gap acceptance theory for German conditions at two-way YIELD-controlled intersections. Therefore, the next German guideline for unsignalized intersections will still be based on gap acceptance theory. Moreover, there seem to be significant differences between behaviour of drivers from different countries. E.g. for roundabouts in Germany a non-linear relation (being of the same type as eq. 1) between capacity and major traffic flow seems to be more realistic (cf. Stuwe's contribution to this book). This may be due to the fact that German drivers are more adjusted to the clear gap-acceptance behaviour.

However, the international discussion is observed carefully. It is expected that - at long terms - also a coincidence of the linear regression techniques and the gap-acceptance approach could be discovered by using flow dependent critical gaps or move-up times (cf. Troutbeck's contribution to this book).

2. Concept and size of critical gaps and move-up times

The only empirical results for representative critical gaps and move-up times under German conditions are those of Harders [18], described in English by Brilon [10] and shown in figure 2.

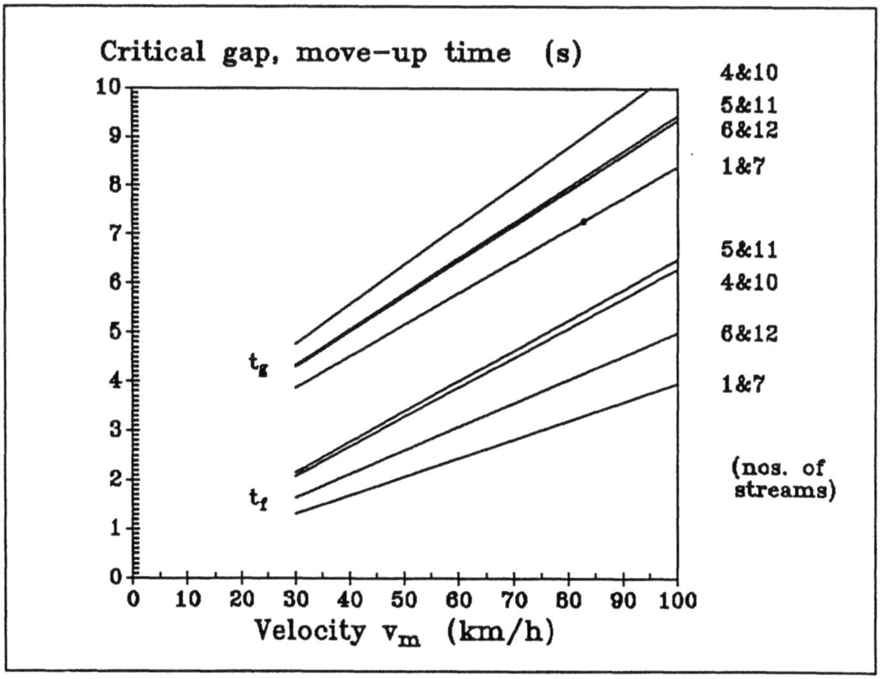

Figure 2. Critical gaps t_g and move-up times t_f for passenger cars in West Germany. (source of data: Harders [18])

One could argue that this data base might have become obsolete and that driver behaviour could have changed in the meantime. Another objection is that the influence of the velocities on the major street might be overestimated in these values. Nevertheless, Harders's results are still the most recent and above all the most comprehensive values for critical gaps and move-up times. During the study 33 intersections were observed for more than 109 hours, resulting in 30,000 measured values.

However, in the research project [11] mentioned above these values were checked with random sampled present test data. The definition of the parameter t_f by Siegloch [5] proves to be the most simple kind for determination of the critical gaps and move-up times. Furthermore, this contains a strong accordance with the derivation of eq. 1. This method allows to determine these values by use of regression techniques. For that, each actual major stream gap has to be evaluated together with the number of vehicles entering it as illustrated by figure 3. However, there is an important condition to use this method: there must be a permanent queue in the observed minor traffic stream.

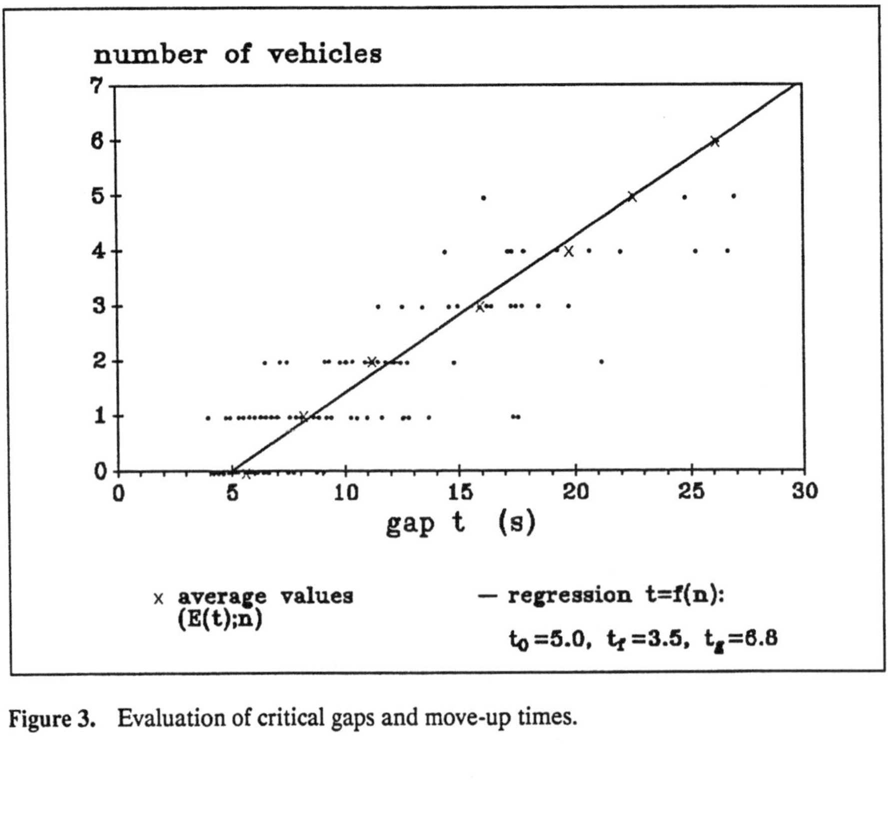

Figure 3. Evaluation of critical gaps and move-up times.

The results of the analysis of 312 actual pairs of values for right turning from minor road also are shown in figure 3. The average velocity on the major road has been observed by 60 km/h. In table 1 the resulting values for critical gaps and move-up times are compared to those from figure 2. The comparison approximately verifies the 1976 values in the range of an important velocity, but this is only *one* value of figure 2. It should, however, be pointed out that these values characterize driver behaviour in West Germany in 1976. Before applying the graphs like figure 1 in other countries, these parameters must certainly be evaluated again. However, for the conditions within one country we must be careful as well, because driver behaviour might be different from one area to another, e.g. it might differ between cities and rural areas.

	from figure 2 (v_m = 60 km/h)	from figure 3 (measured)	from figure 2 (v_m = 70 km/h)
t_0	5.0 s	5.0 s	5.4 s
t_f	3.1 s	3.5 s	3.6 s
t_g	6.5 s	6.8 s	7.2 s

Table 1. Empirical critical gaps t_g, move-up times t_f and the parameter t_0.

Because of these uncertainties another series of research should be started soon covering the whole problem of up-to-date evaluation of driver behaviour at unsignalized intersections. Here all current methods of evaluation including probit-analysis (cf. Hewitt [21] and the paper by Lassarre e.a. in this book) should be employed. It will also be important to see the influence of traffic load on t_g and t_f and to learn more about mechanisms like gap-forcing and priority reversal as it was already pointed out for other countries (cf. papers by Troutbeck, Kyte and Lassarre e.a. in this book). These investigations, however, will consume more time and since there is a practical need for a new procedure the new guideline will be introduced soon, regardless of these still unsolved questions.

3. Impedance Factors

At an intersection, different ranks of priority can be assigned based on highway code regulations. Rank 1 is defined for movements with priority over all other conflicting movements. Each traffic stream (= movement) which has to give priority to another movement of rank r is at least of rank r+1 (see figure 4).

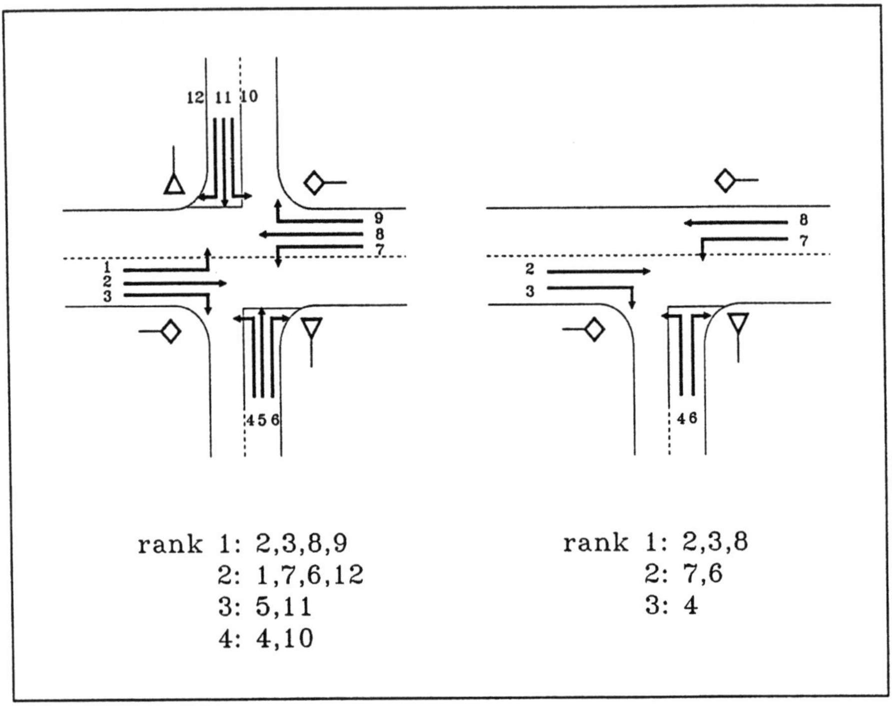

Figure 4. Traffic streams and their rank of priority.

Impedance factors in the former guidelines included this hierarchy for the different movements at an intersection (for more details cf. chapter 10 in HCM [7]). Originally, in the basic queueing model, the impedance factor is the probability that no vehicle is queueing in the minor traffic stream. It is a fundamental property of queueing systems with one service channel and with one uniform service time distribution that

$$p_0 = 1 - a \tag{3}$$

where

p_0 = probability of the empty system

a = utilization factor (= q_n/G)

G = capacity of the minor stream (pcu/h)

q_n = volume of the minor stream (pcu/h)

This regularity has already been pointed out by Siegloch [5]. The original curve for p_0 in the 1972 German procedure [4] and the HCM [7] - which is wrong - was based on approximations from Harders [2]. By the analysis in [11] it was pointed out that for an unsignalized intersection with one major and one minor stream eq. 1 is true with sufficient accuracy, in spite of the fact that two different service time distributions must be applied (cf. Kremser [20], [21] and eq. 16 in [10]). Therefore, eq. 3 can be applied for the probability of no queueing in movements of rank 2. Impedance factors in different streams of rank 2 can be multiplied as proposed by figure 10-4 in chapter 10 of the HCM [7] or table 5 in [10], since queueing in these streams is independent of each other. This has also been proven by extensive computer simulations (cf. Brilon, Großmann [11]).

In former publications, several authors objected that traffic streams of rank 2 and 3 are introduced twice into the calculations according to the German [4] and American [7] procedures. These streams are on the one hand added to the main street volumes (cf. figure 10-2 in the HCM [7]). On the other hand, they are introduced into the impedance factors by which the basic capacities G of rank-3 and rank-4 streams are diminished to calculate the actual capacities L by

$$L = p_0 \cdot G \tag{4}$$

Brilon [10] has discussed arguments which support this double introduction (cf. figures 7 and 8 in [10]).

The reasons for this are as follows:

- During times of queueing in rank-2 streams (e.g. left turners from the priority street) the rank-3 vehicles (e.g. left turners from the minor street at a T-junction) cannot enter the intersection due to traffic regulations by the highway code. Since the portion of time provided for rank-3 vehicles is p_0, the basic capacity calculated from eq. 1 for rank-3 streams has to be diminished by the factor p_0 for the corresponding rank-2 streams (eq. 3 and 4).

- Even if no rank-2 vehicle is queueing, these vehicles influence rank-3 operations, since a rank-2 vehicle approaching the intersection within a time of less than t_g prevents a rank-3 vehicle from entering the intersection.

A recent research project [11] has proven that among the possibilities considered, the existing approach (eq. 4) for rank-3 operations is the best one to account for these relations. For rank 4, however, a minor change of the procedure proved to be necessary, since the probabilities $p_{0,r3}$ (= no vehicle is queuing in rank-3 streams) depend on $p_{0,r2}$ for rank-2 movements. This relationship is expressed by figure 5.

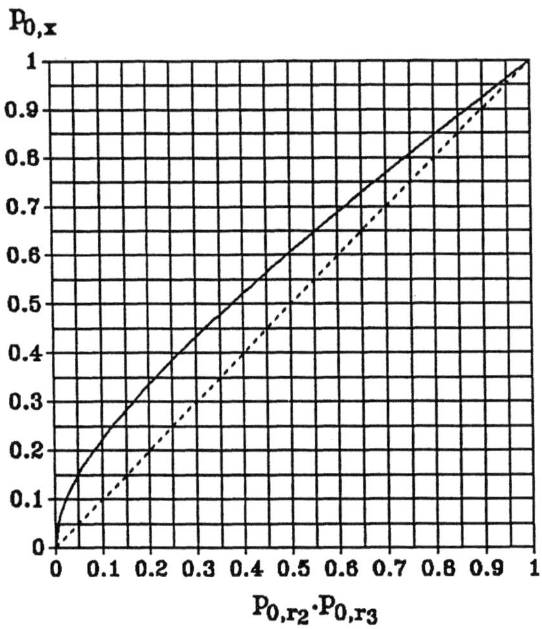

Figure 5. Corrected impedance factor $p_{0,x}$ for evaluating actual capacity of a rank-4 stream.

The diagonal would represent the case of statistical independence between the impedance factors of rank-2 and rank-3 movements. It was found that this assumption of independence, as it was the basis of the procedures in the HCM [7], would underestimate the actual capacity L. To enable an easy consideration of statistical interrelations between $p_{0,r2}$ and $p_{0,r3}$ the solid line in figure 5 has been developed.

$$p_{0,x} = 0.65 \cdot p_{0,y} - \frac{p_{0,y}}{p_{0,y}+3} + 0.6 \cdot \sqrt{p_{0,y}} \qquad (5)$$

where

$p_{0,y} = p_{0,r2} \cdot p_{0,r3}$

It is a representation of simulation results with more than 300 different combinations of traffic flow volumes. The value $p_{0,x}$ represents the impedance factor $p_{0,x}$, which has to be applied instead of the product $p_{0,r2} \cdot p_{0,r3}$. Again, this new factor is only necessary for the computation of rank-4 capacities.

5. Delays, Reserve Capacities and Level of Service

Regarding the concept of level of service (LOS), it is argued that the existing procedures do not clearly distinguish between different levels of service. For the German guideline, this argument is of minor importance, since the German concept is more aimed at capacities and the direct evaluation of traffic quality parameters like delays.

To estimate delays, a series of formulae for the simple case of one priority stream and one minor stream and for steady-state conditions has been developed (e.g. Kremser [22], [23], Poeschl [24]).

Even these formulae, which give the impression of being rather complicated, are only approximations assuming that t_g and t_f are constant values and that the headways are exponentially distributed both in the major and minor stream. Moreover, e.g., Kremser's formula is valid only for the unrealistic case of $t_g = t_f$. Therefore, the delay formula of Harders [2] can be recommended as a sufficient approximation:

$$D = \frac{1 - \gamma}{L - q_n} \cdot 3600 \tag{6}$$

where

$$\gamma = e^{-(p \cdot t_g + n \cdot t_f)}$$

These formulae only apply for the simple queueing system with one major stream (priority stream) and one minor stream (e.g. streams no. 1, 6, 7, 12). A correct formula for other movements of rank 3 and 4 or for more realistic traffic flow assumptions (non-Poisson and/or time-dependent volumes) is not available. For these cases, computer simulation techniques seem to be the only practicable tool.

Irrespective of the delay formula used, one can find that the so-called "Harders's trick" is true. This means that with constant reserve capacity R

$$R = L - q_n \tag{7}$$

(cf. eq. 10-2 in the HCM [7]), average delay D tends to remain on a nearly constant level (see figure 6).

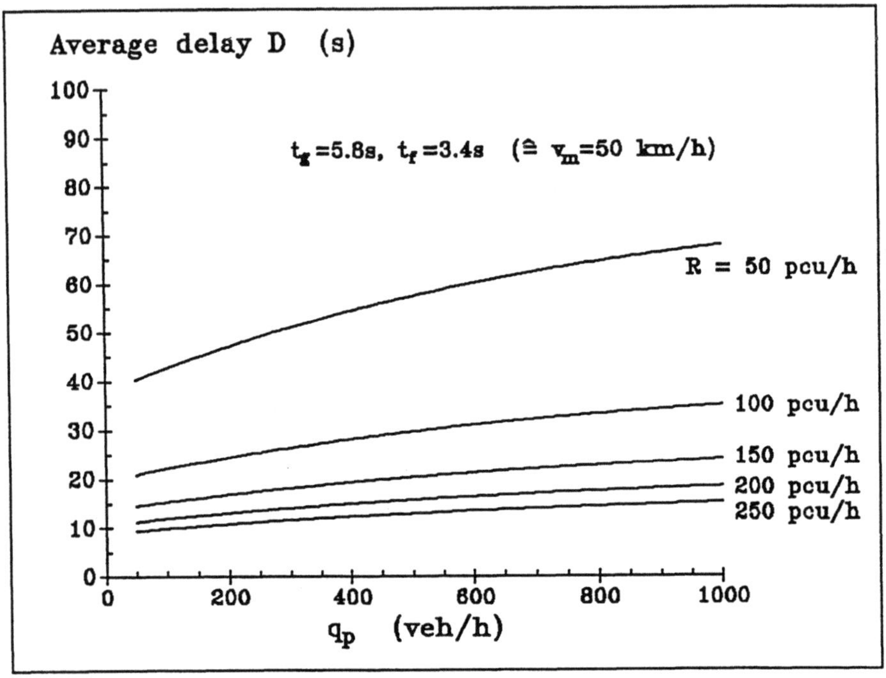

Figure 6. Average delay D for a rank-2 stream of the volume q_n = L - R for different values of q_p.

Hence, it is possible to state that with *R greater than 100 pcu/h* we can expect the average delay of minor vehicles *D to be less than 35 s*. Therefore, the future German guideline will use this limit of R ≥ 100 pcu/h as the limit of practical capacity for minor traffic streams at unsignalized intersections. Recent investigations by Brilon, Großmann [11] also proved that Harders's trick - in principle - is also valid for traffic streams of rank 3 and 4. In these cases, however, the average delay becomes insignificantly higher but remains below 45 s (for R = 100 pcu/h). In the final version of the new German guideline, figures 7 and 8 will be included for a rough estimation of the average delay D from reserve capacities R for rank-2, rank-3 and rank-4 streams.

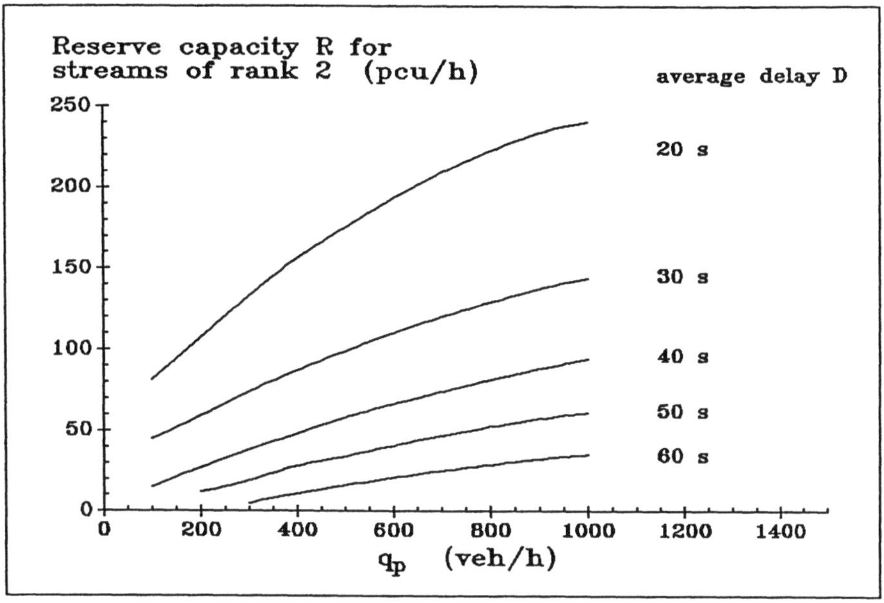

Figure 7. Relation between reserve capacity R and average delay D for rank-2 streams.

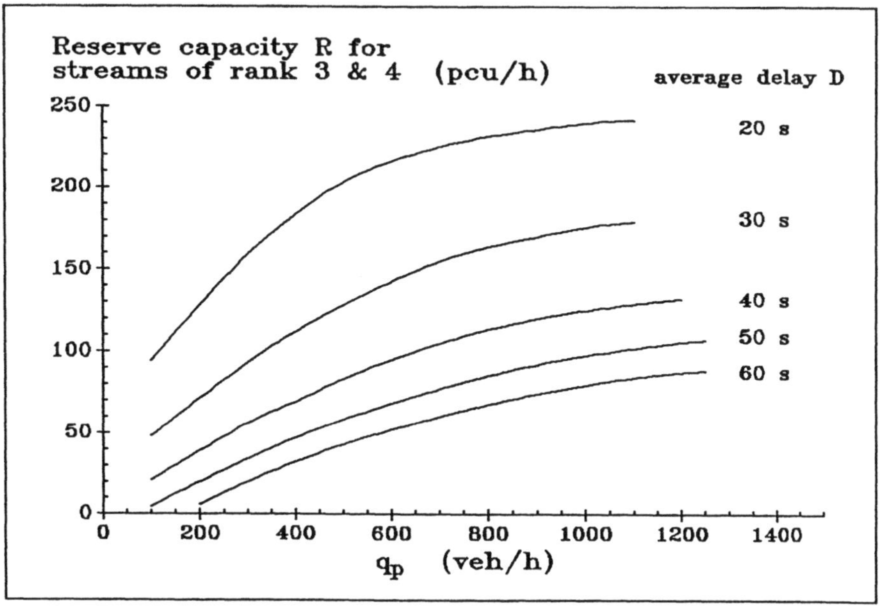

Figure 8. Relation between reserve capacity R and average delay D for rank-3 and rank-4 streams.

Under steady-state conditions delays calculated from eq. 6 tend to infinite values, i.e. they are not realistic, near capacity or with low value of R. A more realistic approach for the estimation of the average delay of vehicles was developed by Kimber, Hollis [23], considering time-dependent demand and capacity of traffic streams.

The average delay D resulting from [23] is illustrated in figure 9 in relation to reserve capacity R for different values of the priority volume q_p. These curves represent the calculated values for the so-called *low-definition* problem [23]. I.e., a rectangular traffic demand flow of volume q and capacity L is assumed in the peak period of duration T = 1 hour. Whereas, $q_0 = 0.8 \cdot q$ and L_0, respectively, are introduced in the adjacent-to-peak periods.

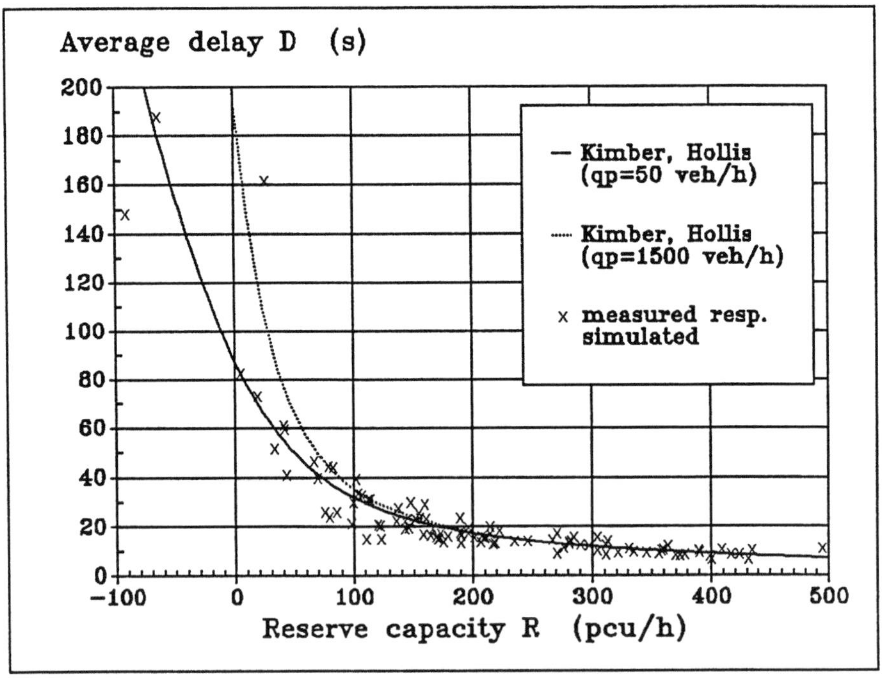

Figure 9. Average delay D in relation to reserve capacity R.

Empirically measured resp. simulated delay data with the corresponding reserve capacities calculated from eq. 7 compared to these curves show that the Kimber, Hollis-formula approximatively enables a quite good representation (see figure 9). This validates the classification mentioned above together with an assignment of delay to calculated reserve capacity, here too. However, a generalization of the relation pointed out in figure 9 needs more investigations. E.g., the definition of the peak period duration T and the relation q_0/q on a generalized basis may cause some problems.

On the basis of the derivation mentioned above, however, the LOS-concept could be either determined from reserve capacities (eq. 7) like in table 10-3 in the HCM [7] or LOS could even be based on average delays, a proposal for which is given by table 2.

Reserve Capacity R (pcu/h)	Average delay for non-priority vehicles	Level of service
> 400	< 10 s	A
300 - 399	15 - 10 s	B
200 - 299	25 - 15 s	C
100 - 199	45 - 25 s	D
0 - 99	> 45 s	E
< 0	---	F

Table 2. Reserve capacities, average delay and level of service.

In future, the queue lengths, aspects of fuel consumption, and emission of pollutants should also be considered. Since each of these aspects differs for unsignalized and signalized intersections, it is not necessary that delays at each level of service correspond at both types of intersections.

6. Proposed Procedure

The new procedure, which has been submitted to the German guideline committees in early 1991, can be characterized by the following features:

- Each of the movements at an intersection has to be evaluated by its traffic volume. There are 12 movements at a 4-way intersection and 6 at a 3-way intersection (see figure 4). Heavy vehicles in the minor streams have to be converted to passenger car units (pcu) using conversion factors (cf. table 10-1 in the HCM [7]).

- The hierarchy of the traffic streams regarding priority leads to the order of evaluating the capacity for each minor stream i. First the basic capacity G_i will be taken from a diagram like figure 1. Parameters are the velocity on the main street and the volume of the priority stream q_p. This priority stream is composed of different streams which are placed hierarchically over the actual minor stream (cf. figure 10-2 in the HCM [7]).

- The actual capacity L_i for rank-2 movements (left turners from the major road and right turners from the minor road) is equivalent to the basic capacity. For the actual capacity of rank-3 or -4 traffic streams, the queueing of privileged streams has to be considered by reducing the basic capacity with the impedance factor p_0 of the corresponding privileged streams (for rank-3 streams: using eq. 3 and 4, for rank-4 streams: using eq. 3, 4 and 5 resp. figure 5).

- If there is a shared lane for more than one movement, the common capacity is calculated from the individual capacities according to the individual volumes using eq. 10-1 in the HCM [7].

- Since the actual capacity describes the situation of "very long" queues and delays, the practical capacity is estimated by the actual capacity L_i minus a reserve capacity R_i (e.g. R_i = 100 pcu/h, cf. paragraph 5.). Depending on this reserve, a specific level of service characterized by average delay D can be guaranteed (see figure 9 and table 2).

- An unsignalized intersection can be called efficient, if the traffic volumes of all minor streams are lower than the corresponding practical capacities: $q_i < L_i - 100$ (or more precisely: if the average delays obtained from figures 7 and 8 are below 45 s).

All calculations can easily be performed using a worksheet which is different for T-junctions and cross intersections.

7. Simulation Model

This new guideline procedure, however, is only capable of estimating reserve capacities and average delays. Observations and simulation runs for realistic situations, however, pointed out that average delay is not a perfect measure for traffic quality. Depending on the rank of the movement, different streams with nearly identical average delay might provide remarkable differences in delay distributions. Therefore, queue length distributions might also differ considerably regardless of average delay.

Thus, for more precise estimations of traffic performance at an unsignalized intersection, the computer model KNOSIMO (cf. Großmann [26]) is recommended for practical use. This model is applied in practice with quite remarkable success. Also foreign traffic engineers are trying to adjust it for their local needs (cf. paper by Teply, Chan in this book). A second edition of the program with a more user-friendly characteristic and with the opportunity to adjust gap-acceptance parameters to local needs will be provided in 1991.

8. Tests for practical applicability

In the research project [11] the developed calculation method was subjected to a test for its practical applicability. For this purpose, data from about 40 intersections could be used which had been prepared by several German authorities or by additionally conducted measurements. Based on these data, for each of the intersections both capacity calculations (according to chapter 6) and simulation runs with the program KNOSIMO [26] have been carried out.

In the first instance, the reserve capacities can be calculated from eq. 7. Then, the delays which have been evaluated empirically resp. by simulation have been attached to these values (cf. figure 9). Qualitatively, this has shown the existing D-R-relation between delays and reserve capacities.

Furthermore, delays were estimated from figures 7 and 8 according to the proposed procedure (cf. chapters 5 and 6) based on the calculated (eq. 7) capacity reserves. Then, the estimated average delays were opposed to the corresponding real (empirical resp. simulated) average delays. These pairs of values are illustrated by figure 10. This comparison shows a considerable conformity. Therefore, the results estimated by the new procedure produce a sufficiently correct description of the real traffic performance.

Finally, table 3 shows a comparison between the former guidelines [4] resp. [7] and the new method by the results of a sample calculation. This again confirms the underestimation of the capacity of unsignalized intersections which has produced the criticism on the ex-procedure.

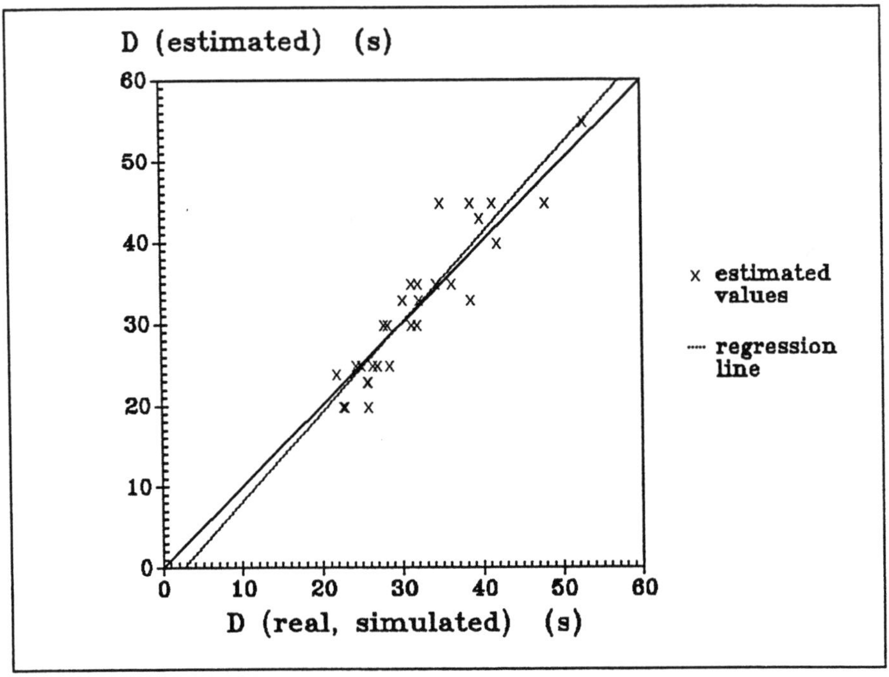

Figure 10. Estimated and real average delays in practice.

Maneuver	Streams	1972 / HCM 85 ("STOP-controlled") L	R	LOS	1991 L	R	LOS	KNOSIMO Delay
Left turns from	1	1030	910	A	1440	1320	A	2.7
the major road	7	825	801	A	1132	1108	A	5.3
Right turns	6	720	706	A	989	975	A	28.7
Through traffic	5	341	277	C	498	434	A	32.3
Left turns	4	133	-1	F	271	137	D	40.1
Shared lane	4+5+6	175	-37	F	332	120	D	36.8
Right turns	12	840	744	A	1146	1050	A	6.3
Through traffic	11	291	177	D	448	334	B	10.5
Left turns	10	192	189	D	343	340	B	16.0
Shared lane	10+11+12	409	196	D	613	400	A	8.8

Table 3. Comparison of calculation results by several methods.

9. Conclusions

On the whole, by the new guideline submitted in this paper capacity and quality of traffic flow, i.e. average delays, can be estimated. The investigations have proved that, on principle, gap-acceptance theory is qualified as theoretical basis of such a calculation method. This contribution briefly presents the intermediate state of the research, not only in Germany. No doubt, there are some requirements for further studies regarding this research field. Especially, this relates to the topic of critical gaps and move-up times regarding their relationships to several parameters, e.g. traffic loads, utilization, time, etc. However, a evident need exists in practice for a calculation method. Therefore, despite all unsolved problems mentioned, the new German guideline will be given to printing, probably even in 1991.

References

1. Grabe, W.: Leistungsermittlung von nicht lichtsignalgeregelten Knotenpunkten des Straßenverkehrs. Forschungsarbeiten aus dem Straßenwesen, Neue Folge, Heft 11, Kirschbaum Verlag, Bielefeld (jetzt Bonn Bad Godesberg): 1954.

2. Harders, J.: Die Leistungsfähigkeit nicht signalgeregelter städtischer Verkehrsknoten. Schriftenreihe Straßenbau und Straßenverkehrstechnik, Heft 76, Bonn: 1968.

3. Jessen, G.D.: Ein Richtlinienvorschlag für die Behandlung der Leistungsfähigkeit von Knotenpunkten ohne Signalregelung. Straßenverkehrstechnik, Nr. 7/8: 1968.

4. FGSV (Ed.): Berechnung der Leistungsfähigkeit von Knotenpunkten ohne Lichtsignalanlagen. Anhang zum Merkblatt für Lichtsignalanlagen an Landstraßen, Forschungsgesellschaft für das Straßenwesen, Köln: 1972.

5. Siegloch, W.: Die Leistungsermittlung an Knotenpunkten ohne Lichtsignalsteuerung. Schriftenreihe Straßenbau und Straßenverkehrstechnik, Heft 154, Bonn: 1973.

6. Brilon, W.: Anmerkungen zur Leistungsermittlung von Knotenpunkten ohne Lichtsignalanlagen. Straßenverkehrstechnik, Heft 1: 1981 und Heft 2: 1982.

7. Transportation Research Board (Ed.): Highway Capacity Manual (HCM). TRB Special Report 209, Washington: 1985.

8. LfV (Ed.): Berechnung der Leistungsfähigkeit für Knotenpunkte ohne Lichtsignalanlagen - Zwischenbericht über den Entwurf eines neuen Merkblattes. Arbeitsblätter Nr. 3, Lehrstuhl für Verkehrswesen, Ruhr-Universität Bochum: 1987.

9. Brilon, W. (Ed.): Intersections without traffic signals. Springer-Verlag, Heidelberg, Berlin, New York: 1988.

10. Brilon, W.: Recent developments in calculation methods for unsignalized intersections in West Germany. In: Brilon, W. (Ed.): Intersections without traffic signals. Springer-Verlag, Heidelberg, Berlin, New York: 1988. pp. 111-153.

11. Brilon, W.; Großmann, M.: Aktualisiertes Berechnungsverfahren für Knotenpunkte ohne Lichtsignalanlagen. Research report for the Federal Minister of Transport, Ruhr-Universität Bochum: 1990.

12. Zhang, X.: Einfluß des Grades der Teilgebundenheit auf die Wartezeit an Knotenpunkten ohne Lichtsignalanlagen. Research report, Technische Hochschule Darmstadt: 1986.

13. Chodur, J.; Gaca, S.; Tracz, M.: Störungen des Verkehrsablaufes an höhengleichen Knotenpunkten ohne LSA-Steuerung bei instationärem Zufluß. Die Straße, Heft 7: 1989.

14. Ashworth, R.: The capacity of priority-type intersections with a non-uniform distribution of critical acceptance gaps. Transportation Research, Vol. 3: 1969.

15. FGSV (Ed.): Richtlinien für die Anlage von Straßen (RAS), Teil: Knotenpunkte (RAS-K), Abschnitt 1: Plangleiche Knotenpunkte. Forschungsgesellschaft für das Straßenwesen, Köln: 1988.

16. FGSV (Ed.): Richtlinien für Lichtsignalanlagen (RiLSA). Forschungsgesellschaft für das Straßenwesen, Köln: 1990.

17. Stamm, J.M.: Der instationäre Verkehrsablauf an Knotenpunkten ohne Lichtsignalanlagen. Dissertation, Technische Hochschule Darmstadt: 1974.

18. Tonke, F.: Wartezeiten bei instationärem Verkehr an Knotenpunkten ohne Lichtsignalanlagen. Schriftenreihe Straßenbau und Straßenverkehrstechnik, Heft 401, Bonn: 1983.

19. Kimber, R.M.: Gap-acceptance and empiricism in capacity prediction. Transportation Science, Vol. 23, No. 2: May 1989.

20. Harders, J.: Grenz- und Folgezeitlücken als Grundlage für die Leistungsfähigkeit von Landstraßen. Schriftenreihe Straßenbau und Straßenverkehrstechnik, Heft 216, Bonn: 1976.

21. Hewitt, R.H.: Using probit analysis with gap acceptance data. Unpublished manuscript, University Glasgow: 1990.

22. Kremser, H.: Ein einfaches Wartezeitproblem bei einem Poisson'schen Verkehrsfluß. Österreichisches Ingenieur-Archiv, Band 16, Heft 1: 1961.

23. Kremser, H.: Ein zusammengesetztes Wartezeitproblem bei Poisson'schen Verkehrsströmen. Österreichisches Ingenieur-Archiv, Band 17, Heft 3: 1962.

24. Poeschl, F.J.: Die nicht signalgesteuerte Nebenstraßenzufahrt als verallgemeinertes M/G/1-Warteschlangensystem. Zeitschrift für Operations Research, Vol. 27 B: 1983.

25. Kimber, R.M.; Hollis, E.M.: Peak-period traffic delays at road junctions and other bottlenecks. Traffic Engineering and Control: October 1978.

26. Großmann, M.: KNOSIMO - A practicable simulation model for unsignalized intersections. In: Brilon, W. (Ed.): Intersections without traffic signals. Springer-Verlag, Heidelberg, Berlin, New York: 1988. pp. 263-273.

A Modular System for Traffic Flow Simulations at Intersections and in Networks

Bernhard Friedrich

Fachgebiet Verkehrsplanung und Verkehrswesen
Technische Universität München

Abstract

Traffic planning and traffic control necessitate a tool to evaluate the performance of the considered traffic system. This requires a multi-purpose calculation method allowing to compute delays, queue lengths etc. for different planning alternatives. These effects can be obtained from the modular system SIMPL for traffic flow simulation described in this paper. Formulating the model components with respect to recent research studies, a representation of traffic flow close to reality can be achieved. On the basis of a microscopic, time and space discrete concept, the simulation model gives the opportunity to integrate additional parameters considering real road user behaviour. In order to avoid the "black box effect" a graphical real time presentation of the simulated traffic provides a valuable tool to detect program inconsistencies and to assess verisimilitude of the model.

1. Research Context

SIMPL is a modular simulation system, designed for the representation and evaluation of the traffic performance at junctions and in networks. Due to the modular approach of SIMPL (the system components can be combined in various ways) priority junctions (according to §§ 8 and 9 of the German traffic regulation - StVO), fixed time controlled intersections, two lane roads and networks containing these components can be simulated.

The goal of this modular concept is to obtain a suitable tool for the evaluation of capacity of priority junctions and signalized intersections as well as to be able to perform direct comparisons between both. Another purpose of SIMPL is to create the possibility to examine the interdependencies of neighbouring junctions and their performance.
Furthermore SIMPL is an attempt to provide a basis for the evaluation of differing traffic control strategies at intersections and in networks.

The results of several studies have been adopted for the formulation of the individual model components. Some parameters that could not be adopted in their precise form were only integrated as estimations and should be subject of further research.

A particular feature of SIMPL is that it takes advantage of a microscopic, increment oriented model. Open problems that can not be solved by traditional methods, due to their simplification of complex interdependencies, can be solved with this simulation approach.

The graphical real-time representation of the vehicle movements in the system provides a way to check program inconsistencies and to assess verisimilitude of the assumptions underlying the model and its parameters.

2. The simulation model

The increment oriented simulation model is based on largely microscopic model assumptions, taking into account the decisions of the drivers in the junction area as well as the interactions of the vehicles in the approach roads. The concept of the model focussed especially on the independence and flexibility of the single system components. Thus the junction area, the entry and exit roads or traffic signals represent individual elements of the system and can, under certain preconditions, be combined in any way. So the system can contain several junctions - which may be signalized or not - connected by links of optional length.

This principle provides the possibility to combine partial networks in a modular way and to simulate the traffic flow of the resulting systems with its network-effects (i.e. congestions of one junction caused by insufficient capacity of the neighbouring junction) (see Fig. 1).

2.1 Junction and road geometry

As the influence of space/time effects should be taken into account by the calculations, junction and road geometry is modeled in an abstracted form (see Fig. 2). The single streams in the junction area are lead in such a way that a very small number of crossings is achieved, thus reducing the number of branchings in the decision logic to a minimum. Left turning lanes of optional length can be provided for all approach roads.

2.2 Generation of driver/vehicle elements

During an initialization procedure preceding the actual simulation the individual driver/vehicle units are supplied with all the information necessary for passing through the system.

Every unit is given the following individual data :

- time of arrival at the system boundary

  Trying to find a distribution for the description of traffic which is made up by "free" as well as "forced" components, DAWSON (1968) examined combined forms of arrival distributions. With the superposition of a shifted negative exponential distrubution and a modified Erlang-distribution DAWSON found a model quite appropriate for describing the headway distribution. Based on these results BRILON/GROSSMANN (1988) showed that empirical distributions fit well with this so-called hyperlang distribution.

  Due to the good applicability of the hyperlang distribution for representing headways it was chosen as the basis for the model described here.

- desired speed

  The applicability of the normal distribution, used in several papers to determine the desired speed (e.g. LEONG, 1968) was confirmed by SCHÄFER (1990).

- gap acceptance

  This is another individual feature affecting performance. For that reason each driver/vehicle unit is given a certain value for its readiness of acceptance. For the distribution of the acceptance of headways a lognormal distribution is used (SOLBERG, OPPENLANDER; 1966)

- route choice

  As traffic observations usually don't yield information about the routes through the network, but provide information about turning percentages, the latter is integrated into the model.

  The route choice of each driver/vehicle unit is determined by equally distributed random numbers.

Apart from permanent information about the identity of each single vehicle the knowledge about its actual status forms the basis for its behaviour within the system.

This status information, which is updated incrementally can be described by the following parameters

- actual position
- actual speed
- persisting reaction-time and other persisting manoevers
- desired speed
- indentity of the preceding vehicle and its status data
- desired turning action at the next junction

Beside these attributes, specific for each vehicle/driver unit, route-related information like conditions of visibility, can be integrated by means of locally valid speed limits.

The individual decisions of vehicles are determined according to this set of internal and external data. So every time-interval includes a test of the intentions of every single vehicle with respect to potential conflicts and a decision about how to react according to given conditions (e.g. reaction time).

2.2 Link model

On the road section the individual reaction is derived from a car following model, containing the distance to the vehicle in front and the speed difference. Starting from the nonlinear approach of GAZIS and ROTHERY (1961) and along the line of the zone-model of WIEDEMANN (1974) a model characterized by the following actions is used:

- free speed
  the vehicle under consideration can realize its desired speed restricted neither by preceding units nor by other constraints. Depending on actual and desired speed the vehicle is accelerated according to empirically validated acceleration values.

- deceleration zone
  When the occurance of a constraint (slower vehicle, traffic lights) is noticed within an area which depends on distance and speed difference, the driver tries to adapt his speed to the car in front or to stop at a certain position.

- car-following zone
  After the speed has been adapted to the preceding vehicle, no more intentional changes are performed. The following car tries to observe certain distance conditions, the speed difference falls within the range of noise.

The change between acceleration and deceleration is modeled as a continuous process.

2.3 Junction model

The decision of a driver/vehicle unit at the junction is determined by the existence of conflicts with vehicles of major streams. The following factors are considered :

- actual position of vehicles having priority and their speed
- speed of the minor stream vehicle
- acceleration and deceleration characteristics of the minor stream vehicle
- braking distances between vehicles, according to the real conflict points
- willingness to take risks

Rendering space and time discrete in a model of traffic flow enables the decision logic to account for dynamic factors as well as spatial dimensions of the junction geometry.

In order to use these additional data for the algorithm, the decision about entering the junction or stopping is based on braking distances which are related to conflict points and expressed as time intervals. In this way for the minor stream vehicle the location of the conflict point and the time of crossing it according to the intended action are determined as well as potential conflicts with vehicles of major streams.

Thus the process "turning left" is differentiated into "crossing" and "merging" depending on the direction of the major streams, where the headway exists. Considerable differences in acceleration conditions demanding different headway lengths exist for both events. HARDERS (1976) mentions this problem and his measurements show that this difference is significant.

Depending on the value for risk-acceptance braking distances of different duration are tolerated. The occurrence of too long or too short braking distances is excluded by introducing a lower and upper threshold for the generation of the normal distributed random numbers.

2.4 Graphical display - Plausibility

The goal to design the model as transparent as possible can be obtained by the graphic representation. Usually result protocols don't provide the information, that can be grasped at once by visual representation.

Graphic representation provides direct check of a program's ability to reproduce adequately the rules and parameters of the model. During the development and implementation of algorithms mistakes may occur (collisions, short distances, endless acceleration), which are not expressed by the result protocols.

Apart from this monitoring of mistakes graphical representation supports in yet another respect the assessment of verisimilitude of the model. For the observer the traffic flow with its platoon-effects braking manoevers etc. is a familiar picture. Thus visualization facilitates judging the plausibility of the model parameters. This is especially true if a real-time representation can be approximated.

Besides that it is possible to omit the graphical component for a confirmed model configuration in order to speed up the simulation.

2.5 Parameters for the assessment of the level of service

In the actual version of the program the following parameters are determined for each junction entry :

- total delay
- longest delay calculated for one vehicle
- average delay
- average congestion length

For the calculation of these values one or more junctions of the given system may be selected. The determination of excess-times follows the definition given by BÖHM (1968). During the simulation in every interval the excess time of each vehicle, resulting from the delay is measured and added successively. From these values the longest, the average and the total delay of each approaching stream is determined. For the average delay and congestion length standard deviations are computed additionally.

Tests with empirically derived data (FRIEDRICH, 1990) indicate, that the program shows a high level of verisimilitude.

3. The simulation program

SIMPL is implemented in FORTRAN on a UNIX workstation which works with a Motorola 68020 Processor and a 68881 Coprocessor. For graphical display a 16" colour-monitor with a resolution of 1024 x 768 pixel is required. The Hewlett Packard specific STARBASE was chosen as graphics tool, as its overall performance is considerably higher than that of the compatible GRAPHIC KERNEL SYSTEM.

By taking time intervals of 0.1 seconds very good temporal resolution can be obtained, giving the impression of a continuous process on the screen.

In its present form the program requires about 500 KByte RAM for the simulation of 10000 vehicles. The question of the program's portability to MS-DOS in case of leaving out the graphic component has not been analysed yet. The input of the required data is done interactively or by an input file. The following information is necessary:

- number and position of junctions
- approach and linking roads
- the lengths of left-turning lanes (expressed by vehicle-lengths)
- turning percentages
- traffic-volumes in the approach roads (patterns of time)
- duration of simulation.

4. Range of application and limits

The program can be used for the evaluation of the performance of junctions, that are controlled by signal or priority rules (according to §§ 8/9 of the German traffic regulations - StVO).

Parameters not covered by SIMPL are:

- different transport modes
- multi-lane roads
- opportunities to overtake
- cyclists and pedestrians
- junction-forms deviating from the given geometry
- public transport vehicles
- parking actions
- junctions with "priority to the right".

5. Conclusion

SIMPL is a successful attempt to model traffic performance integrating in one model several recent results on headway distributions, gap acceptance distributions and car following behavior and to achieve transparency by graphical presentation. Although the model omits omits relevant factors, a representation of traffic flow, close to reality is obtained, as empirical tests show. Due to the modularity of the individual system components the range of applications is wide. For the evaluation of the traffic flow the interdependencies of neighbouring junctions can be considered in particular. As traffic signal control can be included, with/without comparisions for planning-decisions are possible.

At the present time SIMPL exists as a research prototype. The development of a user friendly PC-version for the support of planning-procedures is intended.

References

1. BÖHM, H.: Die Anwendung der Monte-Carlo-Methode in der Straßenverkehrstechnik Teil I: Untersuchungen am ungesteuerten Straßenknoten. Schriftenreihe Straßenbau und Straßenverkehrstechnik, Heft 73, 1968.

2. BRILON, W.; GROSSMANN, M.: Entwicklung eines Simulationsmodells für Knotenpunkte ohne Lichtsignalanlage. Forschungsbericht an den Bundesminister für Verkehr; FE-Nr. 01.1076.85 H, 1988.

3. DAWSON, R.F.: The hyperlang probability distribution - a generalized traffic headway model. Proceedings of Fourth International Symposium on Theory of Traffic Flow, held in Karlsruhe, Germany, June 1968. Published as Beiträge zur Theorie des Verkehrsflusses, Straßenbau und Straßenverkehrstechnik, Vol. 86, S. 30 - 36, 1969.

4. FRIEDRICH, B.: Simulation des Verkehrsablaufs am Knotenpunkt. Diplomarbeit am Fachgebiet Verkehrsplanung und Verkehrswesen, TU München, 1990.

5. GAZIS, D.C.; HERMAN, R.; ROTHERY, R.W.: Nonlinear follow-the-leader model of traffic flow. Operns. Res. 9, S. 545 - 567, 1961.

6. HARDERS, J.: Grenz- und Folgezeitlücken als Grundlage für die Berechnung der Leistungsfähigkeit von Landstraßen. Schriftenreihe Straßenbau und Straßenverkehrstechnik, Heft 216, 1976.

7. LEONG, M.J.W.: The distribution and trend of free speeds on two-lane highways in New South Wales. Proc.Austr.Road Research Board 4, S. 791 - 814, 1968.

8. SCHÄFER, E.: Wunschgeschwindigkeiten auf Außerortsstraßen. Diplomarbeit am Fachgebiet Verkehrsplanung und Verkehrswesen, TU München, 1990.

9. SOLBERG, P.; OPPENLANDER, J.C.: Lag and gap acceptances at stop-controlled intersections. Highw. Res. Rec. 118, pp. 48 - 67, 1966.

10. WIEDEMANN, R.: Simulation des Straßenverkehrsflusses. Schriftenreihe des Instituts für Verkehrswesen, Universität Karlsruhe, Heft 8, 1974.

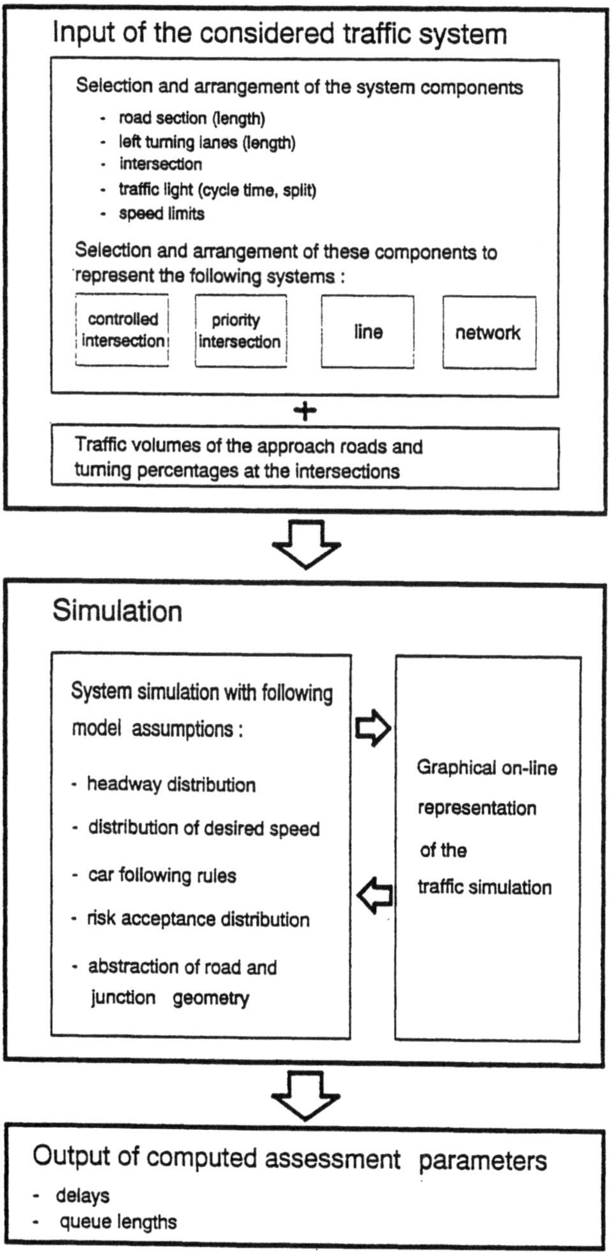

Fig. 1. Structure of the simulation model

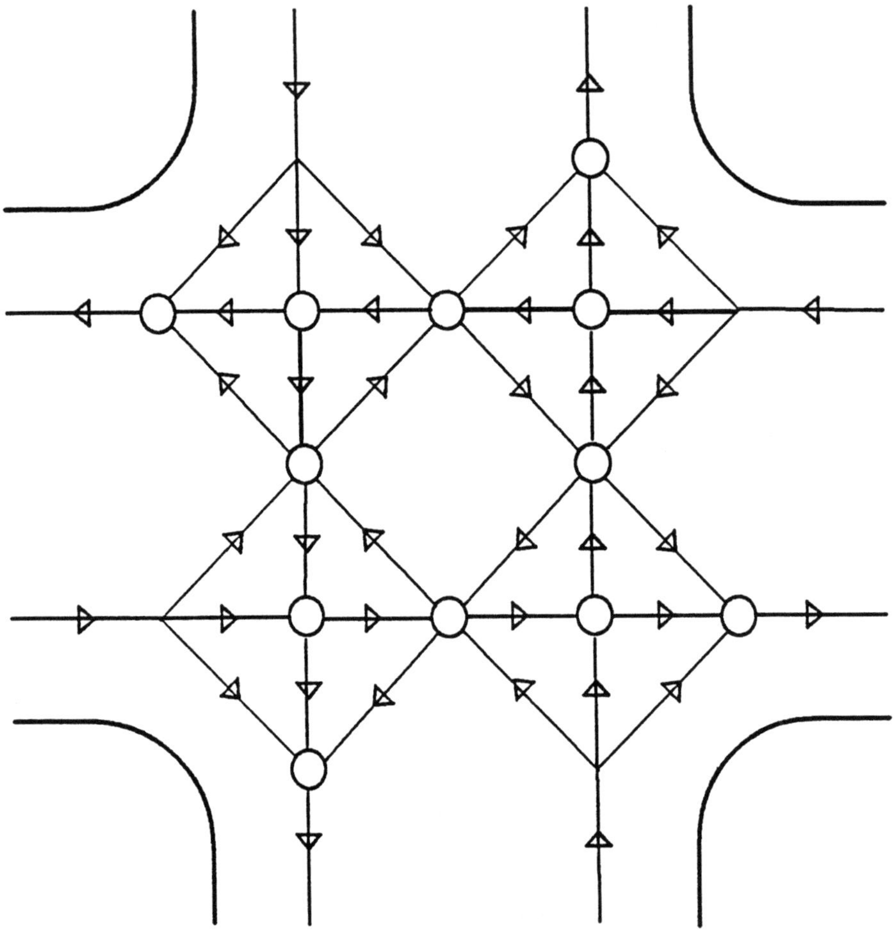

Fig. 2. Junction geometry

A Comparative Study of Roundabout Capacity Procedures

Hakkert A. S., Mahalel D., Asante S. A.
Department of Civil Engineering
Technion- Israel Institute of Technology.

ABSTRACT

This report presents a comparative study on some of the widely used capacity calculation procedures and recently developed models. A comparison of various capacity models for roundabouts is conducted with an independent data set. The main results and conclusions are that the gap acceptance models with fixed or constant parameter values do not provide a complete description of the processes for capacity operation. It is concluded that the move up time should be a function of the circulating flow, and that gap forcing and driver relaxation modify capacity significantly. With the average gap acceptance parameter values from the literature, the gap acceptance models predict capacities fairly well at moderate circulating flows.

INTRODUCTION

High levels of safety and low vehicle operating costs have made roundabouts popular in a number of countries of late and they are increasingly being used on both urban and rural roads as well as in the residential streets for traffic control and aesthetic purposes.

Available theoretical models for capacity analysis of roundabouts are based on the conditions and operating characteristics existing in the country of origin. Human factors (behavior and practices) affecting the driver and the operating characteristics are not the same universally.

The few comparisons of various procedures for capacity analysis (Kimber (2), Schermers (4) Ashworth and Laurence (1) and Troutbeck (6)) have shown that available models tend either to grossly overestimate or underestimate measured capacities. Attempts made at explaining these differences have been attributed to the differences in driver behavior and practices, yet still some of these overestimation and underestimations are so large that it becomes very hard to accept this fact as the sole cause if some of the models themselves are not deficient in some ways. An investigation into the measured and predicted capacities by various models is therefore being carried out as the different approaches and models for predicting capacities have given different solutions to the same problem.

The primary objectives of this research were:

1. to review the approaches to roundabout capacity analysis, in the hope of understanding and reviewing: (a) the theoretical structure of the procedures; (b) the range of applicability of the models; (c) computational procedures, their practicability and accuracy; (d) the investigation of the inherent prediction errors as attributed to the model structure or driver behavior.
2. to compare some of the widely used capacity procedures for roundabouts using independent data.
3. to modify those procedures that fairly accurately predict measured capacities to reflect local conditions for possible adaptation.

The following steps were followed to achieve the above objectives;

1. a literature review and selection of theoretical capacity procedures for comparison;
2. survey and selection of appropriate study intersections;
3. devising a suitable data collection system;
4. determination and selection of the appropriate gap acceptance parameters;
5. analyses of data, capacity calculations and conclusions.

Six models were selected for further investigation and comparison. The first four (Tanner (5), McDonald and Armitage (3), Ashworth and Laurence (1), Troutbeck (6)) are based on the gap acceptance approach. The other two: ARCADY2 by Kimber (2) and the French SETRA model are based on the empirical approach. Tanner initially developed basic capacity formulae for a single lane major stream which he later expanded for multilane major streams. Troutbeck used a dichotomized distribution to develop a capacity formula for single and double circulatory lanes. The McDonald and Armitage model considers roundabouts as a series of linked T-junctions and applies concepts of lost time and saturation flow. The Ashworth and Laurence model is based on Tanner's equation and adapted to flared approaches. The Arcady 2 model was developed by the TRRL and uses an empirical model relating entry capacity to circulatory flow and geometric variables. The Setra model, developed by SETRA in France, uses a statistical approach and expresses capacity in terms similar to the ARCADY2 model.

THE EXPERIMENT

Roundabouts are not very popular in Israel. This, coupled with the following conditions made the identification of suitable sites a difficult task indeed. Since models were being compared here, simple sites satisfying all models selected were therefore the choice.

1. that the roundabout has a simple geometric layout so as to satisfy all the models selected to enable comparison.
2. that the sites have no recent changes in geometric layout so that drivers are not affected by the changes which will cause them to modify their behavior.
3. that there be reasonable flows on all the arms of the roundabout and also a sufficient and stable queue for quite a long period on the study arm so that enough data is collected to justify any statistical conclusions eg a minimum of 60 minutes of saturated entry/circulatory flow observations.
4. good visibility at the roundabouts with no obvious sight restrictions nor significant numbers of pedestrian movements.
5. that all roads meet at approximately level at the roundabout.
6. that the roundabouts are not influenced by traffic signals and are located at least 2 km upstream and downstream from the nearest signalized intersection.
7. that the intersections are free from the interference of other intersections and of parked vehicles.

Two roundabouts were found suitable for the study. Both were four legged and operating under the offside priority rule with single circulatory and entry lanes. Figure 1 describes the basic geometry of the roundabouts.

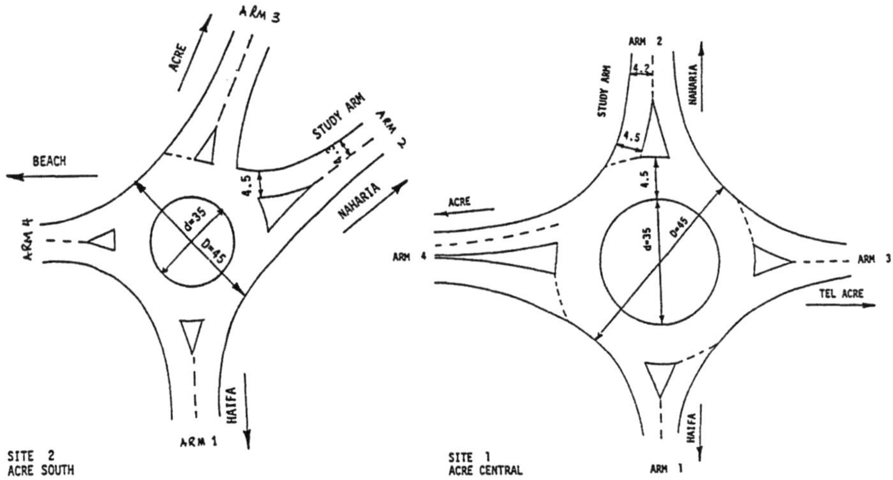

Figure 1: Site plans of the study intersections.

The data collection system used employed both manual and automatic recording equipment. Automatic data collecting systems such as the Archer 6400 and Archer 6200 traffic data loggers by Golden River Ltd. were used for classifications and volume counts. Data was then logged on to an personal computer for analysis. Manual counting was also used for turning movements and actual capacity measurements using manual traffic counters. Geometric measurements were recorded at site and where impossible to measure on site, site maps of scale 1:250 were used. Various separate recordings of traffic flow data were carried out during the afternoon peak periods. Each observation was carried out under similar conditions of weather, daylight, dry road surface, good visibility and when there was stable and sufficient queues on the study arm.

The data were collected at intervals of 2 minutes and included the following; 1. volume of each traffic stream; 2. vehicle class of each traffic stream; 3. turning movements of each stream; 4. needed geometric data; 5. the entry capacity.

PROCEDURE OF ANALYSIS

In this comparative study the models were first divided into two groups according to the modelling approach ie empirical or theoretical gap acceptance approach. Under each approach the models were further subdivided according to the units of measurements:

1. those measuring capacity in vehicles per hour;
2. those measuring capacity in passenger car units.

This is necessary so that within each particular subset/group the models can be compared.

Gap acceptance parameters were estimated in two ways:

1. from the observed capacity data and compared with the parameter estimates as given by each model.
2. from the literature review, average parameter values that have been reported in the past were selected and the various models were tested according to these average values.

Furthermore, a sensitivity analysis was performed on the individual parameters to check the model performance over a reasonable range of values.

Apart from the traffic volumes, the basic parameters used were move up times for entry vehicles β_2 (in seconds), critical gaps T_c (in seconds) and minimum headways for circulating flow vehicles β_1 (in seconds). According to various literature sources, average critical gap values have a range of 3.2 - 5.3 sec. minimum headways range from 0 to 2.3 sec. and move-up times range from 2.0 to 3.3 sec.

A number of methods can be used to examine the results of the various models namely;

1. By comparing the prediction curves across the whole range of observed controlling flows with the central estimate line for the observed capacity data ie the linear regression line through the observations.
2. By fitting a regression line through the calculated capacities to remove systematic errors and finding the standard deviation of the residuals.

3. By calculating the mean and the standard deviation of the differences between the observed and calculated capacities.

Based on the items 1., 2. and 3. described above a model is considered acceptable or good:

- when there is no significant difference between the curves of the models and the central estimate line of the capacity data which represents the measured capacity curve.
- when the standard deviation or the sum squared of the residuals is as small as possible. The closeness of the regression line to the 45 degrees line through the origin also reflects the accuracy of the models.
- when the standard deviation of the differences between the observed and predicted capacities is as small as possible.

The observed capacity curve is taken as the central estimate line of the observations (observed capacity data). This decision means, however, that at 50% of the time capacity may be exceeded. Troutbeck (1984) adopts the upper bound envelope of the observed capacity which, though compatible with the definition of capacity, is not advisable.

At any given value of circulating flow a sample array of maximum outflows is observed due to variations in traffic, vehicle class, drivers and intersection geometry. The capacity value at any given circulating flow is therefore estimated by the mean of the distribution in the corresponding sample array of maximum outflows as shown in figure 2. The mean capacity is recommended as taking account of the sampling errors.

A linear line has been used to represent the capacity curve in this study. To be more certain and conclusive, an exponential curve as assumed by the gap acceptance models was also examined. The exponential and linear curves fitted to the data are shown in figure 2, which indicates that in reality measured capacities vary approximately linearly. It is observed that for most part of the circulating flows, the exponential curve lies within the 95% confidence interval of the regression line through the data sets which is referred to in this as the central estimate line. This is also confirmed by the residual plots for ARCADY2 and Tanners models.

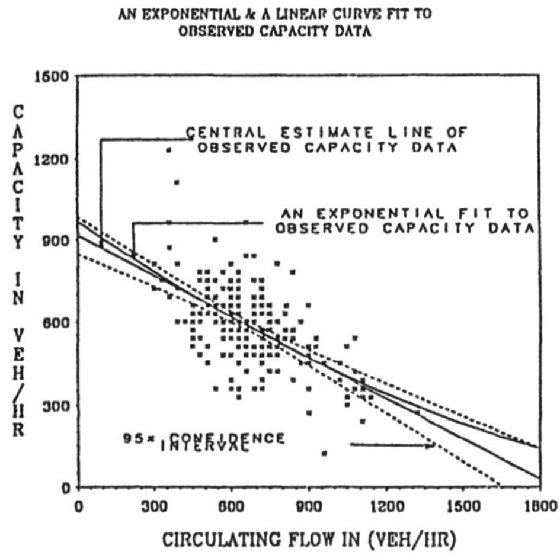

Figure 2: Exponential and Linear curves fitted to measured capacities

The observations from each site were first analyzed statistically for the various streams involved.

The second stage of the analysis involved using standard statistical tests to compare the results from each site as well as the pooled data to check if there were significant differences between the observations from the two sites as well as the pooled data as tabulated in tables 1-3. The results of these tests which were performed on both the means and the variances of the various data sets using the Student t-test and the Fisher F-test respectively showed that there were no significant differences between the various data sets. The pooled data did not therefore differ significantly from the individual data sets.

DATASETS	NUMBER OF OBSERVATIONS	MEAN VALUE	STANDARD DEVIATION	STANDARD ERROR
		veh/hr	veh/hr	veh/hr
SITE 1	100	579	140	14
SITE 2	100	582	156	16
POOLED DATA	199	585	151	11

Table 1: Mean and variance of observations for the entry flow.

DATASETS	NUMBER OF OBSERVATIONS	MEAN VALUE	STANDARD DEVIATION	STANDARD ERROR
		veh/hr	veh/hr	veh/hr
SITE 1	100	647	161	16
SITE 2	100	694	188	19
POOLED DATA	199	670	178	13

Table 2: Mean and variance of observations for the circulating flow.

For the gap acceptance models this part of the analysis focused on a specific example using the average gap acceptance parameter values as reported in the literature (see table 4). This is carried out in order to point out clearly the deficiency in the gap acceptance models considered in the study primarily where these models become inappropriate to represent observed capacities.

DATASETS	NUMBER OF OBSERVATIONS	MEAN VALUE	STANDARD DEVIATION	STANDARD ERROR
		veh/hr	veh/hr	veh/hr
SITE 1	100	557	180	18
SITE 2	100	580	163	16
POOLED DATA	199	572	179	13

Table 3: Mean and variance of observations for the exiting flow.

For the empirical models possible items for modification are; 1. the intercept of the model's prediction line; 2. the slope of the model's prediction line. Using these criteria, the resulting parameter estimates inferred from the observed capacity data for those gap acceptance models based on clearly defined equations for the entry capacity are presented in Table 4.

Figure 3 shows the prediction curves for the empirical models (ARCADY2 and SETRA) using the geometric and traffic data taken from the field across the range of traffic flows measured. A summary of the various items for comparing the models and their performance as described by the observed and predicted capacities

MODEL	MINIMUM HEADWAY	MOVE UP TIME	CRITICAL GAP	RESIDUALS SUM SQUARED
Tanner	1.5s	3.7s	4.2s	3,019,170
Troutbeck	1.1s	3.9s	4.7s	3,020,153
McDonald	1.6s	qs=0.27	L=2.4s	3,017,945
Ashworth	0 s	4.1s	4.1s	3,122,065

Table 4: Parameter estimates for the gap acceptance models for roundabouts using the pooled data.

are also tabulated in Tables 5-6 for the various models used in the study.

Because of the differences in units between some of the models it is not possible to compare the coefficients such as the intercepts and the standard errors for all the models. However, the coefficients such as the slope and the correlation coefficients are directly comparable since they are dimensionless.

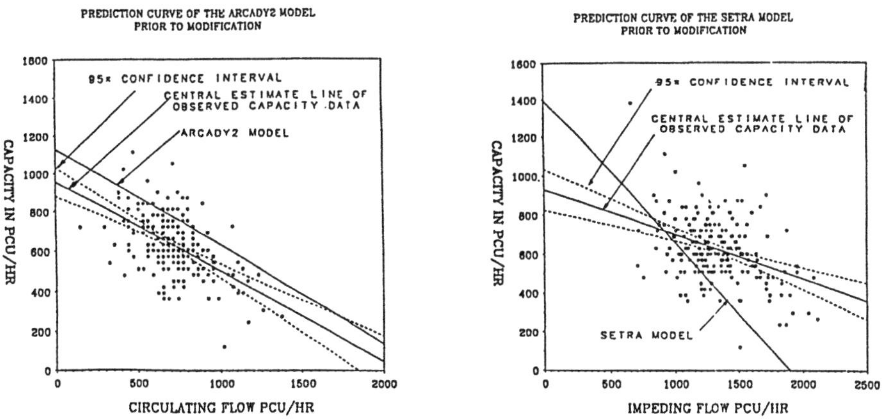

Figure 3: Capacity variation with circulating flow for empirical models ARCADY2 and SETRA using the geometric and traffic data taken from the field

MODEL	INTERCEPT	SLOPE	REGRESSION COEFFICIENT	ROOT MSE
Tanner	389	0.335	0.335	71.60
Troutbeck	398	0.339	0.333	72.89
McDonald	389	0.335	0.335	71.60
Ashworth	441	0.247	0.335	53.0

Table 5: Comparison of observed and predicted capacities for the gap acceptance models using the pooled data.

MODEL	INTERCEPT	SLOPE	REGRESSION COEFFICIENT	ROOT MSE
ARCADY2	573	0.310	0.285	81.30
SETRA	119	0.473	0.147	189.10

Table 6: Comparison of observed and predicted capacities for the empirical models for roundabouts.

Using the capacity data to infer effective parameter estimates the resulting move up times are nearly of the same magnitude as the critical gap (see table 4). However, these move-up times are high for such simple roundabouts used for the study. The difference between these estimates (move-up times) and average values reported in the literature are very substantial. The only reasonable conclusion drawn from this is that the move-up time should be a function of the circulating flow and it increases as the circulating flow decreases. The minimum headway and the critical gap values on their own are very reasonable, however when considered together in relationship with the move up times, they seem artificial.

Using the average values that have been reported in the literature and described earlier the resulting prediction curves are as given in figure 4. It is seen that even with these reasonable parameter values normally applied, the gap acceptance models show substantial overprediction at low flow and underprediction at high flow. At moderate flows these models are seen to perform fairly well with the average parameter values. Further sensitivity analysis on the individual parameters show no improvement in the models.

It is therefore realized that none of the two approaches; either using parameters estimated from the capacity data or values reported in the literature give more insight into the gap acceptance approach. Again this leads to the same conclusion as in the first approach to the problem, that gap forcing and driver relaxation considerably modify capacity. Also the move-up time should be a function of the circulating flow. Thus the failure of the gap acceptance models to incorporate these behaviors

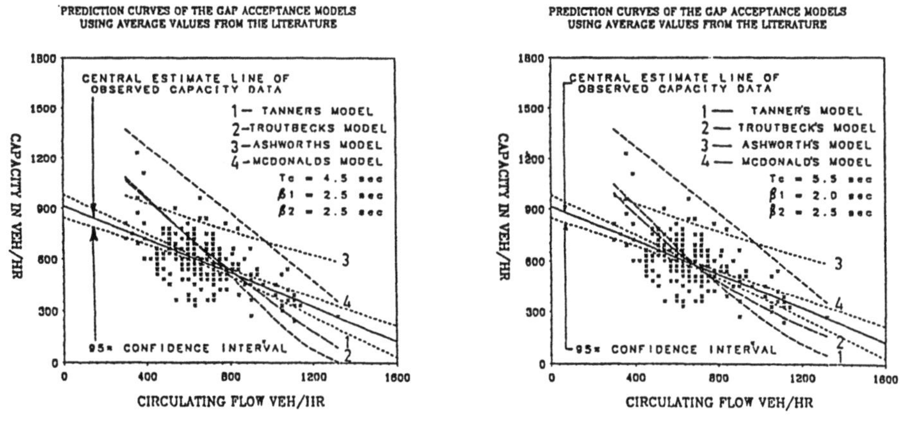

Figure 4: Capacity variation with circulating flow for the gap acceptance models using average parameter values that have been reported in the literature

results in their inability to predict observed capacities in regions of flows where these behaviors are prevalent (the extremities of the priority flows).

For the empirical models, it is seen (Figure 3) that the slope of SETRA differs from those of ARCADY2 and the linear regression other. The slope of ARCADY2 was 0.5065 which lies within the standard error for the slope of the central estimate line through the observed capacity data calculated as 0.4557 with a standard error of 0.051. Furthermore it is observed that the slope of SETRA decreases rapidly with increasing impeding flow compared with the other two curves. Differences in intercepts are also observed. While ARCADY2 gave an intercept of 1128 pcu/hr,the intercept for the central estimate line through the data was 957 pcu/hr with a standard error of 38.5 pcu/hr. The intercept for the SETRA model was 1400 pcu/hr compared with 900 pcu/hr for the central estimate line through the observed capacity with a standard error of 40 pcu/hr using the impeding flow as required by the SETRA model.

It is seen that both models overestimate capacities at low flows. SETRA on the one hand underestimates capacity at moderate and high impeding flows. ARCADY2 on the other hand consistently overestimates capacity by 17% at the mean circulating circulatory flow. For the empirical model ARCADY2, a possible change in the original model is in the intercept with the slope remaining the same. The reduction in intercept, which was put at 120 pcu/hr gave good results and modified the model to local conditions. With this correction the deviation at the mean circulating flow was found to be 6 pcu/hr higher but within the 95% confidence interval calculated as ±20 pcu/hr.

With SETRA both the slope and the intercept of the curve were considered for modifications. The sharp decrease in capacity with increasing impeding flow associated with the model is attributed to the high penalty on the impeding flow. The SETRA model needed major adjustments and recalibration to reflect local capacities, unlike the ARCADY2 model.

CONCLUSIONS

1. Gap forcing and priority sharing significantly modify capacity at high circulating flows. As observed, capacities at high flows are higher than capacities predicted by the gap acceptance models as a result of the gap forcing and

priority sharing which occur at high circulating flows. This behavior of drivers at high flows always increases the capacity of the minor road flow. At low flows, the relaxed driver behavior is also seen to affect capacities. This relaxed driver behavior results in lower capacities than those predicted by the gap acceptance models at low circulating flows. The gap acceptance models were good only at moderate flows. We therefore infer that these models with constant parameter values are limited to the range of flows excluding the extremities. The gap acceptance models with constant parameters applied across the entire range of circulating flow does not give a complete description of observed capacity data. The move up time should be a function of the circulating flow q_c.

2. From the intercept of 957 pcu/hr obtained for the central estimate line of the capacity data, the calculated move up time of 3.75 seconds is arrived at. This is seen to differ substantially from the average 2-3 seconds that have been reported in the literature for the gap acceptance models. Thus, extra caution exercised by drivers and their relaxed behavior leads to increased move up times.
3. The empirical model, ARCADY2 overestimated measured capacity in its original form. This overestimation by ARCADY2 model was expected as roundabouts are not popular in Israel compared to the UK where their popularity has offered drivers good familiarization with their use. A reduction of 120 pcu/hr in the intercept of the ARCADY2 model modified the model to reflect local conditions. The SETRA model performed poorly, it underestimated observed capacities substantially. Major modifications were required both in the intercept and the slope in order to improve the model's performance considerably to reflect local capacities.

REFERENCES

1. Ashworth, R., Laurence, C. J. D., "A New Procedure for Capacity Calculations at Conventional Roundabouts." Proceedings Institution of Civil Engineers, (65) 1978, pp. 1-16.

2. Kimber, R. M., "Gap Acceptance and Empirisicm in Capacity Prediction" Transportation Science, vol. 23 no. 2, 1989 pp. 100-111.

3. McDonald, M, Armitage, D. J., "The Capacity of Roundabouts." Traffic Engineering & Control., 1978.

4. Schermers, G., "Roundabouts - An Alternate Intersection Control", NITRR, CSIR, South Africa., July 1987.

5. Tanner, J. C., "A Theoretical Analysis of Delays at Uncontrolled Intersections.", Biometrika 49, 1962, pp. 163--170.

6. Troutbeck, R. J., "Capacity and Delays at Roundabouts.", ARRB. 1984, 14(4), pp. 205-216.

Roundabouts and Road Safety

State of the Art in France

F. ALPHAND - Engineer with CETUR (Centre d'Etudes des Transports Urbains) - Paris
U. NOELLE - Engineer with CETE Mediterranean (Aix-en-Provence)
B. GUICHET - Higher Technician with CETE West (Nantes)

1 - INTRODUCTION

Built on all types of sites, from open countryside to town centre, via commercial estates, and representing all types of roundabouts, from mini-roundabouts to grade-separated interchanges, roundabouts have made a strong inroad into improving the French road network.

Thus C.E.T.U.R. and S.E.T.R.A., through the intermediary of the C.E.T.E. of Western France, carried out a survey of the 522 roundabouts brought into operation before 1987 in two regions BRITANNY and PAYS DE LA LOIRE, and of the 78 accidents involving personal injury which occurred on them during the year 1988.

To attempt better knowledge of the involvement of pedestrians and drivers of two-wheel vehicles in personal injury accidents, and in view of the small sample represented by these 78 accidents, approximately 175 roundabouts were surveyed, from 20 towns in the West of France. The few 202 accidents involving personal injury which occurred on these roundabouts since their opening, or at least since 1984, were also analysed statistically.
This paper presents the main results of these studies, and the questions that they raise.

2 - GEOMETRY - SAFETY: ANALYSIS OF 522 ROUNDABOUTS GLOBALLY FOR ALL SITES

2.1 Global results

Virtually 80% of the roundabouts are situated in the urban or peri-urban milieu. The remaining 20% are divided over open countryside, industrial or commercial estates.

Site	Number	(%)
Peri-urban	192	37 %
Residential area	131	25 %
Town centre	88	14 %
Industrial estate	22	4 %
Commercial estate	25	5 %
Open countryside	64	12 %
Total :	522	100 %

The rate at which road network managers built roundabouts accelerated from one year to the next : 1984 doubled the number of roundabouts built in 1982; 1987 again doubled that of 1984. A plateau seems to have been reached in 1991, around which the number of roundabouts built is stabilising.

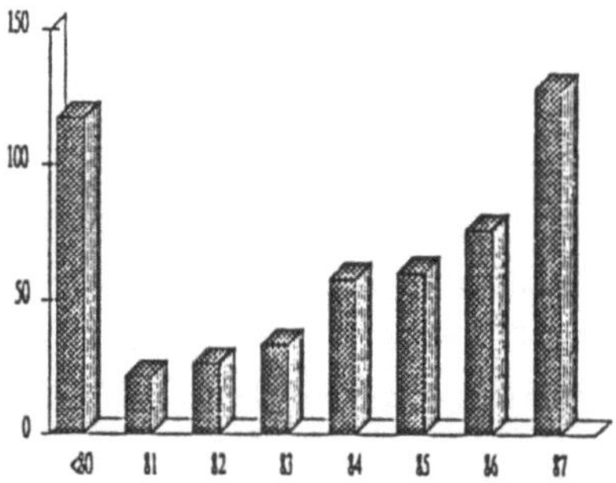

A whole gamut of radii exist, from 2 m to 50m, but most roundabouts are small (average radius < 15 m).

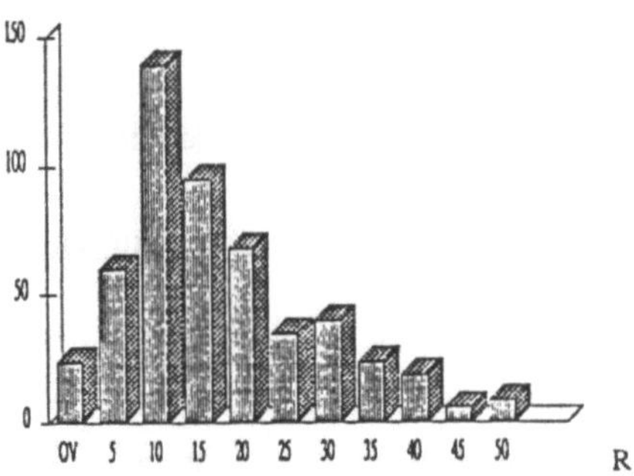

Three-quarters of roundabouts have 3 or 4 branches, approximately 20% have 5 branches or more.

The annual mean daily traffic (AMDT) per roundabout is around
12,500 vehicles per day, and a very distinct drop in number of roundabouts can be seen above 25 000 vehicles per day.

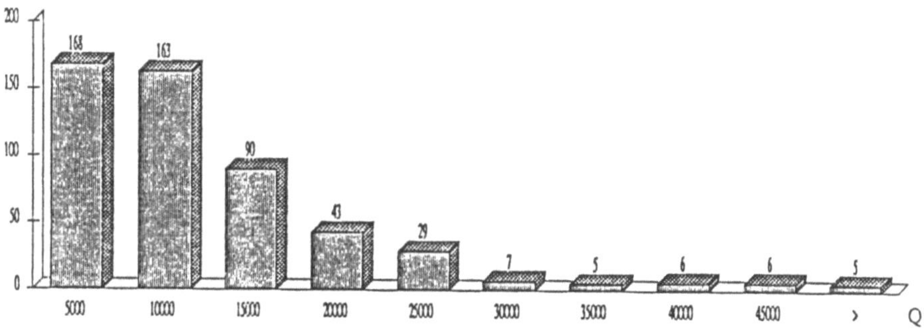

Out of the 522 roundabouts, no personal injuries were recorded on 469 of these in 1988, however 3 roundabouts recorded 4 accidents. Of these latter, there are certainly improvements to be made.

Number of roundabouts	with n accidents	(%)	No of accidents
469	0	90 %	0
36	1	7 %	36
12	2	2 %	24
2	3	-	6
3	4	-	12
Total : 522		100 %	78

As regards seriousness, these 78 accidents caused :
5 dead
26 seriously injured
78 slightly injured

i.e. a gravity rate of (5 + 26)/78 = 0.40

2.2 Evolution per year in operation

Analysis of the time histories highlights the fact that of the roundabouts built prior to 1980 many are "not round" in shape : 10%. During the period 1981 to 1987 the majority of roundabouts (60%) built were of medium or even small size, but 30% of the roundabouts still had an inside radius of over 20 m. Only from 1988 onwards was the use of small radii confirmed : 40% for radii up to 10 m, 30% for radii of 10 to 15 m, the remaining 30% being divided decreasingly over the larger radii.

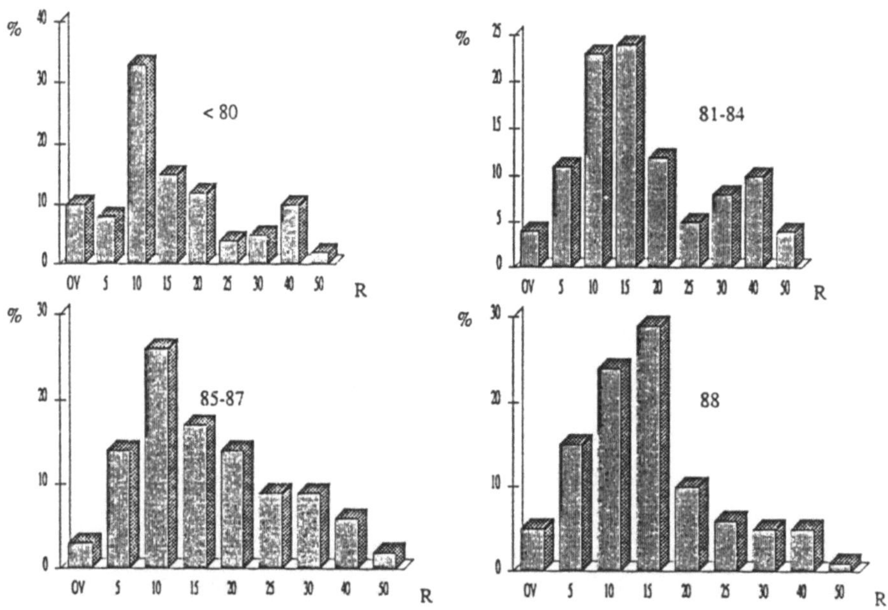

The 78 personal injuries were distributed as follows over the roundabouts of different age :

Sites built in :	**<1980**	**1981**	**1982**	**1983**	**1984**	**1985**	**1986**	**1987**
NO of accidents	29	5	2	4	9	3	4	22
Accidents per 100 roundabouts	24.7	22.7	7.4	11.8	15.5	5.0	5.3	17.2
Average traffic (vehicles/day)	10500	9250	11500	7500	7750	8500	9500	9250

Firstly we note that the number of accidents per 100 roundabouts is distinctly higher on older roundabouts than on others : 23.7 compared to 9.0. However, this number increases for roundabouts in their first year of operation : 17.2. It would have been interesting to check whether there had been any noticeable reduction in 1989.

2.3 Analysis according to location

Analysis according to location provides interesting information on accidents, on the size of the roundabouts and on lighting practices.

2.3.1. Accidents and location

The 78 accidents involving personal injury surveyed in 1988 over the 522 roundabouts are distributed as follows :

Location	**open country-side**	**peri-urban**	**residential area**	**town centre**	**indust. + commerc. estate**
Number of accidents	7	32	18	16	5
Accidents per 100 roundabouts	12.1	18.6	13.7	18.1	10.6
Fatal accidents	2	1	0	2	0
Seriously injured	4	14	4	4	0
Average traffic (vehicles/day)	7500	11500	7000	9000	10000

We note that :

- most accidents occur in the peri-urban area. This is related to the level of traffic and higher speeds of approach than in town, at least on one branch,

- accidents are also numerous in town centres. This is linked to the composition of the traffic and to the greater number of cyclists/motor cyclists and pedestrians.

- accidents are more serious in open countryside. This is related to the higher speeds.

2.3.2. Size of roundabout and location

For open countryside, the average inside radius of a roundabout is 27 m and only 10% of the roundabouts have radii which are either very small (up to 10 m) or very big (more than 40 m).

In the peri-urban milieu, the inside radius of a roundabout is
20 m, and only 15% of roundabouts have radii of more than 30 m.

In town centres and residential areas, we above all (60%) find small roundabouts with an inside radius of up to 10 m. Roundabouts over 20 m are rare.

As regards the industrial and commercial zones, particularly frequented by heavy goods vehicles, the average inside radius is 15 m and 75% of roundabouts have radii up to 20 m.

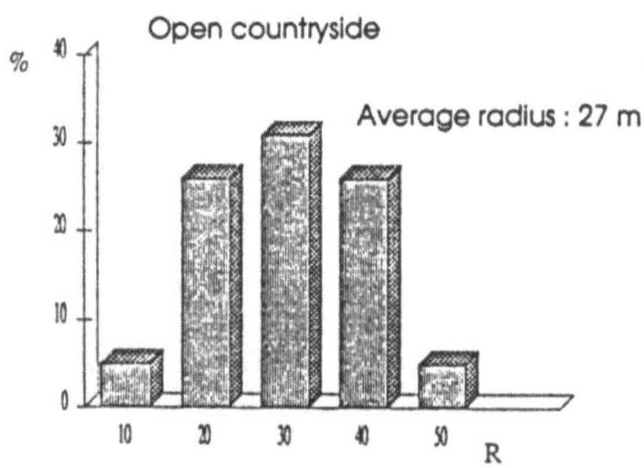

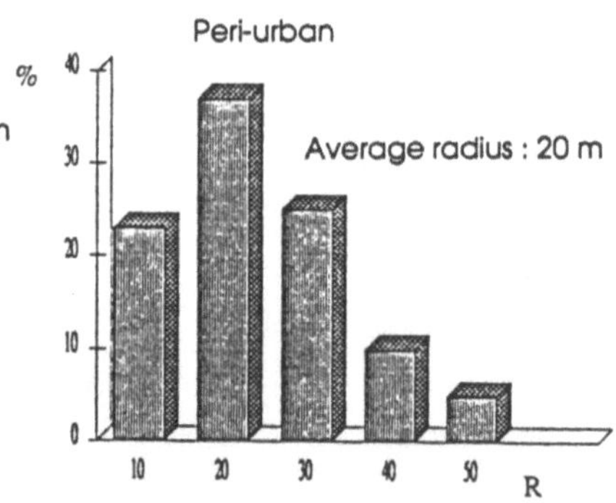

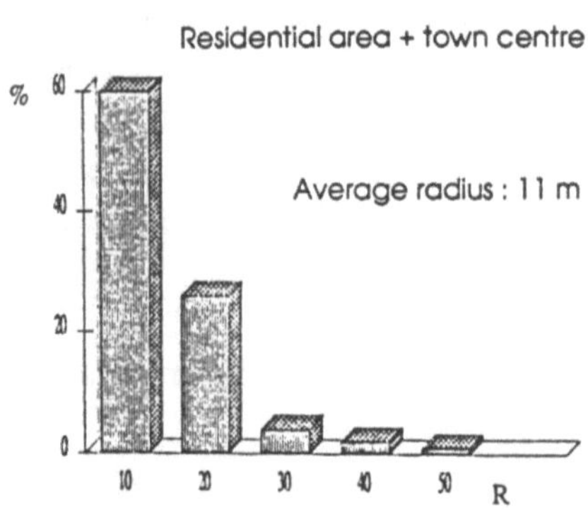

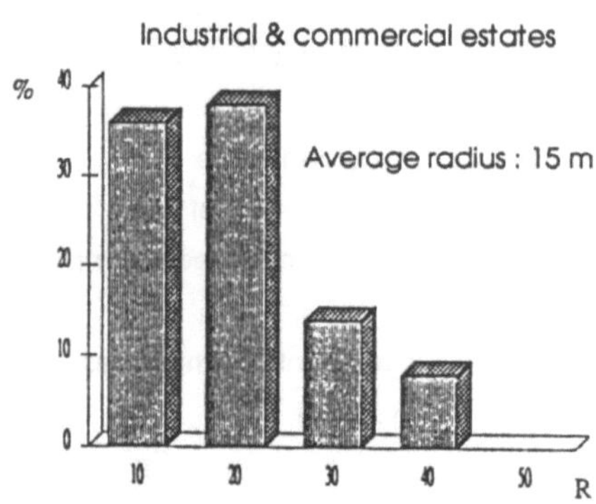

2.3.3 Lighting and site

Over 522 roundabouts, 452 are lighted, i.e. 87%. But although most roundabouts are lighted in towns, 10% are not lighted in the peri-urban area and almost 70% are not in open countryside.

2.4 Analysis according to number of branches

2.4.1 Number of branches and site

In open countryside, industrial and commercial estates, hardly any roundabouts with 3 or 4 branches are found. Roundabouts with 5 or more branches are virtually all urban or peri-urban. The few rare 2-branch roundabouts (public highways) are found in residential areas.

2.4.2 Accidents and number of branches

The 78 accidents surveyed for 1988 over the 522 roundabouts, are distributed as follows :

Number of branches	2	3	4	5	6 and +
Number of accidents	0	14	40	17	7
Accidents for 100 roundabouts	0	12.9	13.3	19.5	25.9
Average traffic (veh/day)	7500	8000	9000	10500	11500

The relative number of accidents increases with the number of branches, but so does AMDT.

2.5 Analysis according to size of radius

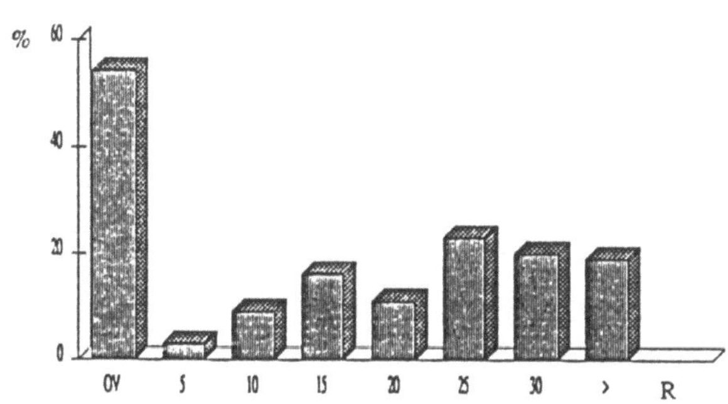

A rate of 54 accidents for 100 "non-round" roundabouts, 13 accidents on average for 100 roundabouts with a regular geometry. The conclusion is obvious : we must avoid building oval-shaped roundabouts and modify these particularly dangerous roundabouts wherever possible.

2.6 Analysis according to traffic

2.6.1 Traffic and locations

In open countryside we find very few roundabouts supporting more than 15 000 vehicles per day and none supporting over 25 000 vehicles per day.

In the peri-urban milieu, many roundabouts support high traffic, and this is where the majority of those catering for a throughput of more than 30 000 vehicles per day are located.

In residential areas, the great majority see less than 10 000 vehicles per day, but a few situated on major traffic routes exceed 20 000 vehicles per day.

In town centres, a distinction should be made between large and small urban centres : traffic depends on the size of the town.

In industrial estates, there is little heavy traffic, but there are specific peak times during the week.

In commercial estates, there is also little heavy traffic, but at specific times in the evenings and above all on Saturdays.

2.6.2 Traffic and accidents

Examination of the time history is clear : there is a direct relation between traffic level and accidents, and this outside of any other data (site, radius, etc).

Traffic and accidents for 100 roundabouts

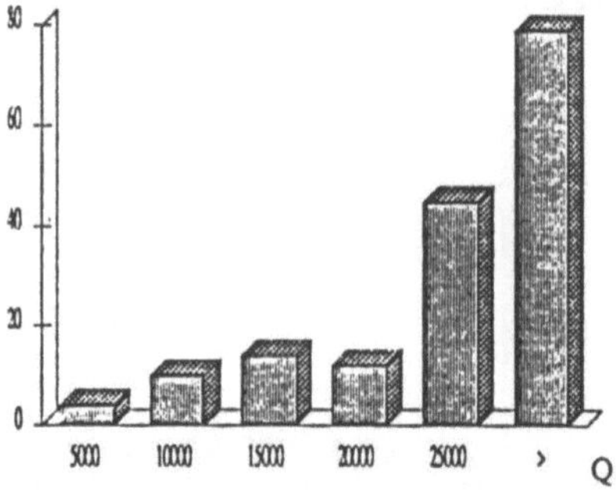

2.7 Summary on accidents

Among the 17 roundabouts where at least two accidents occurred in 1988, we note that :

- 8 were built before or in 1980
- 6 see a throughput of over 25 000 vehicles per day
- 5 are new roundabouts
- 4 are oval in shape
- 2 have 6 branches and are situated in the peri-urban milieu.

The roundabouts where most accidents occurred in 1988 (3 or 4 accidents) cumulate 2 or 3 of these unfavourable elements.

3 - ACCIDENTS ON URBAN ROUNDABOUTS

Based on personal injury accidents in 1988 recorded on the 458 urban roundabouts of the 522 roundabouts studied in all, we found :

. 0.15 accident per year and per crossroads
. 4.45 accidents per 10^8 vehicles throughput
. 0.19 fatal accident per 10^8 vehicles throughput
. 1.57 killed and seriously injured per 10^8 vehicles throughput

No statistic enables comparison with urban crossroads overall, except as regards gravity. Based on the national statistics of SETRA on accidents in 1989, we find on crossroads overall, in the urban milieu :

- 1.8 killed and 19.3 seriously injured for 100 accidents

whereas for the roundabouts analysed within the context of the study, the numbers amounted to :

- 4.3 killed and 31.0 seriously injured for 100 accidents.

These latter figures must be viewed with great caution, the sample of 71 accidents from the study remaining low, but the tendency is confirmed by interpretation of the 202 accidents mentioned in the introduction.

We should not conclude that roundabouts are more dangerous than the other urban crossroads. In fact, personal injuries are less in number on roundabouts, but more serious, for no doubt they more frequently involve that particular risk category : two wheel vehicles.

3.1 Comparison with light-controlled crossroads

The following results come from 15 towns in the West of France, where comparison was made between accidents at light-controlled crossroads and at roundabouts in 1988.

It is true that although traffic statistics, always important in weighting the raw results, are lacking, it may be thought that light-controlled crossroads and roundabouts correspond to similar traffic levels. Indeed, many roundabouts are previous light-controlled crossroads that have been transformed.

	LC crossroads	Roundabouts
Number of crossroads	1 238	179
Number of personal injuries	794	59
Number of accidents involving two-wheel vehicles	278	28
Accidents/year/crossroads	0.64	0.33
2-wheel vehicles/year/crossroads	0.23	0.13
Accidents to 2-wheel vehicles per 100 accidents	35.0	40.7
Serious accidents/year/ crossroads	0.14	0.089
Serious accidents to 2-wheel vehicles/year/crossroads	0.06	0.045
Serious accidents/100 accidents	21.9	27.1
Serious accidents to 2-wheel vehicles/100 accidents to a 2-wheel vehicle	27.0	33.3

The results are clear :

- There are twice as many accidents at light-controlled crossroads than on roundabouts;

- Two-wheel vehicles are involved in accidents more often (+ 77%) at light-controlled crossroads than on a roundabout;
- People are more frequently killed and seriously injured per accident (+ 25%) on roundabouts than on light-controlled crossroads;
- Proportionally, two-wheel vehicle users are more often involved in accidents (+ 16%) on roundabouts than at a light-controlled crossroads and the consequences of such an accident are more serious : 27 seriously injured and killed for 100 accidents on light-controlled crossroads involving a two-wheel vehicle, compared with 33.3 injured and killed per 100 accidents on a roundabout involving a two-wheel vehicle.

3.2 Users involved in personal injury accidents

It was observed that accidents were often more serious on roundabouts than on urban crossroads overall. The following comparison is based on the analysis of 202 urban accidents and provides a few explanatory details.

	Accident rate per type of users	
	All crossroads	**Roundabouts (*)**
Pedestrians	6.3	5.6
Bicycles	3.7	7.3
Mopeds	11.7	16.9
Motor cycles	7.4	4.8
Cars	65.7	61.2
Utility vehicles	2.0	0.6
Heavy goods vehicles	2.0	3.0
Bus/Coach	0.8	0.6
Miscellaneous	0.4	0

(*) Based on the file of the 202 accidents

Analysis of the figures highlights the fact that the percentage of accidents concerning cycles and mopeds increases from 15.4% for urban crossroads overall to 24.2 % for roundabouts, i.e. a relative increase of almost 60%.

3.3 Accident typology

The 202 accidents surveyed can be divided into 16 types which represent virtually all cases imaginable in the urban milieu. These types do not concern the same categories of users in the same way, as we will see below :

Number	Type of accidents	No accidents recorded	(%)
1	Refusal of priority on entry	74	37 %
2	Loss of control on entry	23	11 %
3	Loss of control on ring	33	16 %
4	Loss of control on exit	5	2 %
5	Head-on collision on exit	5	2 %
6	Rear collision on entry	15	7 %
7	Rear collision on ring	1	-
8	Rear collision on exit	2	1 %
9	Shear movement on exit	12	6 %
10	Incorrect overtaking on entry	2	1 %
11	Incorrect crossover on ring	5	2 %
12	Incorrect overtaking on exit	2	1 %
13	Travel in wrong direction on ring	2	1 %
14	Pedestrians on pedestrian crossing	12	6 %
15	Pedestrians on ring	7	3 %
16	Pedestrian not on pedestrian crossing	2	1 %
Total :		202	100 %

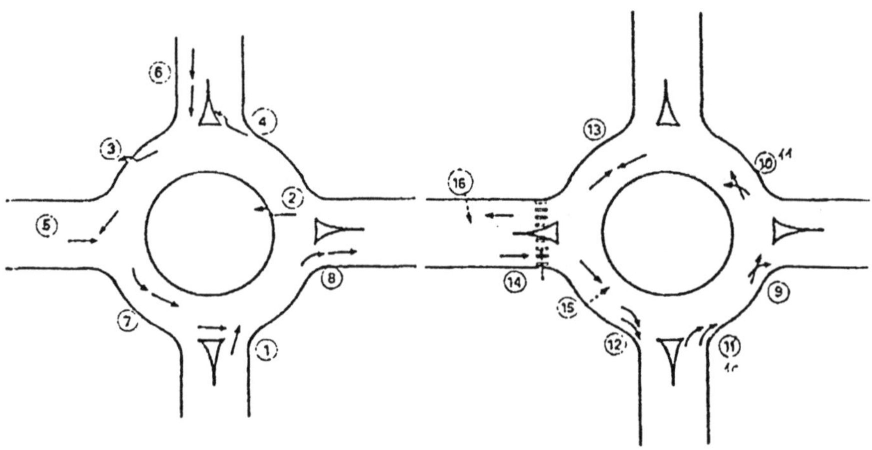

3.4 Distribution per category of users

The table below reads as follows :

- "at least 1 light vehicle is involved in the 66 refusals of priority on entering the roundabout"

Number	Type of accidents	Pedestrians	Bicycle	Mopeds	Motor cycles	Light vehicles	HGV	Public transport
1	Refusal of priority on entry		16	27	5	66	4	2
2	Loss of control on entry				4	18	1	
3	Loss of control of ring			17	5	10	1	
4	Loss of control on exit				1	4		
5	Head-on collision on exit					5	1	
6	Rear collision on entry			1		15	1	
7	Rear collision on ring					1		
8	Rear collision on exit				1	2		
9	Shear movement on exit		3	5	1	12		
10	Incorrect overtaking on entry		1	1		2		
11	Incorrect crossover on ring			3		5		
12	Incorrect overtaking on exit		2				2	
13	Travel in wrong direction on ring			1		1	1	
14	Pedestrians on pedestrian crossing	11	1	1		11		
15	Pedestrians on ring	7				7		5
16	Pedestrian not on pedestrian crossing	2				2		5

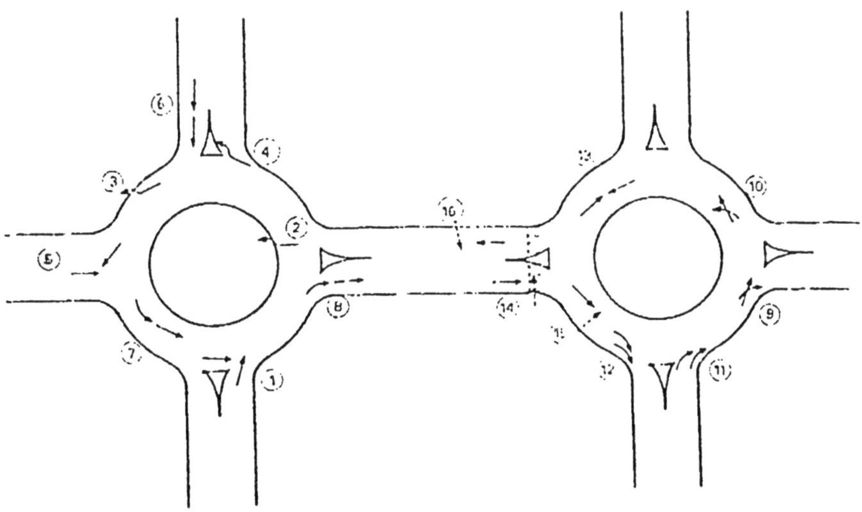

4 - ACCIDENTS INVOLVING BICYCLES AND MOPEDS ON URBAN ROUNDABOUTS

Over the sample of 202 accidents studied, 85 involve cyclists and moped drivers. These accidents were categorised per type of accident, separately for cyclists and mopeds. Elements common to both were also sought.

Number	Different types of accidents	Bicycles	Mopeds	(%)
1	Priority refused to bicycle or moped	16	27	50.6
2	Priority refused by bicycle or moped	2	4	7.0
3	Bicycle or moped hit kerb on entry	1	1	2.4
4	Bicycle or moped hit kerb on exit	2		2.4
5	Shear movement by light vehicle on exit	3		7.0
6	Shear movement by bicycle or moped on exit		2	2.4
7	Crossover on ring		3	3.5
8	Loss of control on ring		16	18.8
9	Miscellaneous	1	4	5.9

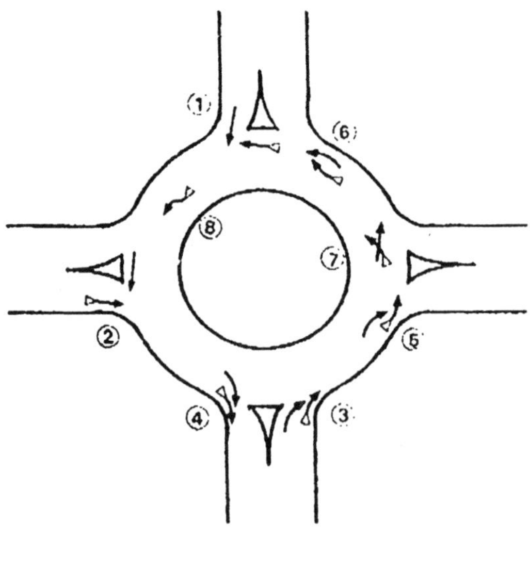

4.1 Priority refused to bicycle or moped

With 43 accidents, this is the most frequent type of accident involving a bicycle or moped. Forty-one cars and 2 heavy goods vehicles refused priority to 16 bicycles and 27 mopeds.

Examination of the distribution per category of radius provided interesting results :

	oval	<5m	5-10m	10-15m	15-20m	20-30m	>30m
(A) % roundabout	4.6	11.6	26.8	18.3	13.1	14.4	11.2
(B) % refusal priority on entry	19.0	4.8	11.9	7.1	14.3	9.5	33.4
Coefficient (B)(A)	4.13	0.41	0.44	0.39	1.09	0.66	2.98

Two categories stand out distinctly : oval shaped roundabouts and the big roundabouts. We know that the first category is the most prone to accidents, partly for two-wheel vehicles. As for the large diameter roundabouts, nothing could have led to expect such a high coefficient. The main common point between these two categories is certainly the possibility of rapid entry, for the first with tangential entries, for the second with a wide entry radius. At least 32 of the 43 accidents occurred on two-lane entries (6 m or more) : this makes us think of entries favouring speed, but also congested situations where 2 vehicles waiting ahead mutually mask the visibility. More precise data than those already collected might perhaps have told us more, however, examination of the traffic may guide our thinking, categorising accidents per traffic section and making a distinction between peak and other periods.

	<10,000 veh/day	10-20,000 veh/day	20-30,000 veh/day	>30,000 veh/day
(A) % roundabouts	63.4	25.5	6.8	4.3
(B) % accidents refusing priority to 2-wheel vehicles	27.9	27.9	13.9	30.2
Coefficient (B)(A)	0.44	1.09	2.04	7.02
Rate during peak periods	66.6 %	50%	50%	61.5%

As could have been expected, the risk of refusal of priority to two-wheel vehicles increases greatly with traffic (coefficient B/A). More than half of these accidents occurred during peak periods.

We point out that 8 accidents occurred at night time, but all roundabouts are lighted in the urban milieu.

4.2 Refusal of priority by two-wheel vehicles

In this fairly rare category of accidents (6 only), only in one case did a two-wheel vehicle truly refuse priority, to a light vehicle travelling round the roundabout. One accident must be discarded from the analysis, the light vehicle having knocked over a two-wheel vehicle trying to escape.

For the remaining accidents, one common point appears instantly : they occurred on roundabouts with very small radii : 3.50 to 7.00 m. It is fully possible that at the time when the two-wheel vehicle decided to enter the roundabout, the vehicle arriving on its left had not yet entered and was possibly even well ahead of its "give way" line.

The opposite is fully possible for the moped accidents recorded on small radius roundabouts in the previous paragraph.

4.3 Two-wheel vehicles wedged against the outside kerbs on entry

Personal injury is fairly rare, yet this type of accident is sufficiently characteristic to mention : a light vehicle and a two wheel vehicle enter the roundabout at the same time to turn right, the trajectory of the light vehicle then hedges the two-wheel vehicle in to the right against the kerbs, causing it to overturn.

On roundabouts with a small entry radius, tyre traces are frequently seen on the outside kerbs at entries : vehicles turning right keep as close in as they can. Two-wheel vehicles should be dissuaded from keeping to these kerbsides alongside of vehicles. To achieve this, the entry widths should be reduced ... to values incompatible with the radius of gyration of heavy goods vehicles.

One solution would consist of isolating the entry trajectory of the two-wheel vehicles tangibly from that of other vehicles, in an overall arrangement with a cycle lane. This would also be beneficial in reabsorbing accidents of the same type on exit.

4.4 Shear movement of two wheel vehicles on leaving the roundabout

This type of accident is well known : a two wheel vehicle travelling on the outside of the ring is knocked over by a vehicle leaving the roundabout, or vice versa. Poor assessment of relative speeds is often incriminated, but let us look closer at this.

Large radii (30 to 40 m), wide rings (3 lanes, marked or not), wide exits (2 lanes), heavy traffic (20 to 45,000 vehicles per day), peak period : these are the characteristics of 7 out of 8 accidents of this type.

4.5 Crossing over on the ring

Three accidents between light vehicles and mopeds crossing over on the ring : the 3 roundabouts concerned had 2 lane entrues and wide rings (11 and 12 m) or variable rings (7 to 9 m).

4.6 Loss of control on the ring

Sixteen accidents to mopeds come into this category. The geometric characteristics of the roundabouts concerned are :

- 3 to 40 m radius
- 5,000 to 45,000 vehicles per day
- interior or exterior slope
- town centre, industrial or peri-urban area

In fact, no element can be highlighted other than the speed of the moped during a turning motion.

4.7 Miscellaneous accidents to two-wheel vehicles

There remain 5 accidents to two-wheel vehicles :

- rear collision of a moped with a light vehicle at a standstill (more frequent accident but often damage only)
- loss of control on cycle lane
- travelling in opposite direction on the ring (in an industrial estate, on a low traffic roundabout)

- moped hit by a cyclist whilst crossing over a pedestrian crossing
- refusal of priority by a cyclist on a cycle lane.

5 - PEDESTRIAN ACCIDENTS

Out of the sample of 202 accidents studied on the urban roundabouts, 20 involve pedestrians. They caused 13 people slight injury and 10 serious injury (no fatal accidents).

It should be specified that this last accident statistic is of no value, since it is not exhaustive. However, it enables a typological study intended to highlight some common elements between roundabout characteristics and types of accidents.

These accidents break down as follows :

- 8 on pedestrian crossings at an entry to a roundabout

- 1 on pedestrian crossing without a central island

- 2 on pedestrian crossings at a roundabout exit (in both cases, a wide, fast exit)

- 7 on the ring :
  of these, a distinction must be made between the pedestrian staggering on the ring, old people who are crossing over to the pedestrian arrangement of the central place, and pedestrians who do not bother to use a pedestrian crossing too far from their route which is greatly channelled by barriers and hedges : the pedestrian walking on the ring could no longer get off it, even on 20 cm of pavement.

- 2 crossings after the roundabout exit.

Two accidents took place on roundabouts with a light-controlled pedestrian crossing, but no relation can be established with this equipment.

5.1 Accidents on entries

If one looks for common points between these 8 accidents, we note that :

- 7 took place on the 2-lane entries,
- 7 occurred on pedestrian crossings adjoining the "give way" line
- 5 (at least) involved children or old people
- 6 took place on roundabouts with more than 20,000 vehicles per day
- 5 occurred in the town centre
- 4 took place in the rush hour

Evolution of Design Rules for Urban Roundabouts in France

F. ALPHAND - Engineer with CETUR (Centre d'Etudes des Transports Urbains), the Urban Transport Study Centre in Paris

U. NOELLE - Engineer with CETE Méditerranée (Aix-en-Provence) - the Engineering Technical Studies Centre for the South of France

B. GUICHET - Higher Technician with CETE de l'Ouest (Nantes) - Engineering Technical Studies Centre for West of France

INTRODUCTION

Roundabouts were introduced in France in 1972 and incorporated into the French highway code in 1984. This defines roundabouts as being the "squares" and intersections satisfying the two following conditions :

- To consist of a central reservation physically impossible to cross, surrounded by a pavement open only to one-way traffic from the right into which traffic enters from different roads.

- To be announced by a specific signpost indicating to the driver that he is coming up to a roundabout and must therefore, by law, whatever the category of road he is just about to leave, give way to users already engaged on the circular pavement.

Advance signing.

Position signal.

After a period of mistrust and observation, roundabouts are currently meeting with increasing success in our towns : they can be found more or less everywhere, in different types of contexts : town centre, suburbs, peri-urban area, industrial and commercial estates, etc.
Thus, the number of roundabouts built over a single year, in 1990, can be estimated at approximately 1000.

This somewhat anarchic proliferation led Ministry of Infrastructure departments such as SETRA (the Roads and Highways Technical Studies department) for open countryside and CETUR (the Urban Transport Study Centre) for the urban milieu, to undertake specific studies which led to the publication of technical guides (in 1984 for inter-urban roundabouts, and 1988 for urban roundabouts).

But changes are occurring very fast in the domain of roundabouts, which has caused CETUR to undertake new works over the past two years, on the basis of observations made on existing intersections, or learnt from the experience of our European neighbours. These studies have essentially covered :

- The definition of a "domain of use" of roundabouts

- Changes in geometric design rules aimed at improving safety

- Research to reduce the size of roundabouts, the ultimate end result being our current experimentation in mini-roundabouts, already in widespread use in England.

We propose, here, to present the main results of these works which will be published in France in the coming months : general guide for urban crossroads, technical guide on mini-roundabouts.

1 - Definition of the domain of use of roundabouts installed in the urban milieu

As we emphasised in the introduction, since 1984 there has been a veritable infatuation for roundabouts in many French towns. Many technicians tend to consider this type of crossroads a panacea to solve all safety, capacity or environmental problems. The forthcoming CETUR publications will attempt therefore to provide a number of criteria to enable determining, according to traffic (vehicles, pedestrians, two-wheel vehicles, public transport) urban context, geometry, the management method and function of the different roadways, whether a roundabout is truly the right solution for the problem posed. We therefore established a list of elements favourable and unfavourable to the use of a roundabout. These elements are not strict rules : it is not because one or other point is favourable that the only envisageable solution is a roundabout. However, an accumulation of unfavourable elements must lead to great caution if building a roundabout is envisaged.

1.1 Favourable elements

• Intersections with more than 4 branches

Immediately there are more than four branches to a crossroads, control by traffic lights becomes problematic. The "roundabout" solution is then particularly well suited to manage the many conflicts. This does not mean roundabouts must be implemented on all crossroads with more than 4 branches : the choice also depends on the geometry (horizontal alignment, vertical alignment, cross-section) of the branches, on whether it is possible to break down the complex crossroads into 2 simple crossroads, on traffic and on pedestrian and two-wheel vehicle-related problems.

• Treatment of transition zones

The roundabout is well suited to the treatment of transition zones : transition between two districts of a different nature or transition between open countryside and urban milieu (roundabouts at town entries). Indeed, its geometry is useful in marking a place where a change of environment occurs, so that users adapt their behaviour accordingly. On entering an urban centre, for example, a roundabout can indicate to drivers that they are entering a built-up area and must therefore reduce speed : it acts as a town gateway.

• Intent to reduce speed along a trunk road

The roundabout can be effective in attempting to reduce speeds on a trunk road. However, it is illusory to think that a single roundabout can reduce speeds other than as a "spot measure" : trunk roads must be treated globally, the roundabout being considered as only one of many tools to reduce speed (reduction of horizontal alignment, barriers, sleeping policemen, visual effects, etc).

• Treatment of accident "black spots"

Generally speaking, personal injuries are much less frequent on roundabouts than on T-junctions or crossroads, whether controlled by lights or not. Roundabouts can therefore constitute an advantageous solution in rearranging crossroads where accidents are found to occur repeatedly.

In some cases, however, when building a roundabout one must at the same time improve conditions on the different approach roads converging upon it. When safety problems are due to insufficient visibility or a poor vertical alignment (for example, crossroads situated at a high point or on too steep a gradient) building a roundabout will not necessarily solve these problems unless accompanied by a modification of the characteristics of the roadway : it can sometimes even worsen the problems.

• Considerable fluctuations in daily or seasonal traffic

In the urban milieu, crossroads are sometimes heavily congested in terms of traffic for very short periods of the day, and little used the rest of the time. In this case, the presence of traffic lights can be too restrictive during off-peak periods and not appear credible to users who therefore infringe regulations, (the French are by nature not always very well disciplined!). Roundabouts are well suited to this type of situation on condition they are capable of absorbing rush hour traffic.

• Many left or right turns

Roundabouts often allow better throughput of left-hand or right-hand turns than crossroads with or without lights, since there is no hierarchiy of traffic flow. Nevertheless, in all cases, it is necessary to calculate the capacity, particularly to ensure that other movements are not over penalised.

• Landscaping

Roundabouts lend themselves well to landscaping or a more general aesthetic treatment of a place. However, this does not mean systematically using a roundabout whenever one wants to improve appearance to mark out an intersection on the road grid!

1.2 Unfavourable elements

* Steep gradients

All types of crossroads pose problems when located mid-way on a steep gradient. However, the most serious accidents observed at roundabouts are loss of control on entry. The frequency of this type of accident will necessarily be increased by the presence of a gradient of more than 4% leading up to the crossroads, particularly on transit routes used by a high number of heavy goods vehicles.

* Many pedestrian crossings and 2-lane entries

Immediately traffic throughput requires entries wider than 4 metres, we consider that pedestrian safety and comfort in crossing the roads can no longer be guaranteed in good conditions on a roundabout. The roundabout solution is therefore questionable in this case.

* Large number of heavy goods or public transport vehicles

Accidents due to heavy goods vehicles overturning are fairly frequent on some types of roundabouts. In addition, the presence of a large number of heavy goods lorries or buses necessitates larger geometric characteristics which may be disfavourable to pedestrians.

In the matter of public transport, roundabouts are source of discomfort for passengers and also, contrary to traffic light controlled crossroads, cannot be used to help bus throughput.

* Two-wheel vehicle traffic in dense traffic comprising a high proportion of HGVs

The combination of these factors is not favourable to roundabouts. Indeed :

- Two-wheel vehicles and HGVs are poor companions around a roundabout.

- The dimensions that have to be provided if the roundabout is to offer good turning conditions for the HGVs, mean prejudicially longer travel distances for cyclists.

• Very poor balance of traffic between a main road and a secondary road

When traffic entering from the most densest secondary road is less than 25% of incoming main road traffic, then the flow of traffic from the secondary road is over-penalised in a roundabout. Another solution must be envisaged.

Whatever this may be, in all cases it is recommended to check capacity, and obviously when a "roundabout" solution proves to be saturated, it must be discarded (except where saturation is only for brief periods, when it can be tolerated).

• Proximity of a level crossing or traffic light controlled crossroads

In this type of case, if there is a risk of queues tailing back to the roundabout, this solution must be discarded in favour of a traffic light controlled crossroad which can be coordinated with the neighbouring crossroads.

• Coordinated trunk road

A roundabout situated on a road regulated by a green wave disturbs the management of the green wave. However, where trunk roads are too long, it can be a useful geometric element to divide the trunk road into two sections.

• Site or layout of branches necessitating an oval shape

On average, there are twice as many accidents on roundabouts of accentuated oval shape than on round shaped roundabouts. The oval type of geometry should therefore be avoided.

• Shops which encourage parking on the roundabout

It is essential that the roundabout be free from all parking. Therefore avoid locating roundabouts in sites where shops are likely to encourage illicit parking around the roundabout.

2 - Evolution of design criteria

The current evolution of the geometric characteristics of urban roundabouts at crossroads in France, characterised by the search for small dimensions, is towards experimentation with "mini-roundabouts" with an uncontrolled central island.

Recent observations indeed show that reducing the characteristics of the diameter of the central island in particular is not prejudicial to its performances, rather the opposite, on condition that certain basic rules are respected, such as :

- ensuring that the roundabout remains easily visible, in particular by visual treatment of the central island and in the design of the longitudinal section of the approaching roadways.

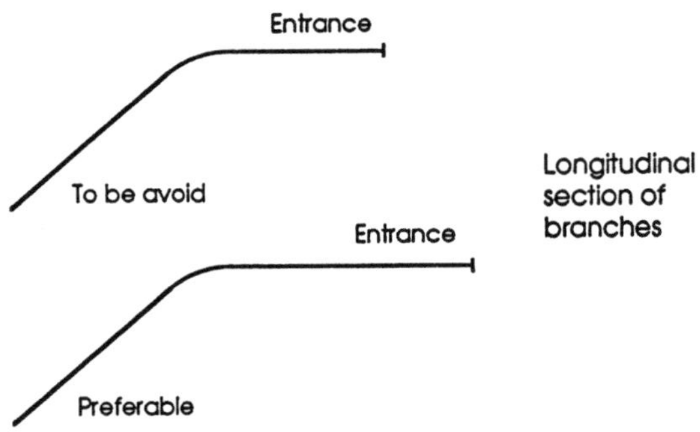

- Unless dictated by the amount of traffic, avoid over-wide entries, avoid over-wide entry radii (> 20m) and avoid trajectories which are too tangential across the roundabout, as they encourage high speeds. The following criteria can be used : to avoid drivers going round the roundabout at a speed of over 50 km/h, preferably none of the trajectories should have a radius over 100 m.

- for small sized roundabouts, check that there is no problem with turning, particularly for heavy goods vehicles and buses. On this subject, CETUR sells a software called "GIRATION" which is used to draft gyration drawings (not only for roundabouts, in fact) and provides precious help in solving this type of problem since it enables constructing curve-against-curve type trajectories.

- study carefully the problems posed by the pedestrian crossings : in our opinion, the positioning of pedestrian crossings can be envisaged in different ways. The best way is to place them 4 to 5 metres before the "give way" line, i.e. behind the vehicle at standstill. But bayonet crossings can also be designed.

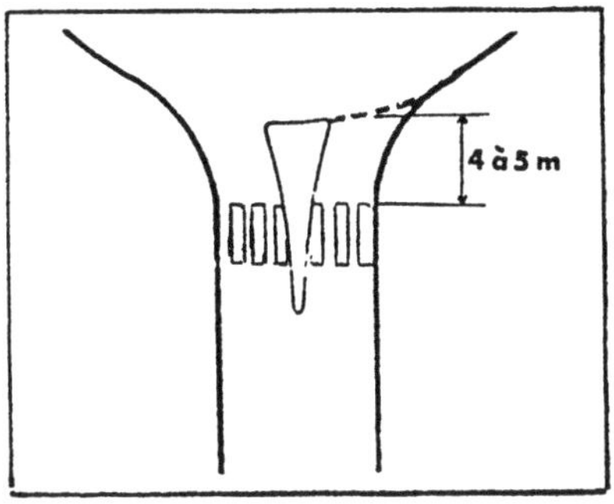

Conventional pedestrian crossing.

Bayonet pedestrian crossing.

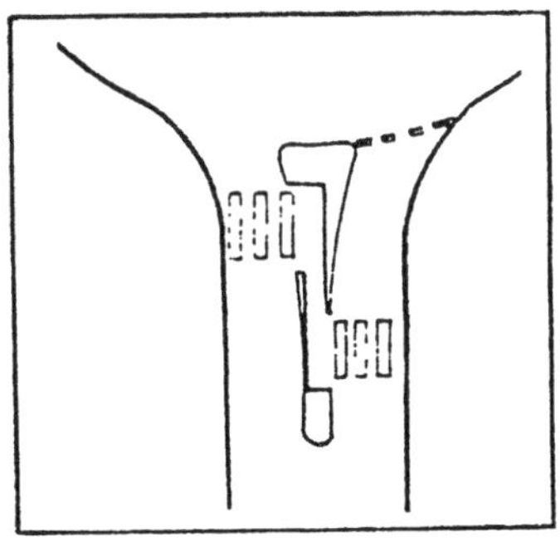

- On the basis of a reliable traffic study, check that the roundabout has sufficient reserve capacity. We have various methods, in France, for making this kind of calculation :

  The GIRCAR software available from CETE Méditerranée is based on the Harders method (the capacity of 2-lane entries is increased by a coefficient of 1.4).

  The GIRABASE software available from the CETE of Nantes also uses the Harders method, but takes geometric characteristics into account with greater accuracy, based on many measurements made on existing crossroads.

The method set out in the CETUR roundabouts guide "Guide carrefours giratoires" is a simplified method, this time based on in situ measurements. It is summarised in the block diagram below.

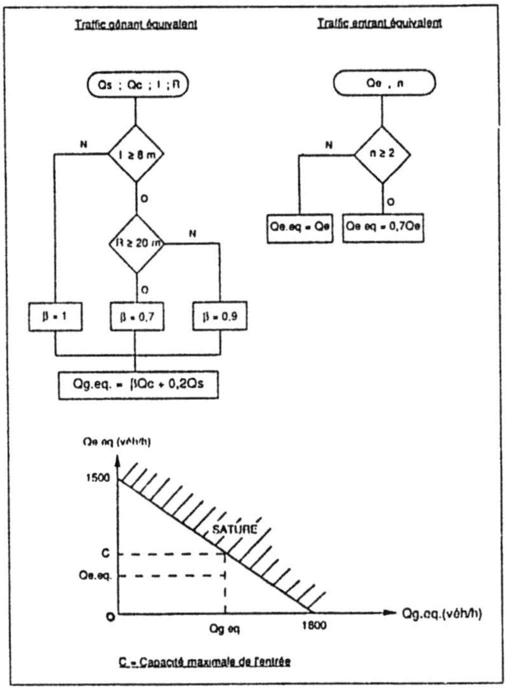

Capacity of a roundabout.

We are currently carrying out observation works intended to improve the accuracy of all of these methods which in some cases appear too pessimistic : perhaps this is due to the improvement in French motorists' behaviour on the roundabouts.

These design rules are common to all types of roundabouts. However we thought it necessary to specify them in relation to the size of the crossroads and to the urban context in which it is found. This led us to develop a classification of the roundabouts as summarised in the table below.

DESIGNATION	CONVENTIONAL URBAN	SEMI-CONTROLLED ISLAND	MINI-ROUNDABOUTS	ENTRIES TO BUILT-UP AREAS
Field of application	All sectors	Urban zones other than main trunk roads	Experimental (with DSCR authorisation)	Crossroads - ring road x axial route
GEOMETRY Radius of hard island Crossing strip Outside radius Width of ring Width of entry Width of traffic island	 7 to 20 m 0 to 2 m 15 to 30 m 7 to 12 m 1 to 3 lanes > or = 2.5 m	 3.5 to 5 m 1.5 to 2 m 11 to 15 m 6 to 8 m 1 lane > or = 1.5 m	 0 0.75 to 2 m 7 to 11 m 6 to 9 m 1 lane 0 to 2.5 m	 10to 14 m 0 to 2 m > or = 18 m > or = 8m 1 to 3 lanes > or = 4 m
Good level of traffic	< or = 2500 PCU*/h (+ if 2-lane entries) *(Passenger car unit)	< or = 2000 PCU/h	< or = 1500 PCU/h	< or = 3000 PCU/h (+ if 2-lane entries)
Pedestrian arrangement	Pedestrian crossings with central islands	Pedestrian crossings with central islands	Pedestrian crossings	Pedestrian crossings with central islands
Two-wheel vehicule arrangements	Continuity of cycle lane or cycle track	None	None	possibly a cycle lane or track continuity
Public transport	Avoid the smallest radii on continuous public transport lines	To be avoid on public transport lines	Not recommended on public transport lines	
Lighting	Peripheral or central post	Peripheral	Peripheral	Peripheral

* Roundabouts at entries into built-up areas (category 4) generally have the function of marking a transition between the open countryside and the urban milieu, and to provide interchanges with trunk roads having dense traffic. In view of the function and the often high speeds seen at approaches to the roundabouts, a fairly big central island is recommended (up to a radius of 40 m).

* Conventional urban roundabouts (category 1) are conventional roundabouts that can be situated in all types of urbanised zones. A few years ago, a minimum central island radius of approximately 15 m was recommended. Experience has shown that these radii can be reduced to 7 m (with outside radius 15 m) without posing a problem of safety or traffic flow. Traffic throughput is ideal when the sum of entering traffic is less than 2500 vehicles per hour. Beyond this threshold, the crossroads can still function if the distribution of turning movements is favourable (capacity must be checked).

* Semi-controlled roundabouts are a new feature : they are as yet few and far between in France, yet they fully conform to existing regulations.

These are very small sized roundabouts which require a site of minimum outside radius 11 m. They have a central island that cannot be crossed, surrounded by a non-controlled surrounding strip intended to facilitate turning for HGVs : the controlled part is round with a surrounding kerb, and its radius can vary from 3.5 to 5 m.

The non-controlled strip must be about 2 m wide and sloping outwards, slope being greater at the superelevation of the pavement ring (approx 4% slope). To ensure that people understand that this strip is only for occasional use, it should preferably be designed in a material different from that of the pavement (paving blocks, for example).

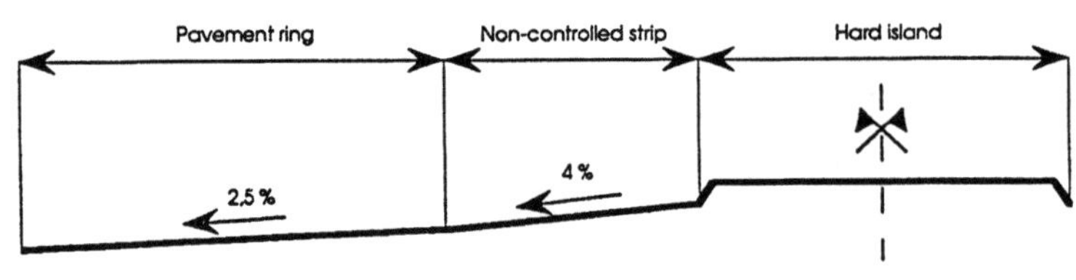

Sectional view of semi-controlled roundabout

This type of crossroads is kept for urban crossroads having a limited site outside of heavy traffic trunk roads : to be avoided when the sum of entering traffic is higher than 2000 PCU/h, when there is a high percentage of HGVs or on regular public transport routes.

For a fairly low cost, it enables solving problems of traffic flow or safety at existing crossroads. In addition, it can be incorporated into operations to enhance the urban environment.

* Mini-roundabouts are, at the current time, an experimental device, which is why we deal with them in a separate chapter.

3 - Experiment with mini-roundabouts

At the current time, the French highway code indeed stipulates that it must be impossible the cross over the central island of a roundabout. Nevertheless, similar to what the English do, we undertook, slightly over a year ago, to experiment with small roundabouts comprising a central island all of which can be driven over, in order to assess what this device could provide (in terms of safety and traffic flow) when used to replace small T or Y junctions whether controlled by lights or not. If the experiments prove conclusive in the short and medium term, we may envisage modifying the Highway Code.

Our procedure was as follows :

- Definition of specifications for experimentation
- Search for four sites that fitted these specifications
- Development of an experiment plan
- Implementation of the experiments
- Analysis of results

3.1 Specifications for experimentation

These specifications set out the criteria the selected sites must satisfy, and the manner of their arrangement.

- The selection criteria are as follows :

  . the total site of the crossroads must be small (diameter of circle inscribed within site less than 21 m)

  . the sum of entering traffic must be less than 1400 vehicles per hour in peak traffic.

  . the lanes must preferably be access lanes or possibly distribution lanes

  . the chosen crossroads must, in the way they operate at present, pose problems of safety, flow, organisation of turning movements, understanding on the behalf of users, or excessive speeds.

- Concerning the arrangement, it was decided in the first stage of experimentation, to modify the geometry at little as possible in order to evaluate the purely intrinsic effects of the mini-roundabout device. The arrangement therefore consisted of :

  . building a dome-shaped central island of 2 to 4.5 m diameter, slightly higher than the pavement (approximately 10 cm) installing signing to indicate to users entering the roundabout that they must give way to users already on the roundabout (traffic signs, road marks)

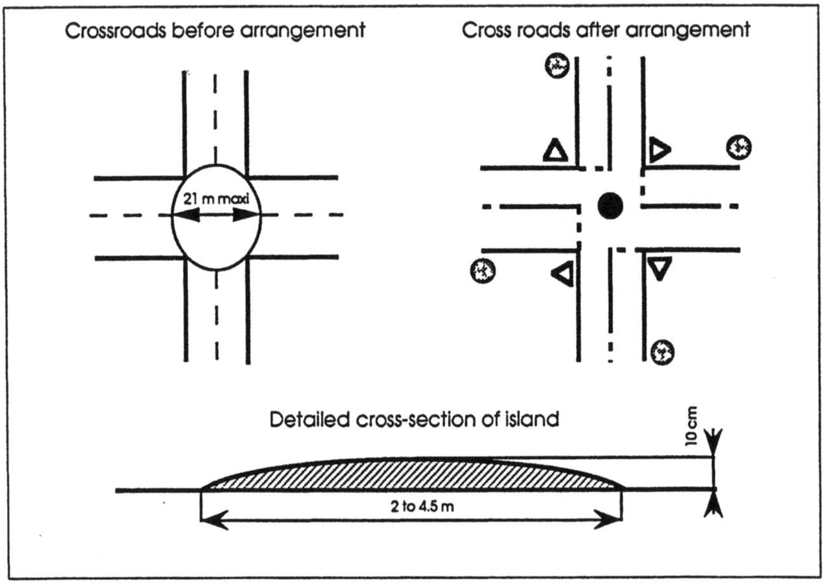

On the basis of the specifications, four sites were selected, one in Grenoble, one in Valence and two in the region of Nantes.

3.2 Evaluation of the arrangement

To set up an evaluation grid for the arrangements, we in advance attributed the mini-roundabouts with a number of objectives :

. to reduce conflicts;
. to reduce speed;
. to improve traffic flow;
. to improve safety and comfort for pedestrians and cyclists/motor cyclists
. to improve the urban environment

The purpose of the experimentation was to check whether these objectives were achieved. The means implemented were :

. Measurements of speed upstream and downstream of the crossroads

. Traffic counts

. Video observations

. In situ observations

. A survey carried out with the help of road managers, users and residents

All of these observations were done using the same methods BEFORE arrangement, TWO MONTHS AFTER arrangement, and SIX MONTHS AFTER, in order to be able to make objective comparison.

The four crossroads were arranged in March 90 and the "after" observations were made in September 90.

3.3 Results

At the time of writing this paper, detailed analysis of the four experiments is not yet complete. However, it will be presented when giving the paper orally.

According to the first results obtained, we can conclude that :

. after a brief period of adaptation, the device seems to be well understood and accepted by drivers,

. the new rule of priority is respected and the non-controlled island is used well by HGVs

. compared with a T-junction or conventional crossroads, the organisation of turning movements is simplified, therefore traffic flow is improved (even if these are not crossroads which see heavy traffic).

. On sites where the speeds were generally high, a noticeable reduction in speed is achieved by the mini-roundabout arrangement.

. As regards increased comfort for pedestrians, the improvement is not obvious, many of them feeling much safer on light-controlled crossroads.

. No personal injuries had been sustained on the four crossroads built in Spring 90.

There is therefore room for a degree of optimism with the regard to the future of mini-roundabouts in France, as this device is also very inexpensive.

Our next work will consist of detailed modifications to the four experimental crossroads (for example minor improvements to the geometry of some entries) and observation of the effects of these modifications on their operation.

In addition, experimentation will be extended to other sites, as there is great demand for this type of roundabout from the towns.

Case Study. Performance Effects of Changing a Traffic Signal Intersection to Roundabout

Paul G Höglund,
Department of Traffic and Transport Planning, Royal Institute of Technology, Stockholm, Sweden

Abstract

In order to study the possible positive effects of changing a traffic signal intersection to a roundabout, a survey was performed during autumn 1989 and spring 1990. The intersection has a central downtown location in the city of Jönköping, Sweden.

The survey was done as a before-after study and is primarily concentrating on differences in the vehicle time consumption when driving through the intersection. The survey is an intergrated part and a case study within a larger research poject aiming to create a Method and Model for Comparative Intersection Research.

The research methodology is based on the use of video pictures taken from the air (helicopter) or from a high located observation point. A special computer programme has been developed for facilitating the data evaluation process, giving a descriptive presentation of relevant traffic flow parameters.

The study gave very positive performance results for the roundabout compared to the traffic signal intersection, and the time saving was remarkable.

SUMMARY

For the studied direction within the intersection and the time of the day, i.e. morning peak hour traffic, the obtained result could be regarded very positive for the vehicle drivers. Also generally for the community the result could be considered positive and is tending to benefit the public at large.

Large time savings as well as a *reduced amount of stopping vehicles* and *reduced speed levels* were obtained. The average time saving per vehicle was 55 %, and the total amount of stopping vehicles was drastically reduced. The speed level was reduced with 13 %, which could mean an *expected reduction of the accidents* and injured persons by an estimated 25 %.

Because of the reduced waiting times and thus reduced driving times and by a significant less amount of stopping vehicles in combination with the the reduced speed level, a *gain in the reduction of detrimental air pollutions could also be expected.*

Finally it could be concluded, that further comparative studies and research within this field could give knowledge about, *which intersection design could be the most suitable for the environment.*

1. Performance Measures

In order to describe the performance of a traffic signal intersection and to make the comparison of posible differences, after the changes of traffic control system to a roundabout has been done, we need some performance measures. We are here using performance measures, which could give essential information regarding the traffic flow characteristics, such as the time consumption (t) (delay), the speed level (sector journey speed) (v) and the flow per time unit (q).Those traffic flow parameters will here be considered as suitable performance measures for this study.

The word time consumption is here used instead of delay for the comparisons. The only difference is, that delay is the extra driving time through the system, needed because of obstruction of the possible optimal driving time. The delay is thus the actual time consumption, driving through a sector or through the system, minus a reference value for driving time at desired speed without stopping or deceleration (retardation).

The variables time consumption, speed and flow, used as performance measures, will in various ways be related to the traffic flow intensity, in order to draw conclusions of possible changes of performance in the before- and after-study of a traffic signal intersection, which is rebuilt to a roundabout.

2. Location of the Site for the Survey

The here studied intersection is located in Jönköping city centre with heavy maximum peak hour traffic in the main directions, see fig. 2.0.1.The figure is showing the traffic streams in the maximum hour. The traffic signal intersection was having rather heavy delays, and the city therefore decided to change the traffic control system to a roundabout in order to try to reduce the unnecessary waiting times in the traffic signal system.

3. Physical Features of the Intersection

The before- and after-design of the studied intersection is shown in fig. 3.0.1 below. In this study only the effects of time consumption reduction is studied for the incoming traffic from the upper left part of the figure. Because of the intensity of the traffic flow in this direction, it could be considered as the main direction. In the before situation the traffic signal had separate phases for turning traffic and four long lane storage bays. One separate bay for respectively left and right turning traffic and two lane bays for straight through traffic.

In the after situation a roundabout with 36 m diameter was built, totally without traffic signal control. All four entrances to the roundabout are with four lanes. The studied entrance in the after situation was physically differently located than in the before situation. Because of a desire of different land use, the road was given a circular form in the direction towards the roundabout. The studied distance along the entrance to the intersection was 180 m in the before situation (to the stop line) and 240 m in the after situation (to the yield line). Thus the vehicles have to travel a 33%

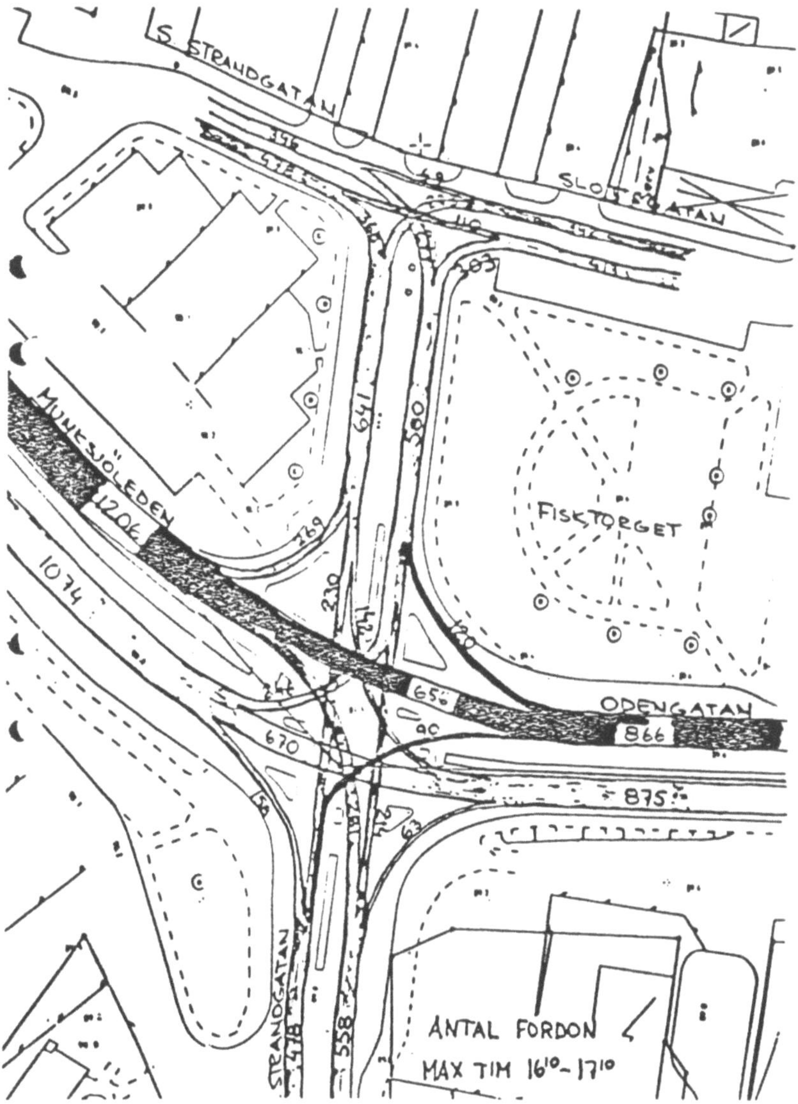

Fig. 2.0.1. Flow intensities during peak hour (16.10-17.10 hrs) in the studied intersection.

longer distance in the after situation from the enterring sector to the departing sector of the studied system. This change of physical design has nothing to do with the intersection design itself; it is purely because of different land use, and the longer travel distances must therefore be taken into consideration when comparing the time consumption results.

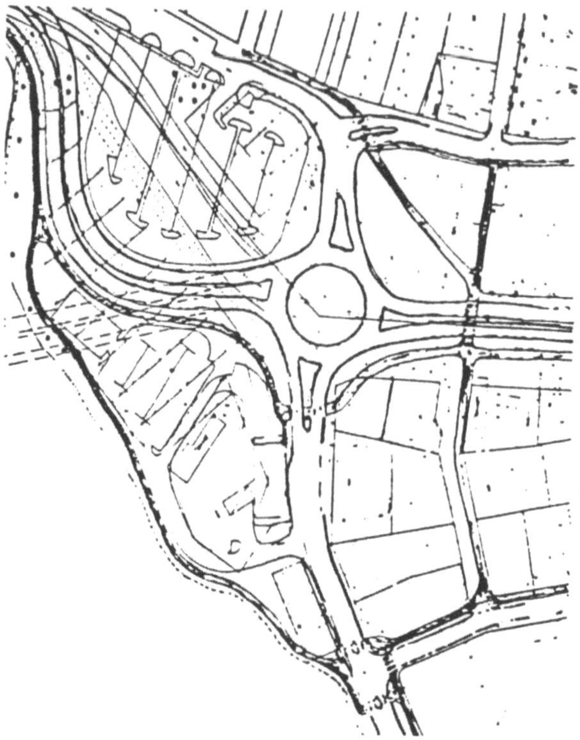

Fig. 3.0.1. Before- and after-design of the studied intersection.

4. Time for the Survey

In order to keep the time depending variables as constant as possible, the before- and after-measurements were tried to be performed at as similar circumstances as possible (wheather, day, time of the day, duration of survey time, contents of traffic flow etc.)

The before study was performed Thursday 89-08-31 in the morning during 07.30 and 08.00. The data used for this survey was collected between 07.37-07.45 hrs, thus during 8 minutes of observation time. The before study was made from a helicopter standing still (hovering) over the intersection.

The after study was performed Friday 90-05-04 during the same time in the morning. Collected survey data was here from 07.00 to 07.30 hrs during 30 minutes of observation time. Because of financial restraints the after study could only be made from a roof based observation of the traffic flow.

The construction of the roundabout was finnished in December 1989, and it had thus been in use five months before the after-study was done. We could therefore assume, that the road users had got accustomed to the roundabout, when the second part of the survey was performed.

5. Data Collection and Descriptive Presentation

5.1. General

All the raw data material and tables are not given within this report text, because of the desire to reduce the volume of the paper. The interested reader can obtain all this information direct from the author!

The extensive amount of data received through this survey technique and data handling could be used for many varying purposes. It enables detailed research of specific intersection design problems and micro studies of the traffic flow. It is also the aim of this paper, mainly describing a case study, to indicate the possibilities of future detailed studies by means of the *Method and Model for Comparative Intersection Research* (see reference /1/), which contains basic research methodology and data collection methods for field study purposes.

5.2. Collected raw data and raw data tables

After data processing and raw data handling, the evaluated raw data is given as tables containing all statistical parameters for the sectors of the studied entrance in the before and after situation. The total raw data tables are not presented in this report for reasons mentioned in 5.1 above.

5.3. Statistical description of raw data

Raw data for the performance parameter **t** is also given as consecutive values according to the incoming vehicles position in the traffic flow. Such data is available for each consecutive section in the studied direction. Fig. 5.3.1 below shows the time consumption values for the first driven sector (AB 0-1) in the studied main direction for the before-situation (with traffic signals). See method and description in reference /1/. In the figures each incoming vehicle time consumption value for the sectors is plotted in consecutive order along a time axis. In those figures it is easy to see the various patterns of time consumptions, and when specific delays occur.

As an example how this method can be used for studying the arrival process at the entrance sector, see fig. 5.3.2 below! This figure shows the arrival times in seconds (after the starting time of the for this method developed Stop Watch Computer Programme) to the section 0 (entrance sector) in consecutive orders. The figure shows a typical systematical disturbance of the traffic flow from a traffic signal placed upstream.

Finally there is also in the raw data material (not presented within this paper) given the regression analysis presentation of the time consumption's (performance parameter **t**) flow dependence by plotting all **t** values against given minute flow values **q** for all sector values. Fig. 5.3.3 below gives an example of the method. The figure shows the time consumption versus the flow for the first driven sector in the main traffic streem (AB 0-1). There is here no particular flow dependence for the driving times in the studied interval.

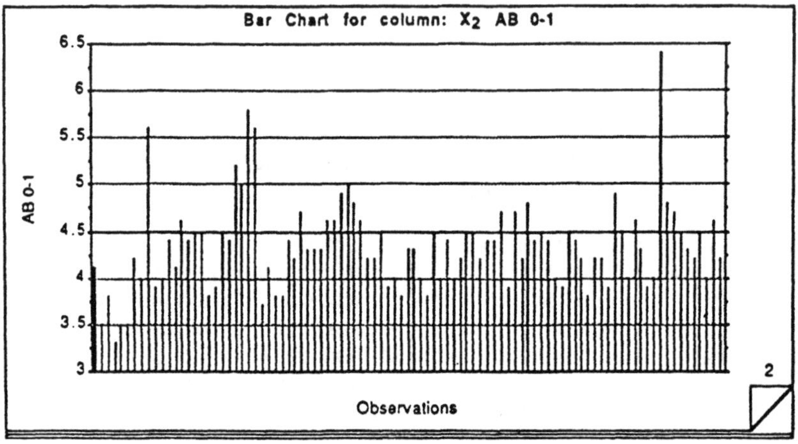

Fig. 5.3.1. Example of sector time consumption for first sector in main direction. Driving time in seconds. Before situation with traffic signals.

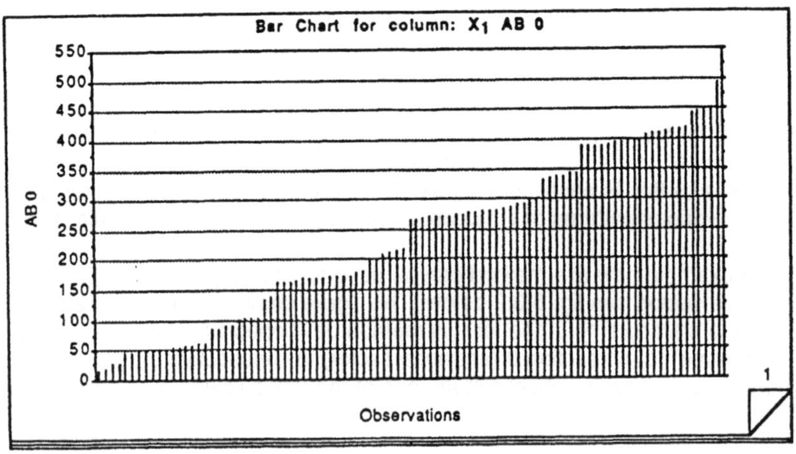

Fig. 5.3.2. Example of studying of the arrival process at the entrance sector. Driving time in seconds for each vehicle from the starting time of the data evaluation computer programme. The figure shows a disturbance of the incoming traffic flow.

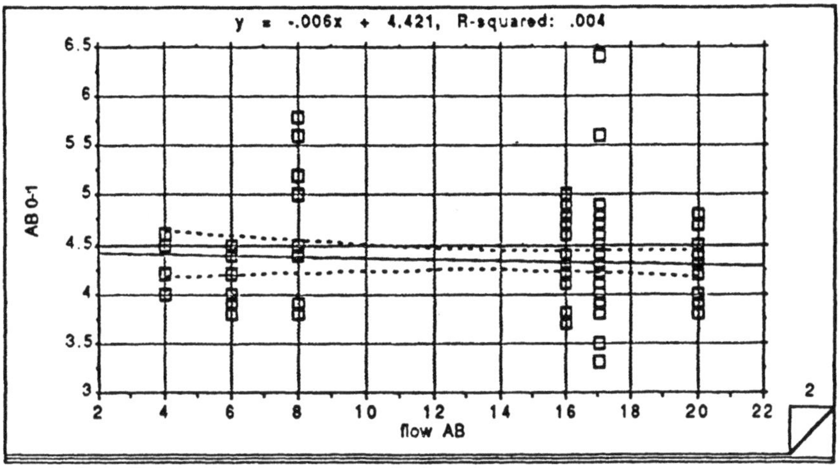

Fig. 5.3.3. Regression analysis of time consumption versus traffic flow for first sector in main direction. x-axis shows flow intensity for the studied flow in vhls/min and y-axis shows driving times in seconds. Before situation with traffic signals.

6. Analysis of the Result

6.1. Problem definition

The specific aim of this survey, apart from the methodological development aspects, is to find out whether or not the *changing of intersection control mode from traffic signal intersection to roundabout has given any positive effects.* Possible quantative measurement parameters for answering such a question could be *time consumption (delay), queue length, amount of stopping vehicles, emission levels* etc.

Due to practical, observation-technical reasons as well as financial limits for this survey, only one incoming direction was studied. By prior experience there were significant delays in this entrance part of the intersection.

Before-after comparisons will be done for the *vehichle time consumption, waiting time at queue end and stop/yield line* and *amount of stopping vehicles per time unit.* From those comparisons conclusions will be drawn concerning differences in performance, when changing from traffic signals to roundabout, for this specific studied intersection.

6.2. Method

A method with comparative regression analysis is used to compare flow related time comparison in different sectors.

Let us first define a certain arrival section called A0 at the beginning of the entrance. This section A0 is physically the same in the before and after situation. The second section A1 is also fixed at a distance of 50 m:s from the arrival sector A0. The physical location is not the same before and after because of the turning road as described in 8.3 above. But as the first sector A 0-1 has the same distance (50 m) in the before-after situation, it could be used for calculating entrance speeds and compare those for the signalled intersection and the roundabout.

After the two in space fixed sections we define a section as a flexible queue end section. The sector A 1-2 will thus consist of the vehicle driving time from section 1 until it stops at the queue end A 2. Sector A 2-3 is defined as the waiting time at the queue end, the sector A 3-4 as the driving time from the vehicle starts again from its position at the queue end until it reaches the stop/yield line. Finally the sector A 4-5 is the vehicle waiting time at the stop/yield line until it finally departs in its departing direction (left, straight or right). The Method and Model here provide for studying of departure sectors as well, but because of that the helicopter could not be used in the after study, we have here observational restraints. Time comparisons in this survey therefore stop with observing differences in waiting time at stop/yield line.

See fig. 6.2.1 below for the usual layout of the sectors in Method and Model for Road Intersection Research. The only modifications for this particular survey is, that the departure sectors 5-6 and 6-7 will not be used. For further information see reference /1/!

6.3. Survey size and sampling

The survey is practically performed as a total survey, where in the before study 98% (187 vehicles of totally 191) and in the after study 94% (528 vehicles of totally 562) are studied. No special statistical sampling technique has been used. The sampling is done by running the Stop Watch Programme (computer evaluation programme) a sufficient amount of times.The *accuracy curves* for the programme runnings in the before-after study are shown in fig. 6.3.1 and fig. 6.3.2. In the before situation it took 27 runs in order to get 98 % of the total amount of the vehicles, and in the after situation it took only 11 runs to get 94 % of the total amount. This because of the difference in the traffic flow situation with reduced waiting times for vehicles in the flow. It is interesting to see, that we have practically a linear situation up to about a 90 % coverage of the total amount of vehicles in both the before- and the after-situation. After this level, and when coming closer to the asymptot (the total count of vehicles), it will require a considerable amount of extra runnings of the computer programme in order to get additional vehicles in the study.

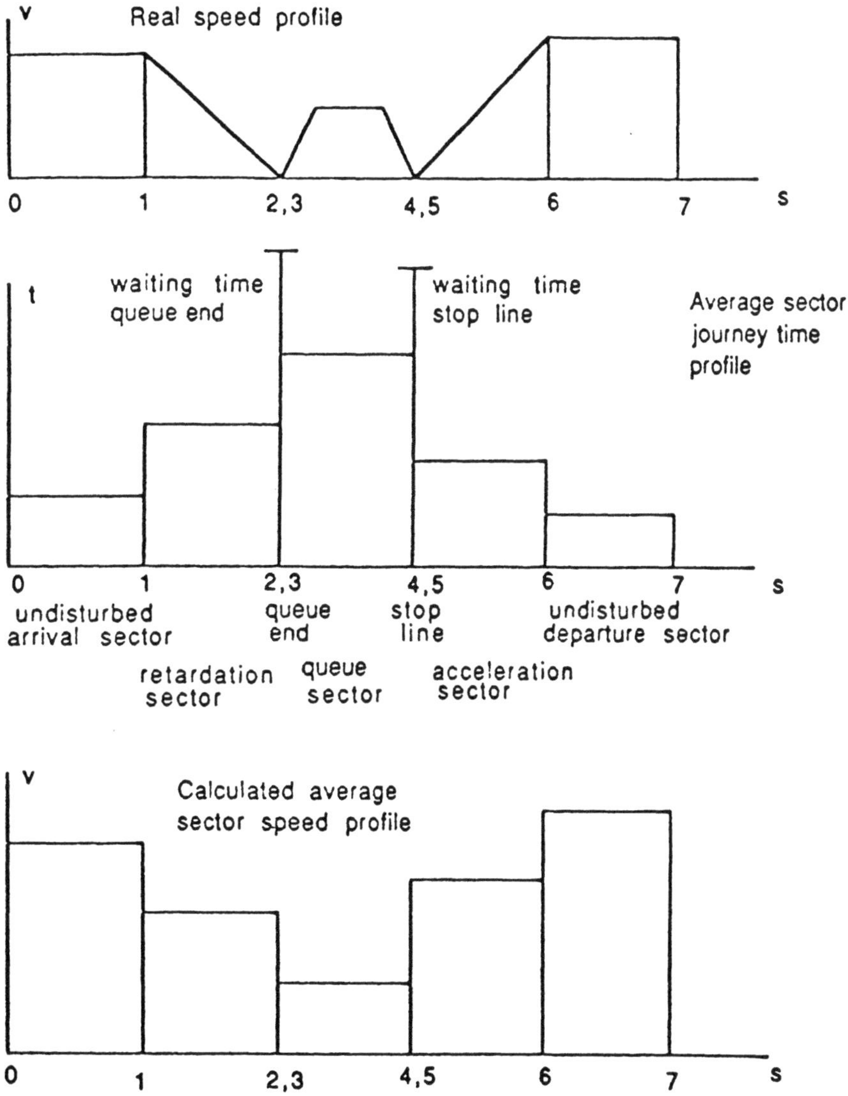

Fig. 6.2.1. Usual-layout of studied sectors in Model for Road Intersection Research. In this particular survey the amount of sectors is limited to 1-5, and ends with the departing from the stop/yield line.

The upper figure shows the *real speed profile*, the middle figure shows the *measured time consumption profile* and the lower figure shows *calculated speed profile.*

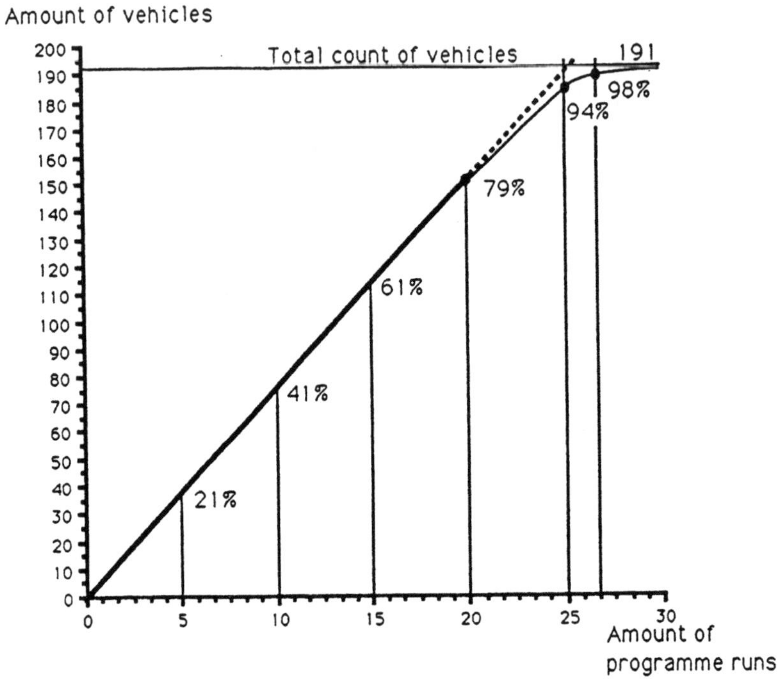

Fig. 6.3.1. Accuracy curve for running of Stop Watch Programme. Before situation with traffic signals and heavy delays and queues.

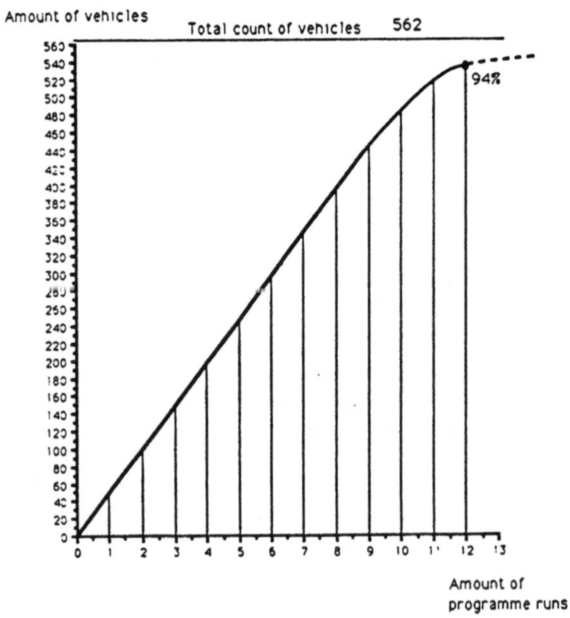

Fig. 6.3.2. Accuracy curve for running of Stop Watch Programme. After situation with roundabout and practically no vehichle waiting times and queues.

6.4. Time consumption and delay

The traffic signal intersection shows a totally different time consumption pattern than the roundabout. Because of the function of the traffic signal, there will practically always be some cars waiting in queues during the red phase. After the roundabout the amount of waiting cars, and the length of the waiting times at the queue end were drastically reduced. Before the avererage waiting time (mean value) at the queue end was 18.14 s per vehicle, which was reduced to 0.53 s after the roundabout. Thus we have here an average delay reduction with 17.6 s per vehicle. In the roundabout situation there were thus practically no queues and car waiting in queues before the yield line.

Before the average waiting time at the stop line (including all passing vehicles during green phase) were 7.22 s compared with the waiting time after at the yield line of 1.30 s. Here is worth noticing, that even though the amount of stopping cars at the yield line were not so few, the length of those stops were so short, that an average time reduction of 5.92 s per vehicle was obtained.

If the driving times through the system before and after are compared, i.e. from arrival section A0 to departure section A5, we have an average of 41.59 s before the roundabout and 18.46 s per vehicle after. When comparing those figures we should also keep in mind, that the driving distance had increased by 33% after the roundabout was built. This means, that the time reduction comparison could have been even more positive, if the driving distance had been kept constant.

6.5. Time consumption comparison by regression analysis

As said earlier, it is necessary to relate a system's parameter value (here time consumption **t**) to the amount of incoming vehicles in the traffic stream, i.e. traffic flow **q**. In such **t-q** diagrams the before and after results could be compared with each other despite of possible differences in minute flows. The diagrams show a slight decreasing tendency of negative flow dependence for the system's parameter **t**. The situations are rather similar in the before and after diagrams. The results of the time consumption comparison are shown in fig. 6.5.1-4 below, where the before and after regression lines are shown in the same figures.

6.6. Speed level at entrance sector

If we use the time consumption values for the entrance sector A 0-1, we can calculate average vehicle speed before and after. As mentioned in 6.2. Method above, the distance for the entrance sector was kept constant at 50 m. For the traffic signal intersection the entrance sector average time consumption was 4.40 s, which gives the average vehicle speed of 40.9 km/h. In the roundabout situation we have an entrance sector time of 5.57 s giving the average vehicle speed of 32.3 km/h.

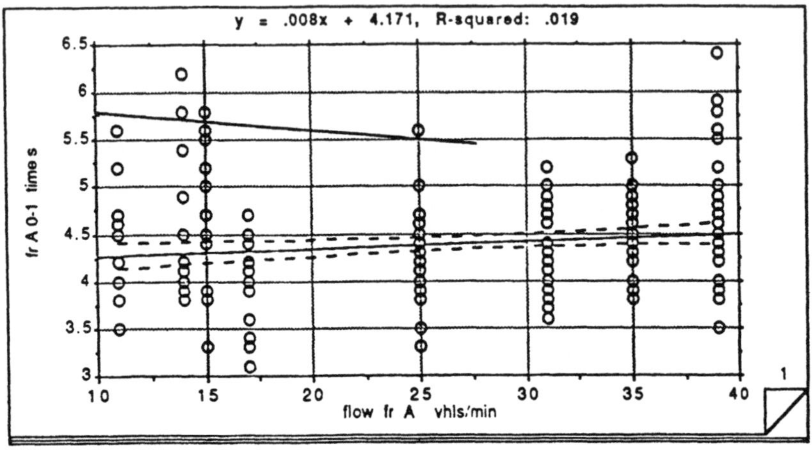

Fig. 6.5.1. Comparison of time consumption for entrance sector (sector 0-1). The upper line shows before situation with traffic signals, and the lower line shows after situation with roundabout. The points (time in sec plotted against the flow in vhls/min) belong to the regression line in the after situation.

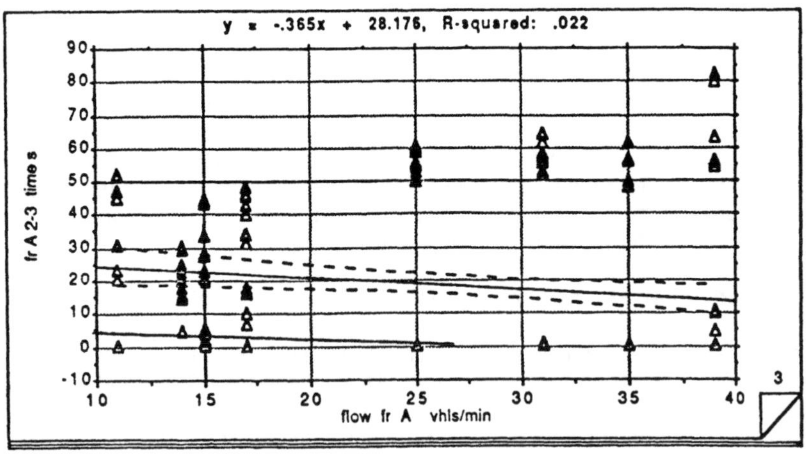

Fig. 6.5.2. Comparison of waiting time at queue end (sector 2-3). The upper line shows the before situation with traffic signals, and the lower line shows the after situation with roundabout. The points (time in sec plotted against the flow in vhls/min) belong to the regression line in the before situation.

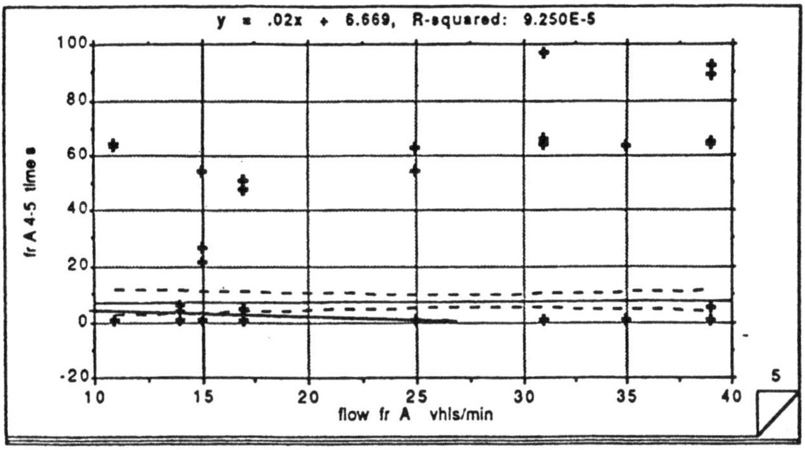

Fig. 6.5.3. Comparison of waiting time at stop/yield line (sector 4-5). The upper line shows before situation with traffic signals, and the lower line shows after situation with roundabout. The points (time in sec plotted against the flow in vehls/min) belong to the regression line in the before situation.

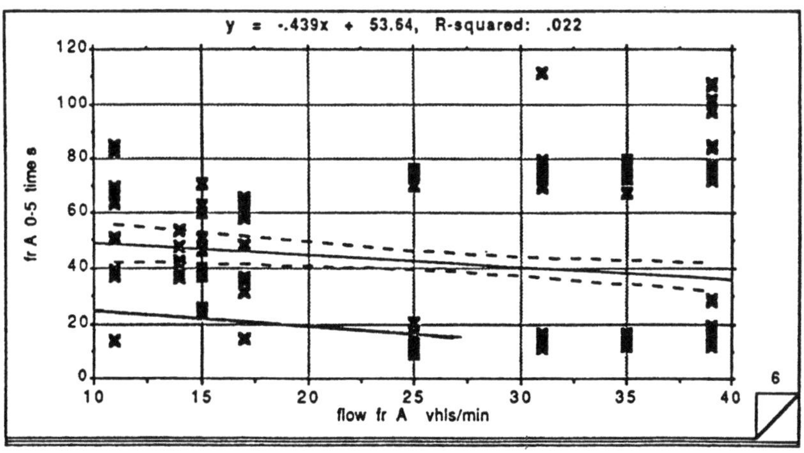

Fig. 6.5.4. Comparison of total driving time through system (sectors 0-5). The upper line shows before situation with traffic signals, and the lower line shows after situation with roundabout. The points (time in sec plotted against the flow in vhls/min) belong to the regression line in the before situation.

This is the same tendency as we received in a comparison between a priority T-junction and a roundabout in another case study. It is naturally not possible to make a general conclusion out of those few values, but the specific survey shows a remarkable reduction of the speed level by 13% during the same time as both the waiting times, driving times and the amount of stopping vehicles were reduced.

6.7. Amount of stopping vehicles

From the raw data tables (not presented here) it is possible to count the amount of stopping vehicles per minute at the queue end and at the stop/yield line. The amount of vehicles, which had to stop at a queue end in the before situation, was an average of 8.75 vehicles per minute, and the amount stopping at the stop line for red light was 2.88 vehicles/min. In some heavy traffic flow situations a car stopping at the queue end could have to stop once more at the stop line, but the traffic signal system mostly eliminates the need for stopping twice.

After the roundabout there were practically no vehicles stopping at a the queue end, and the average figure was 0.67 vhls/min. On the other hand the amount of vehicles stopping at the yield line increased to 4.20 vhls/min. Those stops are inevitable for giving priority to traffic flow within the roundabout. However the average stopping time for those vehicles was much shorter, and the total waiting time at the stop/yield line was reduced in a great extent, see above under 6.4!

7. Conclusions

7.1. General

In this chapter some conclusions are drawn based on the ongoing development of the Method and Model for Comparative Intersection Research (see reference /1/) and some obtained survey results, which are presented in the previous chapters. The essence and the most essential parts of the discussions and conclusions are very briefly presented in the abstract and the summary of this paper.

7.2. Summary of Survey Goal

This research work, survey and paper has some basic goals, namely:

--- **1.** *Present a case study as an example of the positive effects, which could be obtained by changing from a traffic signal intersection to a roundabout*

--- **2.** *Compare the performance between two types of road intersections (here a traffic signal intersection and a roundabout), by means of certain performance measures and descriptive traffic flow parameters*

--- **3.** *Inform and give a general background for the overall project approach concerning Intersection Performance and Traffic Control, and then especially the work of creating a Method and Model for Comparative Intersection Research, which could be used for data collection and analysis of various aspects of Capacity, Delay, Safety and Environment for road and street intersections*

7.3. Performance Comparison. Traffic signal intersection to roundabout

7.3.1. Time consumption and delay

The conclusions are here drawn from a survey with a before- and after-study regarding a traffic signal intersection and a roundabout. The traffic signal intersection was changed to a roundabout with yield priority at all legs, i.e. the traffic flow within the roundabout circle has priority. At the before situation there were heavy delays at peak hour periods and thus time consuming stopping in queues. The time saving results obtained here are typical for this particular surveyed intersection type, and therefore it is very plausible, that significantly *large time savings could generally occur when changing from traffic signal control to roundabout.*

When a traffic signal intersection is changed to a roundabout, the *driving times will be more equally distributed.* For the direction with heavy queues and stopping vehicles (before the change) the gain in driving time in peak hour was in our case study approximately an average of 23 seconds per vehicle (55 %) for a distance of 240 m through the intersection. Here should also be taken into consideration, that the physical road legth had been increased by 33 % after the design change.

The average waiting times before the stop line in the same direction were reduced from 7.2 sec to only 1.3 sec before the yield line, an average gain of 5.9 sec per vehicle during peak hour periods. But the heavy waiting times at the queue end (before the change) were reduced from 18.1 sec to only 0.53 sec (average value per vehicle). The waiting time reduction was thus 17,6 seconds per vehicle.

The studied direction of the intersection is passed by 12480 vehicles per day. This should mean a total time saving for the drivers of c:a 80 hrs a day. If as an economical estimate we calculate with a (low) time value of SEK 25:- (USD 4.5:- January 1991) we would obtain quantified time savings of SEK 720.000:- (c:a USD 140.000:-) per year only for one studied direction through the intersection.

Consequently *there are certainly many traffic signal intersections, which could give substantial economical profit for the community when being changed to roundabouts.*

7.3.2. Traffic safety

There is the common belief, that lower speed levels also mean a lower risk level. If we accept that statement, the *changing to roundabout has* also *increased the level of safety* at the studied intersection. The speed level in the beginning of the influence area before the intersection (180 m respectively 240 m before the stop/yield line) was 49.9 km/h for the traffic signal intersection and 32.3 km/h for the roundabout. The changing to roundabout thus gave a reduction of the speed level by 8.6 km/h or 13 %.

Refering to some research information from the Swedish Road Research Institute and Lund University, Sweden, a 13 % reduction of the speed level could give an estimated reduction in the amount of accidents with human injuries by about 25 %.

7.3.3. Environmental aspects

Less driving time in an intersection and reduced waiting times at stop or yield lines are reducing the total emissions from the traffic in the intersection. In a research report from General Motors Research Laboratory, it is stated, that *"hydrocarbon emissions can be expressed as a linear function of average trip time per unit distance for urban driving at low speeds (less than 60 km/h)"*, see reference /28/!.

In this particular studied case with a changing of a traffic signal intersection to a roundabout, the major directions will have a significantly reduced driving time through the system. It is here evident, *that a changing to a roundabout should give a total saving of the amount of vehicle seconds spent in the system.* In such cases we should also get savings in emitted pollutions from the traffic.

The surveying of detrimental pollutions at intersection and its relation to intersection design and choice of intersection type is a very difficult research task, because of the lack of data from vehicle driving patterns through the system. It is therefore essential to put *more attention to the finding of suitable emission models and methods of estimating levels of emission at intersections.* Before those methods are ready and developed, it is very difficult to do any comparative research, regarding *which intersection design could be the most suitable for the environment.*

8. Methodological and Accuracy Aspects

In this presented paper the general methodological aspects are only dicussed briefly in relation to the creation af a Method and Model possible to use for collecting survey data. Particularly the measurement accuracy and data quality aspects are not covered in any higher extent here. For further studies of those problems and also for the questions of validity and reliability aspects, see reference /1/, where some of these aspects are discussed. The author also intend to do a lot more research within this field of comparative intersection research.

List of References

1. Royal Institute of Technology, Stockholm, Sweden, 1990. Department of Traffic and Transportation Planning. Paul G Höglund. Method and Model for Comparative Intersection Research (Intersection Performance and Traffic Control; Capacity, Delay, Safety and Environment).

2. Royal Institute of Technology, Stockholm, Sweden, 1990. Department of Traffic and Transportation Planning. Paul G Höglund. T-junction Performance. Comparison between Priority Junction and Roundabout.

3. Royal Institute of Technology, Stockholm, Sweden. Department of Traffic and Transportation Planning. Ann Storkitt. Korsningsutformning och trafikreglering. Litteraturstudie (Intersection Design and Traffic Control. Literature study.

4. Royal Institute of Technology, Stockholm, Sweden, 1989. Department of Traffic and Transportation Planning. Gösta Gynnerstedt.Trafiktekniskt underlag för utformning av korsningar och trafikplatser (Traffic engineering basis for the design of intersections).

5. Royal Institute of Technology, Stockholm, Sweden, 1987. Department of Traffic and Transportation Planning. Gösta Gynnerstedt. Road Traffic Simulation Programmes on a Micro-level in the Frame of the Prometheus Programme.

6. Royal Institute of Technology, Stockholm, Sweden, 1989. Instrumentation Laboratory. Andrej Bielek, Tomasz Abramczuk. Real-time Wide Area Traffic Monitoring by use of Image Processing.

7. Royal Institute of Technology, Stockholm, Sweden, 1990. Department of Traffic and Transportation Planning. Andrzej Sobzcak. The Stop Watch Programme.

8. Swedish Road and Traffic Research Institute (VTI), Report 361, 1983. Hans Åke Cedersund. Cirkulationsplatser (Roundabouts).

9. Swedish Road Administration (VV), VV 1977-11 TV 137. Cirkulationsplatser (Roundabouts).

10. Discussions conducted with Hans A Cedersund at Swedish Road and Traffic Research Institute, Linkoping, Sweden.

11. University of Lund (LTH), Sweden. Traffic engineering. Sverker Almqvist. Försök med minicirkulationsplats i Växjö (Experiment with miniroundabout at Växjö, Sweden).

12. University of Lund. Arne Hanson. Studies in driver behaviour with applications in traffic and planning. Two examples. Appendix 2. Traffic flow at uncontrolled intersections.

13. Intersections without Traffic Signals, Springer Verlag, 1988. Werner Brilon et al.

14. Intersections without Traffic Signals, Springer Verlag, 1988. R.J. Troutbeck, ARRB, Australia. Current and Future Australian Practices for the Design of Unsignalized Intersections.

15. Intersections without Traffic Signals, Springer Verlag, 1988. Frank Poppe, SWOV, Netherlands. Intersection Layout, Priority Rules and Traffic Safety.

16. Intersections without Traffic Signals, Springer Verlag, 1988. J M Gambard, SETRA, France. Safety and Design of Unsignalized Intersections in France.

17. Werner Brilon. Der Kreisverkehr; Eine vergessene Knotenpunktform.

18. SWOV, Holland. Jaap van Minnen. Roundabouts. Some remarks on the safety and capacity of roundabouts and roundabout priority systems.

19. TRRL Annual Report 1988, Great Britain. Director's forword.

20. SETRA, France, 1983. Carrefours Giratoires--Analyse de la Sécurité.

21. CETUR, France, 1988. Conception des Carrefours a Sens Giratoire Implantes en Milieu Urbain.

22. SINTEF, Norway. Rundkjöringer (Roundabouts).

23. Seattle Engineering Department, Transportation Services Division, USA. Edwin W von Borstel. Traffic Circles, Seattle's Experience.

24. ITE Journal, USA. November 1988. Russi P Bhesania. A Procedure for Identifying Needed Improvements in a Traffic Network.

25. ITE Conference, San Francisco, USA, 1984. Marl L Schlappi. Traffic Signal Coordination as its affects Carbon Monboxide Concentrations.

26. SNV Report 3276, Sweden, 1987. Karl Erik Egebäck. Hastighet, Föroreningsutsläpp. Bensidrivna bilar (Speed, Pollution. Gasoline driven cars).

27. VTI, Sweden, 1989. Harald Perby. Beräkningar av koloxidproduktion vid bensin och dieselförbränning (Calculation of Carbon Dioxid Production Consumption of Gasoline and Diesel).

28. General Motors Research Laboraties, Warren Michigan, 1976. Leonard Evans. Exhaust Emissions, Fuel Consumption and Trafic: Relations Derived from Urban Schedule Data.

A Microscopic Simulation Model of Roundabout Entry Operations

-the influence of flows leaving by previous exit on entry capacity
-the determination of entry capacity formula in Switzerland

TAN Jian-an

Institute of Transport and Planning
Swiss Federal Institute of Technology at Lausanne
CH-1015 Lausanne, Switzerland.

Abstract

A microscopic computer simulation model of roundabout entry operation, which can simulate the phenomena that vehicles leaving by the previous exit disturb entry vehicles, is built for study the influences of leaving vehicles on entry capacity and the determination of entry capacity formula in Switzerland. The comparisons between simulation results and field observations indicated that the simulation model can reproduce the real situations adequately in terms of maximum entry flow. The model is applied to study the influences of leaving vehicles on entry capacity and to determine the entry capacity formula. The results show that the larger the travel distance between the conflict point of exit and conflict point of entry, the fewer the disturbance of leaving vehicles on entry vehicles. If the distance is large enough, then the disturbance will disappear. The entry capacity formula obtained from the simulation study can take into account of the influence. Field observations made in Switzerland have confirmed this formula gives better predictions than the formulae proposed by Kimber, CETUR and Harders. This new entry capacity formula has been adopted in the new Swiss Guide On Roundabout Design.

1. Introduction

After the off-side priority was introduced, roundabout as a special intersection has demonstrated a lot of advantages over other type of junctions in certain cases, such as higher capacity, less delay, more safety and lower operation costs, so more and more roundabouts are being and will be constructed in Switzerland.

Although a lot of research works about roundabout operations have been done since 1960, it is not assured that the research results obtained can be applied to Swiss traffic conditions. In order to provide a work sketch which can ensure the quality and the coherence of the roundabout design in Switzerland, a new Swiss guide on roundabout design is being prepared and a series of researches is needed. One of the important questions is how to predict entry capacity relative accurately in Swiss traffic conditions.

Because there are many random factors affecting roundabout operations, such as vehicles arriving and gap acceptance, statistical methods can be used. In order to obtain reliable results by this method, a lots of well-chosen field observations on various geometric conditions are needed. So this method is costly and time consuming, furthermore the results can only be applied in the similar geometrical conditions and traffic conditions.

Roundabout can also be thought as a random arrived and random service queueing system, so the related mathematical model can be applied. But roundabout operations are more complex than that described by the mathematical models, so the results given by the model is often unsatisfactory in practice. With the development of computer technology, microscopic simulation has become a favourite method for the study of roundabouts. This paper describes the development of a simulation model, and its application to study the influence of flows leaving by previous exit on entry capacity and to determine an entry capacity formula.

2. The mechanism of vehicles entering the roundabout

Generally, a driver of a vehicle at the yield line of a roundabout will observe the conflict traffic, if there is no conflict traffic, the vehicle will enter into the roundabout, if there exists conflict traffic then it will try to find a gap sufficient large in the conflict stream and then enter the roundabout by using this gap. The minimum gap (in unit of time) needed for entering a roundabout is dependent on the roundabout geometries, characteristics of vehicles, human factors and especially the state of vehicles at the yield line (in moving or stop).

Because a vehicle travelling in the circulating carriageway, if it doesn't reach the diverging point of leaving the roundabout and going ahead (named as the conflict point of exit), can be thought by entry vehicle to travel in front of entry. The possible situation that the entry flow is disturbed by leaving vehicles is described as follows.

As shown in Fig.1, point b is the conflict point of exit, vehicle V2 hasn't reached the point b, the travel distance between point 'b' and point 'a' is relatively small, usually the driver of vehicle V4 don't know if it leaves or not, so he will judge the gap between V1 and V2, if the acceptable minimum gap for the driver of vehicle V4 (named as Gc(V4)) is larger than the time predicted for vehicle V2 travelling to conflict point a (named as Time(v2, a)), then V4 will wait for another gap even if V2 will turn to right (Fig.1), therefore traffic leaving the roundabout disturbs the traffic flow at the entry. But if the acceptable minimum gap Gc(V4) < Time(V2,a) then in this case leaving traffic flow does not disturb the entry

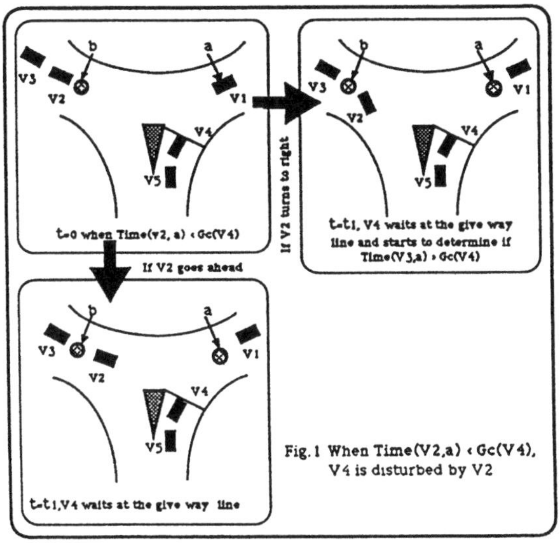

Fig.1 When Time(V2,a) < Gc(V4), V4 is disturbed by V2

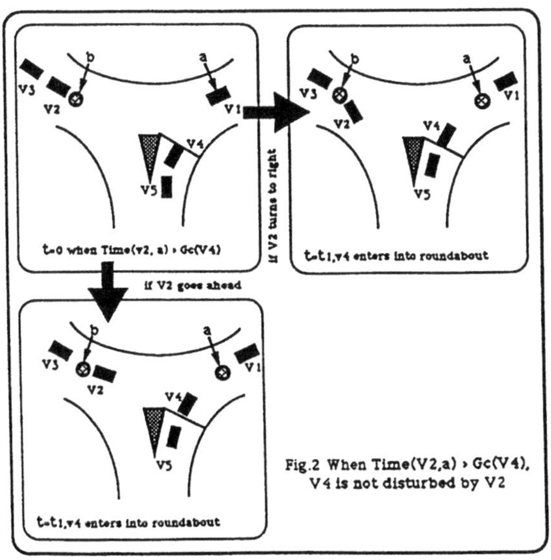

Fig.2 When Time(V2,a) > Gc(V4), V4 is not disturbed by V2

vehicle (Fig.2). Therefore whether vehicles leaving the roundabout disturb the entry stream, it depends on the minimum gap accepted by vehicles at the entry and travel time needed from conflict point of exit to conflict point of entry (i.e. point a in the Fig.).

In the case that the entry vehicle is disturbed by leaving vehicles, is one vehicle leaving the roundabout equivalent to one vehicle going ahead? To answer this question, see Fig.3, vehicle V2 hasn't reached the conflict point b, and Time(V2,a)< Gc(V4), then V4 will wait at the stop line (Fig.3 (1)). If V2 goes ahead ,V4 will wait for another gap after V3 (Fig.3 (2)). If V2 leaves the roundabout, as soon as the driver find that, he will quickly judge whether the gap is large enough for entering, then there are two situations which may happen: (1) if Time(V3,a) > Gc(V4) then V4 will immediately enter into roundabout (Fig.3 (3)), suppose the time needed for finding that V2 turns to right is t_r, then in this case, as existing of V2, the time wasted is equal to t_r , only a part of the gap between V1 and V3 . (2) If (V3,a) < Gc(V4), then V4 will wait for another gap after V3 (Fig.3 (4)), in this case the gap between V1 and V3 is wasted, evidently V2 leaving the roundabout is equivalent to V2 going ahead.

Because the arrival of vehicles is random, the gaps of conflicting flow is a random variable. In addition to that, as conflicting flow includes vehicles leaving the roundabout, the gaps can be changed when a vehicle turns to right and leaves the roundabout. It's too complex to determining the real distribution of gaps mathematically in this case. This results in that it is nearly impossible to apply the classical analytical method in this study.

Simulation method can be used for this study. the idea is to construct a simulation model which can simulate microscopically the traffic operations described above. By applying the calibrated simulation model, the influence of leaving vehicles on the entry flow and entry capacity taking account of this influence can be studied.

3. Simulation model of roundabouts

Now there exist several simulation models of roundabouts, such as SIMRO, developed at University of Southampton of England. But in all these simulation models, conflicting flow is thought as circulating vehicles travelling in front of entry, so they can't be applied to this study.

A new simulation model is constructed, Fig.4 shows the outlines of the model. The simulation model includes the following mechanisms:

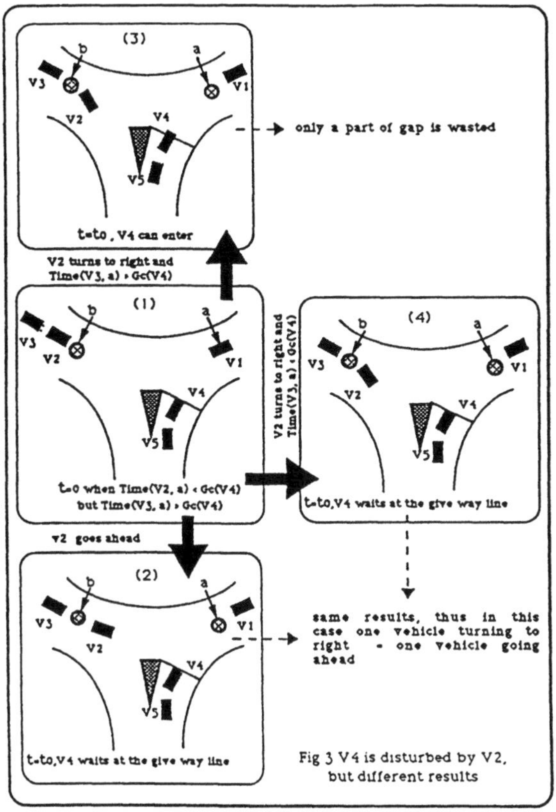

Fig 3 V4 is disturbed by V2, but different results

a) vehicle kinematics;
b) vehicle generation (both for entry flow and conflicting flow);
c) vehicle leaving generation;
d) gap determination and gap acceptance.

1) Vehicle kinematics

The microscopic representation of vehicle movement needs to define the position and other kinematic characteristics of each vehicle throughout the simulation process. The positions of vehicles moving along the approaches of the roundabout are represented in a one-dimension co-ordinates system with reference to the give way line at entry (named system E in the model). The positions of circulating vehicles and the positions of vehicles at exit are also represented by two one-dimension co-ordinates (named system C and B respectively). Fig.5 shows these three co-ordinates systems. The position of each vehicle is updated at regular simulation time intervals. In each system the position of exit is defined by boundary values, if the coordinate of vehicle is larger than the

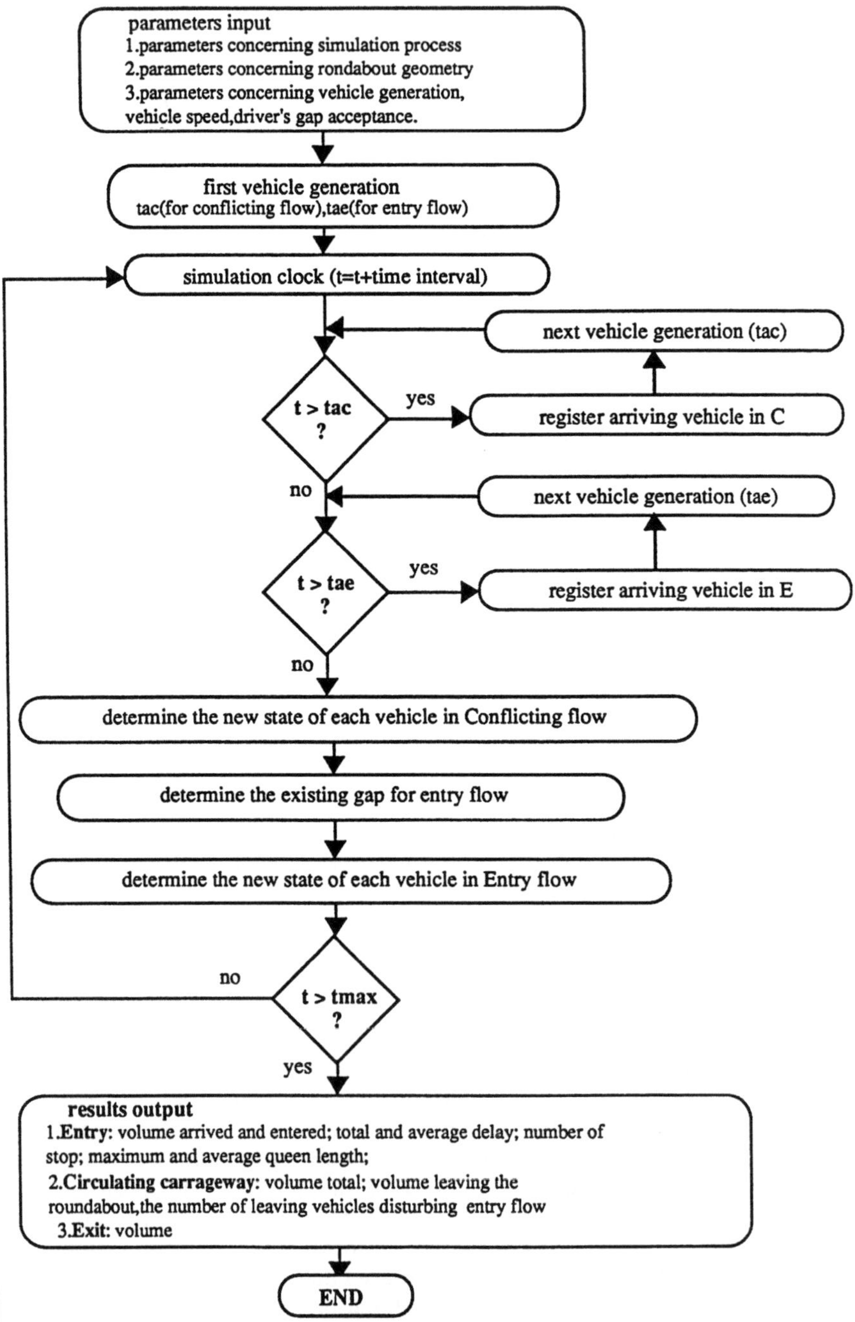

Fig.4 Outlines of computer simulation model of roundabout entry operation

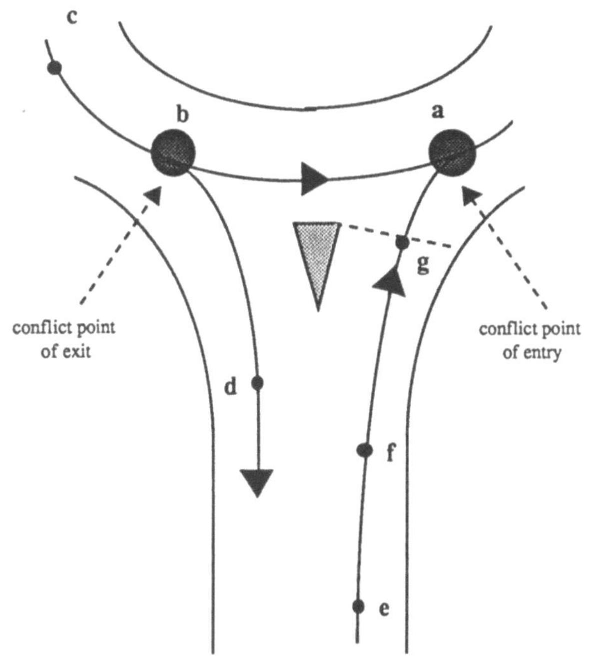

Fig.5 Three co-ordinate systems defined in the simulation model

boundary value, the vehicle is considered leaving the system. Points **a**, **g**, and **d** are the boundary values of system C, E, and B respectively.

2) Vehicle generation

Vehicle arriving at entry is generated at point **e**, original point of system E. Vehicle arriving at circulating carriageway is generated at point **c**, the original point of system C, see Fig 5. Following 4 types of distribution can be chosen:

a) negative exponential distribution;
b) shifted exponential distribution;
c) composite exponential distribution;
d) observed distribution.

Negative exponential distribution is the simplest distribution and often used to describe the traffic flow in many studies. Shifted exponential distribution can assure the headway is always larger than a certain value, this is the case of one lane traffic flow. Composite exponential distribution can be used to represent the vehicle arrival in platoon, it's the case where upstream there is a signalized intersection. The model also provides a possibility to use observed headway distribution.

3) Vehicle leaving generation

Vehicles in system C turning to right (to exit) can also be generated randomly according the percentage of turns. A turning index generated randomly is assigned to each vehicle of the system C at the time it is registered in the system. When leaving vehicle arrives at conflict point of exit, it leaves system C and enters system B.

4) Gap determination

Gap is defined as time difference between leading vehicle and following vehicle passing the conflict point of entry. At the time t, the gap is the time needed for next vehicle to pass through the conflict point of entry. In order to model the situation that vehicles in the entry don't know exactly if circulating vehicle will leave the roundabout or not, each circulating vehicle including leaving vehicle, if it hasn't reached the conflict point of exit (point b), will be thought to travel in front of entry. This is the important feature different to other simulation models of roundabouts, and with this feature the effect of leaving vehicle on the entry vehicles can be studied.

5) Gap acceptance

Field observations have shown that a longer gap is needed for the vehicle entering into roundabout when it has stopped at yield line, and a smaller gap for the vehicle without complete stop. In each time interval, the model estimates if the first vehicle in the entry will stop or not when it arrives the yield line. The minimum gap needed for vehicle stopped is the critical gap, and for vehicle without stop is the move up time. The minimum gap needed is not fixed for each vehicle, for example, the following vehicle is one without stop when it reaches the yield line, so minimum gap is the move up time, but when it can't enter into roundabout, it will stop at yield line, so at next time interval, the minimum gap is the critical gap. 4 - 4.5 sec can be used as the critical gap, 2 - 2.5 sec as the move up time.

6) Model input and output

Fig.4 shows the model input. The parameters concerning the simulation process include:

a. the simulation period (exclude warm up time)
b. the warm up time;
c. the random seeds.

The parameters of roundabout geometry include:

a. the travel distance from conflict point of exit to conflict point of entry (named as L_{ba});
b. the number of lanes in the circulating carriageways *;
c. the number of lanes in the entry and exit*.

(* only one lane is considered now)

The parameters about vehicle generation and operation are:

a. the traffic demand;
b. the minimum following time for vehicle in the entry to pass the conflict point a;
c. the critical gap and move up time;
d. the minimum headway (if shifted exponential or composite exponential distribution is used);
e. the mean headway between restrained vehicles (platoon), the mean headway between free flowing vehicles and the proportion of vehicles in platoon (if composite exponential distribution is used);
f. the observed headway distribution (if observed distribution is used);
g. the average travelling speed of vehicles in each system.

The model output includes following:

1) Entry:

a. the volume generated, registered and left;
b. the time of arrival at point **e** and the time entering into roundabout of each vehicle;
c. the total and average delays;
d. the maximum and average queue length.

2) Circulating carriageway:

a. the volume generated, registered and left;
b. the volume turning to right;
c. the time arriving at **c** and leaving at **a** of each vehicle;
d. the number of leaving vehicles disturbing the entry vehicles.

3) Exit:

a. the volume registered and left;
b. the time arriving at **b** and leaving at **d** of each vehicle.

4. Model validation

Before applying the simulation model, it must be validated to ensure the real situation can be adequately simulated. This is done by comparing the results predicted by the simulation with that observed from the field. Because the objective of the model is to evaluate the influence of leaving vehicle on the entry capacity and to study entry capacity, the measure to be compared is maximum entry flow Q_e (v/h).

Two roundabouts are chosen to validate the model. One is located near the centre of a small town of Payerne (the arm of Avenue Général Jomini, L_{ba}= 14m), another is near the train station of the town of Pully (the arm of Avenue C.F Ramus, L_{ba}= 12m). Table.1 gives the observed and simulated data for these two roundabouts in the situations that the entries are saturated.

For each roundabout, the observed values of Q_{cir} (circulating flow across the entry concerned) and Q_s (flow leaving by the previous exit) are used as inputs of the simulation model. In order to produce saturated situation at entry, high traffic demand is used as the entry demand data (900 v/h for the roundabout of Payerne and 1290 v/h for the roundabout of Pully) and 2 minutes as the warm up time, this results in that at the beginning of the sampling period, there is a queue in the entry of roundabout. The simulation period is 20 minutes.

A	Q_{cir} (v/h)	Q_s (v/h)	Q_e (v/h)
observed	914	395	528
simulated	919	396	530

A. roundabout of Payerne

B	Q_{cir} (v/h)	Q_s (v/h)	Q_e (v/h)
observed	363	197	1113
simulated	360	212	1107

B. roundabout of Pully

Table.1 comparisons between the average simulation results and the average observed data

After correcting errors and adjusting several parameters in the model, the model can reproduce the real situation satisfactorily. Fig.6 gives the comparisons between observed data and simulated data. Table.1 summarizes the comparisons between average observed data and average simulation results (5 runs, where the average observed values of Q_{cir} and Q_s are used as the inputs) for these two roundabouts. The comparisons also indicate that the general agreements between the traffic volumes observed and the traffic volume simulated have been achieved.

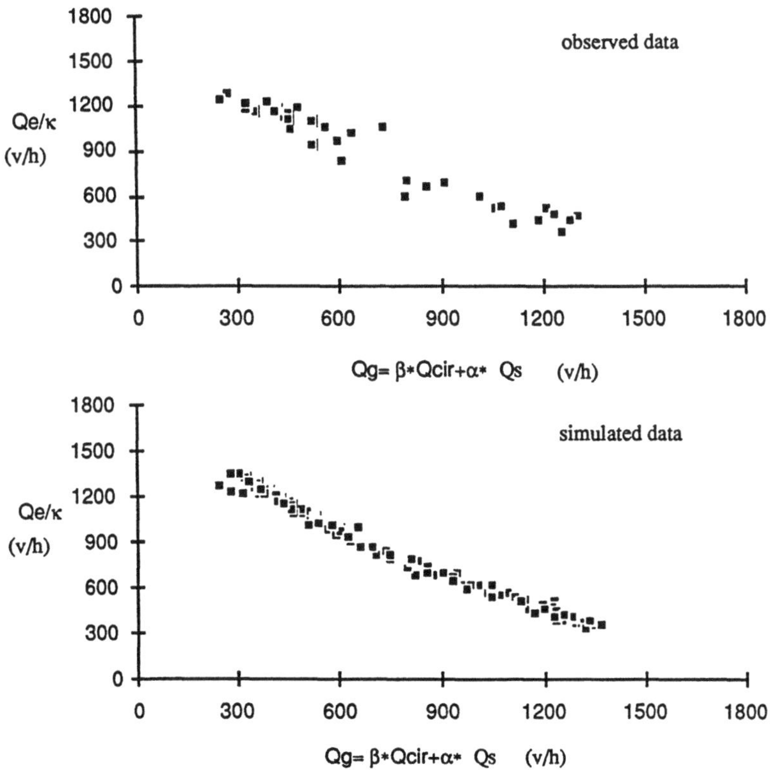

Fig.6 observed and simulated data at two roundabouts (Payerne & Pully)

5. The influence of leaving vehicles on entry capacity and the determination of an entry capacity formula

After validation process, the model is used as a tool to study the influence of turning vehicles on entry capacity under different roundabout geometries.

The model's inputs remain the same except the travel distance from conflict of exit to conflict point of entry. The distances introduced in the simulation are 9m, 10m, ...,28m, 29m. The outputs are the maximum entry volumes, the leaving volumes and the percentage of leaving vehicles disturbing the entry flow (case 1 and case 2). Fig.7 shows the simulation results.

The results indicate that the larger the travel distance between the conflict point of exit and the conflict point of entry, the smaller the percentage of disturbance resulting from leaving vehicles. When travel distance is larger than 28m, then the leaving vehicles don't disturb the entry vehicles. Also it indicates that when L_{ba} is smaller, the value in the case 1 (entry vehicle is distuebed so that it can't enter) is relatively higher, on contrary if L_{ba} is relatively higher, the value in the case 2 (entry vehicle is disturbed, but it can enter and only a part of gap is wasted, see Fig. 3 (3)) is higher.

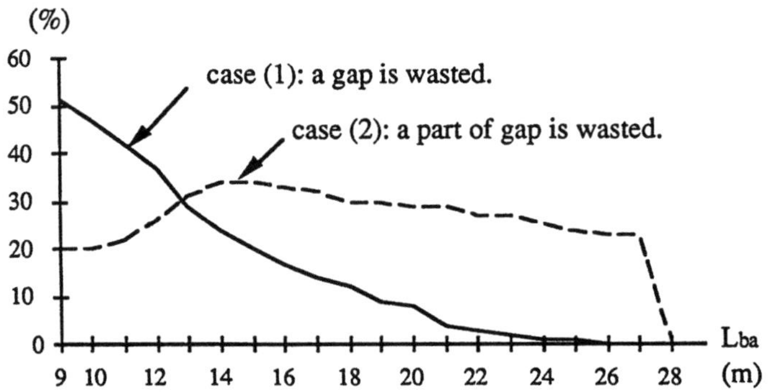

Fig.7 percentage of leaving vehicles disturbing entry vehicle

Investigations and researches by Mr. R. M. Kimber (Transport and Road Research Laboratory, England) and CETUR (Centre d'Etudes des Transports URbains, France) have been shown that

$$Q_e = F - f_g * Q_g$$

where F and f_g are the coefficient determined by the geometries of roundabouts. But in the formula of TRRL, Q_g is considered as those circulating flow crossing the entry (Q_{cir}), and in the formula of CETUR, $Q_g = Q_{cir} + \alpha * Q_s$.

As considering the French type of formula can take account of the influence of leaving vehicles by the previous entry, therefore the study is concentrated on this type of formula. In order to reasonably determine the coefficient of α in the formula, the parameters F and f_g should be determined first. Based on field observations in Switzerland, F=1500 is reasonable. From Fig.7, when $L_{ba} >= 28$, leaving vehicles don't disturb entry vehicles, therefore $\alpha=0$, then

$$f_g = (F - Q_e) / Q_{cir}$$

By using the input data Q_{cir} and output data Q_e in the simulations, $f_g = 0.893 = 8/9$.

Therefore

$$Q_e = 1500 - 0.893*(Q_{cir} + \alpha * Q_s).$$

From this equation, the coefficient of α can be calculated as follows

$$\alpha = \{[(1500 - Q_e) / 0.893] - Q_{cir}\} / Q_s$$

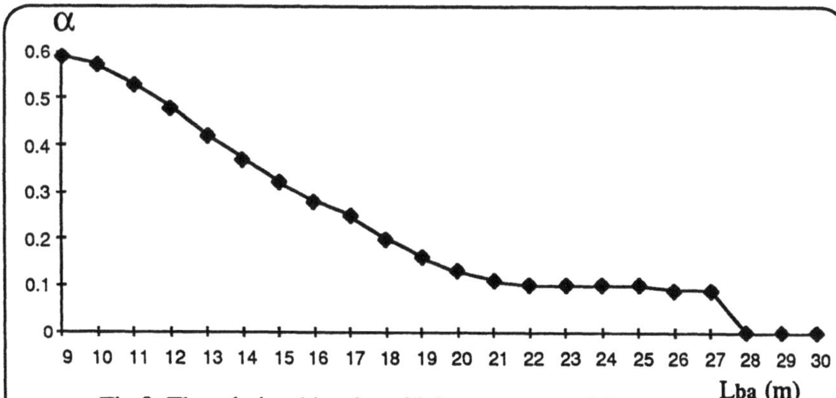

Fig.8 The relationship of coefficient α and Lba

by applying simulation input data Q_{cir}, Q_s and output data Q_e. The calculation results are shown in Fig.8. It shows that the larger the travel distance between the conflict point of exit and the conflict point of entry, the smaller the coefficient of α. If the distance is larger enough, the influence of leaving vehicles by the previous entry disappears.

It should be noted that here α is determined under the assumption that entry vehicles can't know whether circulating vehicles will turn to right or continue to travel if they don't reach the diverging point of exit. This results in over-estimating the value of α for the roundabouts with three branches and some roundabouts with special geometries, because the entry vehicles in this case can often easily make correct judgement and take their decisions even before circulating vehicles have not reached diverging points. In order to correctly determine α for this type of roundabout, it's suggested to use the half value of α determined. Another method, proposed by Mr. Ph Glayre (TRANSITEC Ingénieurs-Conseils SA, Lausanne), is that only the leaving flow from its left branch is taken as the leaving volume Qs when applying the formula.

In the simulation study, the circulating speed is fixed at 20 - 25 km/h. Evidently if the speed is higher than this value, more disturbance can be resulted, so in this case the value of α determined here should be adjusted to take account of it (generally adjust in the interval $[\alpha, \alpha+0.1]$). On contrary, if the speed is lower, α should adjusted in $[\alpha\text{-}0.1, \alpha]$.

It should be noted that here only roundabouts with one lane in the entry and one lane in the circulating carriageway are taken into account. Based on some observations made in Switzerland (need more field observations), the French method to take account of multilanes situation is accepted. That is to use parameter κ and β to take account of multilanes in entries and in circulating carriageways respectively. Therefore the new formula is

$$Q_e = \kappa(1500 - 0.893 * Q_g) \qquad (1)$$

where $Q_g=(\beta * Q_{cir} + \alpha * Q_s)$, α is determined based on Fig.8 and κ and β are determined as follows (need further investigations):

1 lane in the entry: $\kappa = 1$;
2 lanes in the entry $\kappa = 1.4 - 1.6$;
3 lanes in the entry $\kappa = 2$;
1 lane in the circulating carriageway: $\beta = 0.9 - 1.0$;
2 lanes in the circulating carriageway: $\beta = 0.6 - 0.8$;
3 lanes in the circulating carriageway: $\beta = 0.5 - 0.6$.

6. Further comparisons with field observations

A series of field observations have been made at 12 entries of 7 roundabouts in Switzerland. The video recording method was used in the investigation, and the Q_{cir}, Q_s and Q_e are registered when the entries concerned are saturated. Fig.9 shows the comparisons of field observation results and theoretical prediction by the new formula. It indicates that this formula gives good predictions.

According to observed average values of Q_{cir} and Q_s as well as geometric parameters, the predicted entry capacities are calculated by applying this new formula and the formulae of Kimber [8],

CETUR [9] and Harders [11] (in the calculations by the formula of Harders, t_c=4.2 s and t_f=2.2 s are used). The calculated results are compared with field observed values. The detailed calculation and comparisons for each roundabout are presented in [3]. Table.2 shows its comparison results. It has confirmed that this new formula gives better predictions than other formulae.

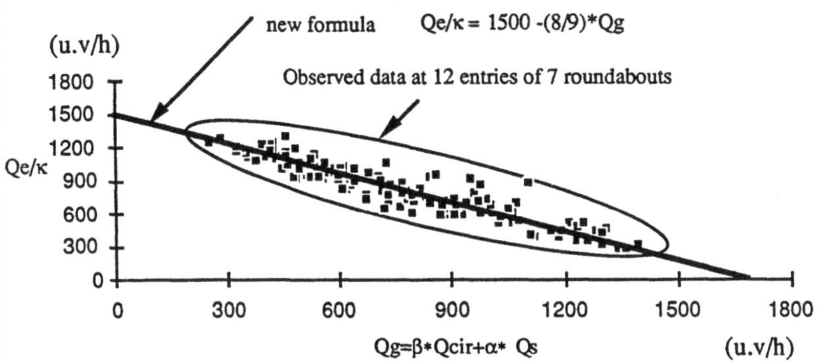

Fig.9 comparison of theoretic predictions and field observations in Switzerland

Another comparison has been made between the new formula and the linear regression formula Based on the observed data Q_{cir} and Q_e (without taking account Qs). The following expression

$$E=\Sigma(Q_{ep}(i)-Q_{eo}(i))^2/Q_{eo}(i)$$

has been used in the comparison, where Q_{ep} is the predicted entry capacity and Q_{eo} observed entry capacity. The calculation results show that E=1619 for the new formula and E=2186 for the regression formula, therefore the new formula gives better predictions than the regression formula without taking account of Q_s.

7. Conclusions and further researches suggested

A microscopic computer simulation model of roundabout entry operations was built. One of the important features, distinguished with other existing simulation models, is that the phenomenon that leaving vehicles disturb entry vehicles can be simulated.

In order to ensure the model can predict the entry capacity adequately, the model is validated by comparing the simulated results and field observations. The comparisons show that the model can reproduce the real situations quite well in terms of the entry capacity. The simulation model then is used as a tool to study the influences of leaving vehicles on entry capacity and determine an entry capacity formula.

Branch (roundabout)		new Formula	Kimber	CETUR	Harders
		formula			
Viaduc (Moutier)		1.6%	72.6%	30.4%	28.2%
Industrielle (Moutier)		4.8%	34.0%	11.5%	4.8%
General Jomi (Payerne)		4.7%	66.1%	27.4%	17.1%
C.F. Ramus (Pully)		-2.0%	-23.1%	4.7%	2.1%
Schwäbis (Steffisburg)		7.6%	-4.5%	15.2%	17.0%
Thun (Steffisburg)		2.0%	8.5%	12.0%	2.9%
Frutigen (Thun)		12.7%	14.4%	24.1%	13.5%
Chablais (Mally)		1.7%	67.6%	13.6%	3.8%
Provence (Mally)		-12.3%	28.0%	2.5%	-5.6%
Ponte (Losone)		2.0%	-13.5%	-1.2%	-7.2%
Losone (Losone)		-0.6%	-16.6%	-6.2%	-12.3%
Ascona (Losone)		-2.8%	-29.0%	-4.0%	-9.7%
Error limit	-	**-12.3%**	**-29.0%**	**-6.2%**	**-12.3%**
	+	**12.7%**	**72.6%**	**30.4%**	**28.2%**

Table.2 Prediction errors of the formulae by comparing field observations

The results from simulation runs indicates that the larger the travel distance between the conflict point of exit and conflict point of entry, the fewer the disturbances of leaving vehicles on entry vehicles, and the smaller the coefficient of α, the higher the entry capacity. Field observations at 12 entries of 7 roundabouts in Switzerland indicate that the new formula gives good predictions, and gives better predictions than the other formulae mentioned in this paper.

The further effort is needed to study the cases where there are more than one lanes in entry and/or in circulating carriageways. On the other hand, the simulation model is needed to be improved in order that the delays and queue lengths can be studied as well.

The simulation model can be further developed to study some special problems concerning roundabout operations, such as roundabout with pedestrian crossings, twin roundabouts, roundabout near a signal intersection.

Acknowledgements

The author is grateful for the support and encouragement from Professor Ph. Bovy, P. De Aragao and F. Molina (Institute of transport and planning, Swiss Federal Institute of Technology at Lausanne). The autor also thanks M. Simon (Emch+Berger AG) for providing the video-tapes of the roundabouts of Steffisburg, Thun and Losone.

References

1. P. de Aragao: Les carrefours giratoires. Institut des transports et de planification, Ecole polytechnique federale de Lausanne, Suisse. 1988

2. Tan Jian-an: Influences of leaving vehicles on entry capacity of roundabout. Institute of transport and planning, Swiss Federal Institute of Technology at Lausanne, 1990 (unpublished).

3. Tan Jian-an: Investigation of entry capacity of roundabouts in Switzerland. Institute of transport and planning, Swiss Federal Institute of Technology at Lausanne, August 1990 (unpublished).

4. H. C. Chin: SIMRO: a model to simulate traffic at roundabouts. Traffic Engineering & Control, March 1985, 109 - 113.

5. Ph. Bovy et P. De Aragao: Vers une moderation intelligente de la circulation: les nouveaux giratoires. Ingénieurs et architectes suisses n.24, Lausanne, 16 novembre 1988.

6. Fonds de la recherche routière et Fonds de la sécurité routière: Guide Suisse des giratoire. Document préparé en vue de la consultation de mai 1990, Lausanne, Suisse.

7. P. de Aragao: Off-side priority roundabouts as a traffic and town planning tool. Proceeding of European transport and planning 18th summer annual meeting, 1990.

8. R. M. Kimber: The traffic capacity of roundabouts. TRRL Report 942, Crowthorne 1980

9. Center d'Etudes des Transports Urbains: Conception des carrefours a sens giratoire implantés en milieu urbain. France, Juillet 1988.

10. W. Brilon, M. GroBmann, B. Stuwe: Towards a new German guideline for capacity of unsignalized intersections. Ruhr-Universität Bochum, Lehrstuhl für Verkehrswesen, Universitätsstrasse 150, 4630 Bochum 1, West Germany, 1990.

11. CETE Médiderranée: Carrousel-gir: calcul de capacité d'un carrefour giratoire, guide de l'utilisateur. France, mars 1989.

12. M. Simon, O. Hintermeister, U. Lehmann: Empirische ermittlung der leistungsfähigkeit von innerörtlichen kreisverkehrsplätzen. Emch+Berger Ingenieurgemeinschaft, Schweiz, Oktober 1990.

A General Capacity Formula for Unsignalized Intersections

Helmut Wegmann

Technische Hochschule Darmstadt

Abstract

A capacity formula for unsignalized intersections is derived that is general enough to be applicable to intersections with non Poissonian traffic streams, with nonhomogeneous and inconsistent behavior of minor road drivers and to allow taking into account impedance effects. It generalizes most known capacity formulae. The relation to these formulae is cleared up. The dependance of the capacity on gap distributions and bunching of major road traffic, on impedance, and on the parameters describing the behavior of minor road drivers is investigated.

1 Introduction

The capacity of an unsignalized intersection for a minor traffic stream crossing or merging with one or more major streams is defined as the maximal possible flow rate from the minor road queue. The classical formula for capacity was developed by Tanner [11] and has been extended and generalized by many authors. We mention only a small selection in order to show that the problem has regained interest in the last years: Catchpole and Plank [8], [3], Troutbeck [12], [13], Brilon [2], Jacobs et al. [7], and Wegmann [14]. Many more are referenced in these papers. Capacity estimates depend on

a) the flow rates and
b) the structure of the major road traffic streams

and on the characteristics of drivers on the minor road such as

c) gap acceptance and
d) move–up behavior

In this paper a capacity estimate is presented, that is based on rather general assumptions. The stochastic model used for the major road traffic is a gap–block model that does not demand independent headways and exponential gaps. The estimate can be applied to multilane traffic and to traffic streams of different priority (impedance) on the major

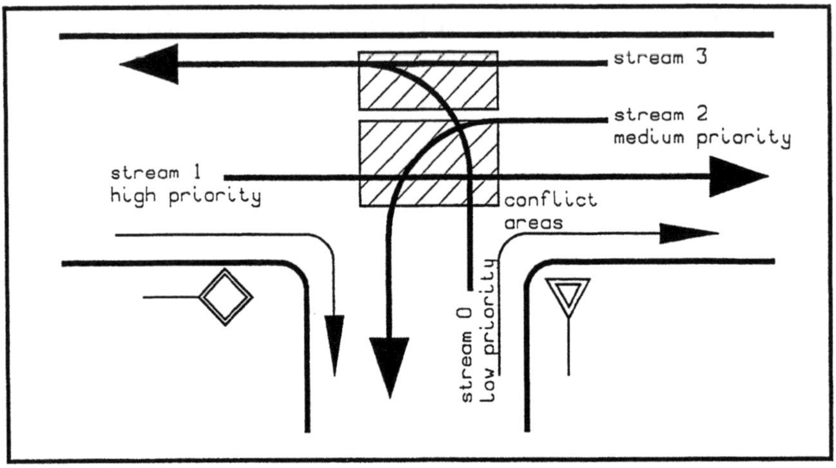

Figure 1: Traffic streams of different priorities

road. It is also applicable to inhomogeneous and inconsistent behavior of drivers on the minor road, i.e. to a stream with different types of vehicles (passenger cars, trucks, busses) with different intended maneuvers (crossing, turning left).

Most known capacity formulae can be rederived as special cases on the basis of weaker assumptions. Thereby it becomes plausible, why some capacity estimates coincide, though they have been developed in the framework of different models.

Our approach utilizes mainly ideas of Tanner [10], [11] and Catchpole, Plank [3].

In section 2 of this paper the model is introduced, and in the following section the capacity formula (12) derived. The proof of the main theorem is given in the appendix. The quantities appearing in (12) are not the same as used in other capacity formulae but closely related. Section 4 serves to present the link, to show how known capacity estimates come out as special cases and to discuss the influence of model assumptions on the estimate. For instance it will turn out, that inhomogeneity decreases capacity while inconsistency has an increasing effect. Finally in section 5 the dependence of the capacity on the gap distribution and on the amount of bunching on the major road will be analysed.

2 Assumptions and notation

On the minor road we assume an infinite queue of vehicles waiting for an opportunity to cross or merge with the major road streams. For an interval $I \subset [0, \infty)$ denote by $N(I)$ the number of departures from the minor road queue during I. The capacity is defined by

$$\lambda_c = \lim_{t \to \infty} \frac{N([0,t])}{t} \tag{1}$$

provided this limit exists a.s.

The model to determine the capacity is given by

a) a description of the major road traffic as seen from the leading vehicle on the minor road and

b) the behavior of minor road vehicles such as

b1) the gap acceptance behavior and

b2) the move–up behavior

a) The driver of the leading vehicle on the minor road observes an alternating process $G_1, B_1, G_2, B_2, \ldots$ of "gaps" and "blocks". We describe a gap and the following block by random variables G and B resp., which may be dependent, but assume that the sequence of pairs (G_i, B_i) is an i.i.d. sequence. For capacity estimation we need only *the distribution of* G and the *mean length* EC of a "major road cycle"

$$C = G + B \tag{2}$$

b) The characteristics of minor road drivers are assumed to be independent of each other and of the major road process.

b1) The gap acceptance behavior of a minor road driver/vehicle is determined by its "type" τ and its "critical gap" A_τ. If a vehicle of type τ leads the queue during a block, the driver accepts the following gap G, if G exceeds the critical gap A_τ. The parameter τ is defined as the probability of acceptance:

$$\tau = \text{prob } (G > A_\tau). \tag{3}$$

Different drivers may belong to different types. We therefore interpret τ as a random variable taking values in some parameter set Γ. In applications Γ will be a finite set (one parameter value each for passenger cars, trucks, busses etc.), but this restriction is not essential for the theory. We only need the assumption that all acceptance probabilities are positive, more precisely

$$\Gamma \subset [\tau_0, 1] \quad \text{for some} \quad \tau_0 > 0.$$

If Γ contains only one element, i.e. if τ is degenerate, we speak of the homogeneous case. Otherwise we have inhomogeneous gap acceptance behavior. The *distribution of* τ is given by its distribution function F. For fixed $\tau \in \Gamma$ the *critical gap* A_τ is a random variable, which may be degenerate. In this case a driver of type τ uses the same critical gap for each acceptance decision (consistent behavior). Otherwise he samples his critical gaps from the distribution of A_τ (inconsistent behavior).

b2) The move–up behavior is determined by a *move–up function* $z(t)$, $t \geq 0$, which is defined as follows: Assume there is a departure from the minor road queue at t_0. Let $Z(t)$ be the (random) number of the further departures within the interval $(t_0, t_0 + t]$ given there is no impediment from the major road traffic during this interval. Then

$$z(t) = EZ(t), \qquad t \geq 0$$

is the expected number of departures.

By these assumptions the model is completely determined. The quantities typed in italics in the preceding text are needed for capacity estimation. Examples for a proper choice are given in section 4.

We close this section by collecting some notation used in the following.

For a nonnegative random variable Y we denote by

$$\begin{aligned} Y(x) &= \text{prob}(Y \le x) && \text{its distribution function} \\ \bar{Y}(x) &= \text{prob}(Y > x) && \text{its survival function} \\ Y^*(s) &= E(e^{-sY}) && \text{its Laplace transform} \\ EY &= \int_0^\infty \bar{Y}(y)\,dy && \text{its expectation.} \end{aligned}$$

By A we denote the marginal critical gap, i.e. a random variable with distribution function

$$A(x) = \int_\Gamma A_\tau(x) F(d\tau). \tag{4}$$

The random variable G_τ is distributed according to the conditional distribution of the gap G given it exceeds the critical gap A_τ:

$$G_\tau(x) = \text{prob}(G \le x \mid G > A_\tau). \tag{5}$$

Furthermore we define

$$\begin{aligned} \lambda &= \text{total flow rate of the major road traffic} \\ \lambda_i &= \text{flow rate of the } i\text{'th stream } (\lambda = \lambda_1 + \cdots + \lambda_n) \\ f_i &= \text{mean occupation time of the conflict area} \\ &\quad\ \text{by a single vehicle of the } i\text{'th stream} \\ m &= \text{mean move-up time for minor road traffic} \\ s &= \text{mean bunch size of the major road traffic.} \end{aligned}$$

3 The capacity formula

From the assumptions formulated in the previous section we obtain, that the departure process from the minor road queue is a regenerative process. Regeneration points are those endpoints of major road cycles that contain an accepted gap. At those instants a new cycle starts, and we have a new vehicle at the head of the minor road queue. This vehicle initiates the following regeneration epoch. It is plausible that vehicles with small acceptance probability τ will more likely reject gaps and thus initiate regeneration epochs than those with large τ. Therefore it is expected that the types of vehicles initiating regeneration epochs will not be distributed according to the distribution function F. This is in fact true as shown in the following theorem.

Theorem 1 *The types $\tau_1, \tau_2, \ldots$ of vehicles heading the minor road queue at successive regeneration points form a Markov chain with state space Γ and transition density*

$$p(\tau', \tau) = 1 + Ez(G_{\tau'} - A) - Ez(G_{\tau'} - A_\tau) \tag{6}$$

with respect to $F(d\tau)$.

Given a regeneration epoch R is initiated by a vehicle of type τ the expected length and the expected number of departures within R are given by

$$ER = EC/\tau \quad \text{and} \quad EN(R) = 1 + Ez(G_\tau - A) \tag{7}$$

The distributions of G_τ and A are defined in (4) and (5).

The proof will be given in the appendix.

The Markov chain (τ_n) possesses a steady state distribution $\tilde{F}(d\tau)$ provided the density (6) is positive and continuous on Γ (see Feller [5, p. 272]). In all cases of practical importance (for instance if Γ is finite) this assumption is satisfied. Let $\tilde{\tau}$ be distributed according to the steady state distribution. Then by the strong law for Markov chains (see Révész [9, p. 130]) we obtain from (1) and (7) the capacity formula

$$\lambda_c = \frac{1 + Ez(G_{\tilde{\tau}} - A)}{EC \cdot E(1/\tilde{\tau})} \,. \tag{8}$$

The steady state distribution $\tilde{F}(d\tau)$, which is needed to calculate the expectations in (8) satisfies the equation

$$\tilde{F}(d\tau) = F(d\tau) \cdot [1 + Ez(G_{\tilde{\tau}} - A) - Ez(G_{\tilde{\tau}} - A_\tau)], \quad \tau \in \Gamma \tag{9}$$

as follows immediately from (6). For finite Γ this is a finite system of linear equations, which can be solved to determine $\tilde{F}(d\tau)$, $\tau \in \Gamma$. This can be avoided by the following assumption on the gap acceptance and move–up parameters: If

$$z(t) = 0 \quad \text{for} \quad t \leq \sup_{\tau, \tau' \in \Gamma} (A_\tau - A_{\tau'}) \tag{10}$$

(see Catchpole, Plank [3], where an equivalent condition is formulated), then

$$z(G - A_{\tau'}) = 0 \quad \text{whenever} \quad G \leq A_\tau, \quad \tau, \tau' \in \Gamma.$$

Therefore

$$Ez(G_\tau - A_{\tau'}) = \frac{1}{\tau} Ez(G - A_{\tau'}), \quad \tau, \tau' \in \Gamma. \tag{11}$$

From (8), (9) and (11) the final capacity formula

$$\lambda_c = \frac{1 + E(z(G - A_\tau)/\tau)}{EC \cdot E(1/\tau)} \tag{12}$$

can easily be deduced.

In the homogeneous case, where

$$A_\tau = A, \quad \tau = P(G > A)$$

we obtain

$$\lambda_c = \frac{P(G > A) \cdot (1 + E(z(G - A) \mid G > A))}{EC} \,. \tag{13}$$

4 Specialized models

In this section we will introduce more detailed assumptions for the model. We want to show how the quantities appearing in formula (12) are related to the parameters commonly used in capacity estimates and thus rederive these estimates as special cases of (12). At the same time we will study the influence of model assumptions onto the capacity estimate.

4.1 Major road cycles

Most capacity estimates are based on *independent headways* (in our reference list all with exception of Tanner's results). In our model this case is obtained by taking $B = 0$. Then necessarily

$$EC = EG = 1/\lambda \tag{14}$$

with λ = total flow rate on the major road.

If we assume that each major road vehicle occupies the conflict area for a (random) time interval of length F with $f = EF$ and that a block is generated by (say) S *vehicles in a bunch*, they block the conflict area during an interval of length

$$B = F_1 + \cdots + F_S \, .$$

Let S be independent of $F_1, F_2, \ldots$ then

$$EB = ES \cdot EF = s \cdot f \quad \text{and} \quad s = ES = \lambda \cdot EC.$$

Therefore

$$EC = s/\lambda \quad \text{and} \quad EG = s(1 - \lambda f)/\lambda \tag{15}$$

Tanner [10] considers blocks, which are the busy periods of a M/G/1-queue with deterministic service time of length f. In this model S obeys the Borel–Tanner distribution and has expectation $s = 1/(1 - \lambda f)$. Thus

$$EC = 1/\lambda(1 - \lambda f) \quad \text{and} \quad EG = 1/\lambda.$$

Exactly the same capacity estimate as Tanner's will be obtained, whenever S obeys any other distribution with the same expected value, for instance a geometric distribution (see Plank, Catchpole [8]).

Assume n *lanes on the major road* with flow rates λ_i, mean occupation times f_i and mean bunch sizes s_i. Then as in (15) we have on the i'th lane

$$EC_i = s_i/\lambda_i \quad \text{and} \quad EG_i = s_i(1 - \lambda_i f_i)/\lambda_i.$$

Assume furthermore that the gaps are exponentially distributed and the processes on the n lanes are independent and in steady state. Then a driver on the minor road viewing all lanes together observes exponential gaps with mean

$$EG = \Big(\sum_{i=1}^{n} \frac{\lambda_i}{s_i(1 - \lambda_i f_i)}\Big)^{-1} . \tag{16}$$

The probability to observe at some instant t a gap on the i'th lane is $p_i = EG_i/EC_i = 1 - \lambda_i f_i$. Therefore we obtain for the probability to observe a gap common to all n lanes

$$p = \prod_{i=1}^{n}(1 - \lambda_i f_i) = EG/EC.$$

Thus

$$EC = EG/\prod_{i=1}^{n}(1 - \lambda_i f_i) \tag{17}$$

with EG given by (16). Tanner's capacity formula in [11] is obtained by this specification (see also Troutbeck [12]).

If we have on the major road several traffic streams of different priorities using the same conflict area *(impedance)*, then a block will be the busy period of a queueing system with priorities. It is well known from queueing theory that there is no chance to obtain a simple explicit formula for the expected length of this period, even if all input streams are Poissonian. In an earlier paper [14] we analysed the problem in detail: It is assumed that there are n Poissonian streams of decreasing priority with flow rates $\lambda_1, \ldots, \lambda_n$. The vehicles of the i'th stream block the conflict area during their approach for vehicles of lower priority (the (random) blocking time is denoted by A_i) and during their crossing maneuver for all vehicles (this blocking time is denoted by F_i). Let $\Lambda_i = \lambda_1 + \cdots + \lambda_i$ be the total flow rate of the first i streams, let B_i be the length of the busy period generated by these streams and denote by $b_i = EB_i$ its expectation. Then as shown in [14]

$$b_1 = \frac{1}{\lambda_1}\left(-1 + \frac{A_1^*(-\lambda_1)}{1 - \lambda_1 f_1}\right) \tag{18}$$

and for $i = 2, \ldots, n$

$$b_i = \frac{1}{\Lambda_i}\left[-1 + \frac{1 + \Lambda_{i-1}b_{i-1}}{1 - \lambda_i f_i(1 + \Lambda_{i-1}b_{i-1})}\left(1 + \lambda_i \frac{A_i^*(-\beta_i) - 1}{\beta_i}\right)\right] \tag{19}$$

with

$$\beta_i = \Lambda_i - \Lambda_{i-1}B_{i-1}^*(\lambda_i). \tag{20}$$

These equations allow in principle to determine recursively $b_2, \ldots, b_n$ and thus

$$EC = 1/\Lambda_n + b_n. \tag{21}$$

The main problem is to determine β_i from (20) because this involves the evaluation of the Laplace transform $B_{i-1}^*(s)$ at $s = \lambda_i$, which is possible only after solution of a complicated nonlinear equation. Expanding this transform up to the linear term we obtain the approximation

$$\beta_i = \lambda_i(1 + \Lambda_{i-1}b_{i-1}) \tag{22}$$

which simplifies (19) to

$$b_i = \frac{1}{\Lambda_i}\Big(- 1 + \frac{A_i^*(-\beta_i) + \Lambda_{i-1}b_{i-1}}{1 - \lambda_i f_i(1 + \Lambda_{i-1}b_{i-1}}\Big). \tag{23}$$

With equations (18), (22) and (23) the recursion to determine b_n and finally EC in (21) can now easily be carried out.

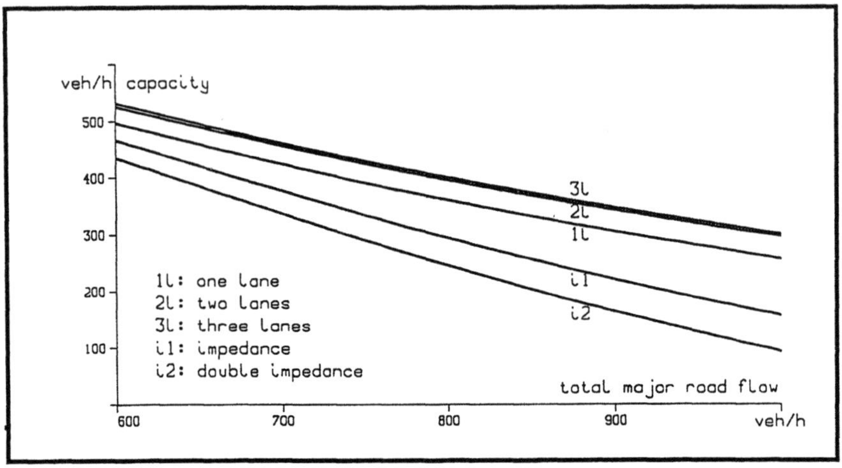

Figure 2: Capacity estimates for multilane traffic and impedance as functions of the total flow rate

In Fig. 2 capacity estimates derived under different model assumptions for the major road traffic are compared. In all examples the same total major flow rate λ is assumed. The mean cycle length EC is calculated by equation

(i) (15) for a single lane

(ii) (17) for 2 or 3 nonconflicting lanes

(iii) (18), (21)–(23) for 2 or 3 conflicting streams

The parameters were chosen as follows

(i) $f = 1.5$ sec, $\quad s = 1/(1 - \lambda f)$

(ii) $\lambda_i = \lambda/n, \quad f_i = 1.5$ sec, $\quad s_i = 1/(1 - \lambda_i f_i) \quad$ for $\quad i \leq n, \; n = 2, 3$

(iii) $\lambda_i = \lambda/n, \quad f_i = 1.5$ sec, $\quad A_i = 5$ sec $\qquad$ for $\quad i \leq n, \; n = 2, 3.$

Homogeneous and consistant gap acceptance behavior and exponential gaps were assumed such that the capacity estimate (29) could be applied with parameters $\alpha = 1$, $A = 5$ sec, $m = 3$ sec and $\gamma = \lambda$. In the model for impedance the blocking time during the approach of the major road vehicles is incorporated into the block B. Therefore the remaining gaps are accepted with probability 1. Consequently in case (iii) the critical gap A was set to 0.

4.2 Distribution of gaps

Most capacity estimates in literature are based on *exponentially* distributed gaps (see for instance [2], [6], [10] and [11]). This assumption is criticized frequently because headways never fall below some minimum value t_0. Therefore *shifted exponential* or more generally *mixtures* of deterministic and shifted exponential distributions are considered (see for instance [2], [3], [7] , [8] and [12]). These mixtures seem to have been introduced by Buckley (see [8]) but are used independently by Jacobs et al. ([6] and earlier papers).

They allow the modelling of bunching: Assume headways are independent and take the constant value t_0 with probability p and are shifted exponential with probability $1-p$. If we interpret vehicles with headways t_0 to form a bunch, we obtain a traffic flow with bunches of expected length t_0/p separated by exponential gaps. Therefore these models are in fact special gap–block models. It is not surprising that the associated capacity estimates are equal to estimates obtained by other models with the same expected block length (see Plank, Catchpole [8] compared with Tanner [10] and Troutbeck [12] compared with Tanner [11]).

The common characteristic of exponential, shifted exponential distributions and mixtures as described above is that they have exponential tails, i.e. that for some α, γ and t_0

$$P(G > t) = \alpha \cdot e^{-\gamma t}, \qquad t \geq t_0 .$$

If $A_\tau \geq t_0,\ \tau \in \Gamma$, we obtain

$$\tau = P(G > A_\tau) = \alpha \cdot E(e^{-\gamma A_\tau}) = \alpha \cdot A_\tau^*(\gamma) \tag{24}$$

and

$$E(z(G - A_\tau)/\tau) = \int_0^\infty z(t)\gamma e^{-\gamma t} dt = z^*(\gamma) . \tag{25}$$

Substituting (24) and (25) into (12) leads to the capacity estimate

$$\lambda_C = \frac{1 + z^*(\gamma)}{EC \cdot E(1/\tau)} . \tag{26}$$

The distribution of types τ enters this formula only through the expectation $E(1/\tau)$ in the denominator. This is no longer true for *other gap distributions.* We will discuss the influence of gap distributions on the capacity estimate in section 5.

4.3 Characteristics of minor road drivers

The *gap acceptance probabilities* τ are uniquely determined by the distributions of gaps G and critical gaps A_τ (see (3)). This has already been utilized in the estimate (13) for the homogeneous case. *Inhomogeneous gap acceptance behavior* is discussed by Catchpole, Plank [3] and Jacobs et al. [7] (see also Harders [6] and Brilon [2]). The effect of departures from the homogeneous case is complicated to analyse unless gap distributions with exponential tails are assumed. Then estimate (26) is applicable.

Denote by λ_C^{inhom} the estimate (26) and by λ_C^{hom} the capacity in the homogeneous case with acceptance probability $\tau = \tau_0 = E(\tau)$. Then

$$\lambda_C^{\text{inhom}} / \lambda_C^{\text{hom}} = \frac{1}{E(1/\tau)} \cdot \frac{1}{E(\tau)} \ .$$

As the harmonic mean $1/E(1/\tau)$ is always less than or equal to the mean $E(\tau)$ we observe a decrease of capacity for departures from the homogeneous case. This fact has already been noted by Catchpole, Plank [3].

The relation between consistent and inconsistent behavior, i.e. the influence of the *critical gap distribution* on the capacity estimate will again be discussed only in the case of gaps with exponential tails. From (13), (24) and (25) it can be concluded that the replacement of the (random) critical gap A by its expected value EA changes the capacity estimate by the factor

$$E(e^{\gamma(A-EA)}) \cong 1 + \tfrac{1}{2}\gamma^2 \text{Var}\ (A)$$

for $\gamma(A - EA)$ small. Because of

$$\lambda_C^{\text{incon}} / \lambda_C^{\text{con}} \cong 1/(1 + \tfrac{1}{2}\gamma^2 \text{Var}\ (A))$$

at least for low flow rates and/or small departures from consistency a decrease of capacity is to be expected (see Plank, Catchpole [8]).

To obtain a reasonable model for the *move-up function* $z(t)$ we assume that minor road vehicles depart at instants $t_0, t_0 + M_1, t_0 + M_1 + M_2, \ldots$ as long as possible concerning the major road traffic, where $M_1, M_2, \ldots$ is an *i.i.d.* sequence with $m = EM$. Then the number of departures until $t_0 + t$ equals

$$Z(t) = \max\{k : M_1 + \cdots + M_k \leq t\} \ .$$

Therefore

$$z(t) = EZ(t) = \sum_{k=1}^{\infty} \text{prob}\ (M_1 + \cdots + M_k \leq t) =: U_M(t) \tag{27}$$

is the renewal function associated with the random variable M. For constant $M = m$ this function equals

$$z(t) = [t/m], \tag{28}$$

the integer part of t/m. This function is widely used for capacity estimation (see [2], [3], [4], [10], [12]). For random M see Plank, Catchpole [8]. It is easily verified that with $z(t)$ as defined in (27)

$$1 + z^*(t) = \frac{1}{1 - M^*(\gamma)}$$

which equals $1/(1 - e^{-\gamma m})$ in case (28). For gaps with exponential tails we therefore obtain in the homogeneous case from (26)

$$\lambda_C = \frac{\alpha \cdot A^*(\gamma)}{EC \cdot (1 - M^*(\gamma))} \tag{29}$$

(see Plank, Catchpole [8]).

An argument as used above to study the influence of inconsistency on capacity shows that random move–up times increase the capacity compared to constant $M = m$. For general gap distributions the numerator in (12)

$$1 + E(z(G - A_\tau)/\tau) = 1 + E(U_M(G - A_\tau)/\tau)$$

can not be evaluated explicitly. Therefore we use a linear approximation of the renewal function $U_M(t)$:

$$U_M(t) \cong \tfrac{t}{m} - \tfrac{1}{2} + \tfrac{v}{2}, \qquad t \geq 0$$

(see Feller [5, p. 366]), where $v = \text{Var}\,(M)/m^2$ is the square of the coefficient of variation of M. This approximation suggests the move–up function

$$z(t) = \left(\tfrac{t}{m} - \tfrac{1}{2} + \tfrac{v}{2}\right)^+, \qquad t \geq 0. \tag{30}$$

The +–sign indicates that $z(t) = 0$ whenever the term in parentheses is negative. With this choice we obtain from (12) the capacity estimate

$$\lambda_C = \frac{1 + E((G - \bar{A}_\tau)^+/\tau \cdot m)}{EC \cdot E(1/\tau)}$$

and in the homogeneous case from (13)

$$\lambda_C = \frac{P(G > A)}{EC}(1 + \tfrac{1}{m}E((G - \bar{A})^+ \mid G > A)) \tag{31}$$

with $\bar{A}_\tau = A_\tau + \frac{m}{2}(1 - v)$ and $\bar{A} = A + \frac{m}{2}(1 - v)$.

Again an increase of capacity with increasing variance of M is obvious.

5 The dependence of capacity from gap distributions and bunching

In this section homogeneous and consistent gap acceptance behavior is assumed. We use the estimate (31) with

$$EC = s/\lambda \quad \text{and} \quad EG = s(1 - \lambda f)/\lambda$$

(see (15)), and compare the estimates obtained for different distributions of G as functions of the mean bunch size s. From (31) we obtain

$$\lambda_C = \frac{1 - \lambda f}{m} - \frac{\lambda}{s}\Big(\frac{1}{m}\int_0^{\bar{A}} \bar{G}(t)\, dt - \bar{G}(A)\Big). \tag{32}$$

As $\bar{G}(t)$ is a decreasing function with $\bar{G}(0) = 1$

$$\bar{A} \cdot \bar{G}(\bar{A}) \leq \int_0^{\bar{A}} \bar{G}(t)\, dt \leq \bar{A}\,.$$

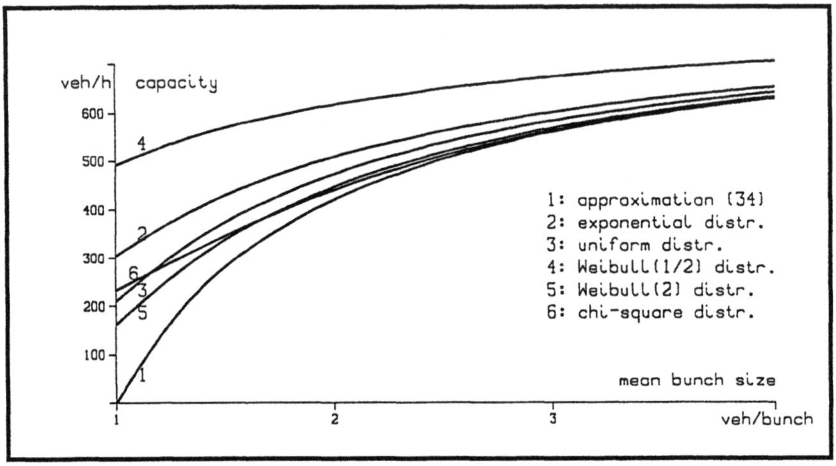

Figure 3: Capacity estimates for different gap distributions

Therefore

$$\frac{1-\lambda f}{m} - \frac{\lambda}{s}\Big(\frac{\bar{A}}{m} - \bar{G}(A)\Big) \leq \lambda_C \leq \frac{1-\lambda f}{m} - \frac{\lambda}{s}\Big(\frac{\bar{A}\cdot\bar{G}(\bar{A})}{m} - \bar{G}(A)\Big)\,. \tag{33}$$

For $\bar{G}(\bar{A}) = 1$ both bounds coincide and we get the estimate

$$\lambda_C = \frac{1-\lambda f}{m} - \frac{\lambda}{s}\Big(\frac{A}{m} - \frac{1+v}{2}\Big)\,. \tag{34}$$

For increasing bunches the gaps will get longer and $\bar{G}(\bar{A})$ can be expected to be close to 1. Therefore the estimate (34), which does not depend on the distribution of G, should approximate the "true" capacity for increasing s.

The following numerical examples indicate that this is in fact true.

In Fig. 3 the capacity estimates (32) for different gap distributions and the approximation (34) as functions of the mean bunch size on the major road are plotted. The parameters are chosen as follows:

$$\lambda = 600 \text{ veh/h}, \quad f = 1.5 \text{ sec}, \quad A = 5 \text{ sec}, \quad m = 3 \text{ sec} \quad \text{and} \quad v = 0.$$

Then $\bar{A} = 6.5$ sec and $g = EG = 4.5 \cdot s$. The capacities have been calculated for the following distributions of gaps:

Exponential distribution with parameter $\gamma = 1/g$:

$$\bar{G}(t) = e^{-\gamma t}, \quad EG = g, \quad \text{Var}\,(G) = g^2$$

Uniform distribution on the interval $[0, 2\cdot g]$

$$\bar{G}(t) = 1 - t/2g, \quad 0 \leq t \leq 2g, \quad EG = g, \quad \text{Var}\,(G) = g^2/3$$

Weibull distributions with parameters $\alpha = (\Gamma(1+1/\beta)/g)^\beta$, $\beta = \frac{1}{2}$ and $\beta = 2$.

$$\bar{G}(t) = e^{-\alpha t^\beta}, \quad EG = g,$$
$$\text{Var}\,(G) = 5g^2 \text{ for } \beta = \tfrac{1}{2} \quad \text{and} \quad \text{Var}\,(G) = 0.27g^2 \text{ for } \beta = 2.$$

Chi square distribution with g degrees of freedom

$$\begin{aligned} \bar{G}(t) &= \int_t^\infty (2^{g/2}\Gamma(g/2))^{-1} t^{g/2-1} e^{-t/2} dt \\ EG &= g, \quad \text{Var}\,(G) = 2g. \end{aligned}$$

The plot shows that there are considerable differences between capacity estimates for different gap distributions. With increasing amount of bunching the differences disappear and the approximation (34) becomes a realistic estimate. The plot indicates that the capacity increases with increasing variance of the gaps. An extreme example is the Weibull (1/2)–distribution with variance $5g^2$.

Appendix

In this section we present the proof of the theorem in section 3. The essential idea of the proof is due to Catchpole, Plank [3].

Assume at time t_0 a regeneration epoch starts that is initiated by a vehicle of type τ'. Let this vehicle be given number 0. The vehicles queueing behind it have types $\tau_1, \tau_2, \ldots$. The driver of vehicle 0 will eventually accept some gap and depart. The accepted gap is distributed like $G_{\tau'}$ (see (5)); let its actual length equal g. A random number $N(g)$ of queuers are able to follow until vehicle number $N(g)+1$ is stopped, because the next major road block comes up. It will initiate the following regeneration epoch. Given $N(g) = k$ the type of this vehicle is τ_{k+1}. We calculate the probability $G(\tau)$ of the event $(\tau_{k+1} \le \tau)$:

$$\begin{aligned} G(\tau) &= \sum_{k=1}^\infty P(N(g) = k,\ \tau_{k+1} \le \tau) \\ &= \sum_{k=1}^\infty P(N(g) \ge k,\ \tau_{k+1} \le \tau) - \sum_{k=1}^\infty P(N(g) \ge k+1,\ \tau_{k+1} \le \tau). \end{aligned}$$

The events $(N(g) \ge k)$ and $(\tau_{k+1} \le \tau)$ are independent and $(N(g) \ge k+1)$ occurs if and only if $Z(g - A_{\tau_k}) \ge k$. Therefore

$$G(\tau) = \sum_{k=1}^\infty \int_0^1 P(Z(g-A_t) \ge k-1)\, F(dt) \cdot F(\tau) - \sum_{k=1}^\infty \int_0^\tau P(Z(g-A_t) \ge k)\, F(dt)$$

and after interchanging the order of summation and integration

$$G(\tau) = \int_0^1 (1 + Ez(g-A_t))F(dt) \cdot F(\tau) - \int_0^\tau Ez(g-A_t)F(dt).$$

Equation (6) now follows by integration with respect to $G_{\tau'}(dg)$.

To prove (7) observe that a regeneration epoch R contains a random number K of major road cycles with rejected gaps and a last one with an accepted gap. Departures are only possible during the gap of the last cycle. Given the epoch is initiated by a vehicle of type τ the number K is geometrically distributed:

$$P(K = k) = (1-\tau)^k \cdot \tau, \qquad k = 0, 1, \dots .$$

Therefore

$$\begin{aligned} ER &= EK \cdot E(C \mid G \leq A_\tau) + E(C \mid G > A_\tau) \\ &= \frac{1-\tau}{\tau} \cdot E(C \mid G \leq A_\tau) + E(C \mid G > A_\tau) = \frac{1}{\tau} EC \end{aligned}$$

and

$$\begin{aligned} EN(R) &= E(N(G) \mid G > A_\tau) \\ &= 1 + \sum_{k=1}^{\infty} \int_0^1 P(Z(G - A_{\tau'}) \geq k \mid G > A_\tau)\, F(d\tau') \\ &= 1 + \int_0^1 Ez(G_\tau - A_{\tau'})\, F(d\tau') = 1 + Ez(G_\tau - A). \end{aligned}$$

These equations complete the proof.

References

[1] Brilon, W.(Ed.): Intersections without traffic signals. Springer Berlin 1988.

[2] Brilon, W.: Recent developments in calculation methods for unsignalized intersections in West Germany. In [1], 111–153.

[3] Catchpole, E.A.; Plank, A.W.: The capacity of a priority intersection. Trans. Res. B 20B(6), 441–456(1986)

[4] Dunne, M.C.; Buckley, D.J.: Delays and capacities at unsignalized intersections. Proc. 6th ARRB Conf. 6, 345–362(1975)

[5] Feller, W.: An introduction to probability theory and its applications II. Wiley, N.Y. 1971

[6] Harders, J.: Grenz- und Folgezeitlücken als Grundlage für die Leistungsfähigkeit von Landstraßen. Schriftenreihe Straßenbau und Verkehrstechnik 216(1976)

[7] Jacobs, F.; Anduszies, H.; Hertkorn, W.: Leistungsfähigkeitsberechnung von Knotenpunkten ohne Lichtsignalanlagen. Veröff. a.d. Inst. f. Straßen- und Verkehrswesen 1-29, Univ. Stuttgart 1989

[8] Plank, A.W.; Catchpole, E.A.: A general capacity formula for an uncontrolled intersection. Traffic Eng. and Control 25, 327–329(1984)

[9] Révész,P.: The laws of large numbers. Academic Press N.Y. 1968

[10] Tanner, J.C.: A theoretical analysis of delays at an uncontrolled intersection. Biometrica 49, 163–170(1962)

[11] Tanner, J.C.: The capacity of an uncontrolled intersection. Biometrika 54, 657–658(1967)

[12] Troutbeck, R.J.: Average delay at an unsignalized intersection with two major streams each having a dichotomised headway distribution. Trans.Sci. 20(4), 272–286(1986)

[13] Troutbeck, R.J.: Current and future Australian practices for the design of unsignalized intersections. In [1], 1–19.

[14] Wegmann, H.: Impedance effects on capacity estimation. ZOR (1991), to appear.

Intersections Without Traffic Signals - Swedish Experience on Capacity and Traffic Safety

T. BERGH

National Swedish Road Administration

Summary

The objectives of this paper are twofolded. The first is to summarize the main results of Swedish research on capacity and traffic safety at intersections as they have been presented to practicing traffic engineers. The second is to describe the principles of the modern Swedish traffic engineering guidelines on choice of intersection type and control, on evaluation of alternative intersection types and on geometric design.

The paper also gives a background to the guidelines and the capacity and traffic safety research program they rely on.

First Generation Swedish Traffic Engineering Guidelines

The first Swedish traffic engineering guidelines on intersections were published in 1967 for rural areas (ref 1) and in 1973 for urban areas (ref 2). Both defined a number of standard intersection types including traffic control, gave recommendations at which traffic flows they were to be used and how they were to be designed.

Choice of Intersection Type and Traffic Control

The criteria for choice of intersection type and traffic control were based on level-of-service considerations for a future design hour to be decided, all stemming from the HCM 50. In rural areas, for instance, left turn lanes was recommended if through traffic on the primary road in the intersection otherwise would be as impeded as traffic on a two-lane highway alone, at a traffic flow requesting motorway at that time.

In urban areas traffic signals or roundabouts (if fairly equally distributed traffic) were recommended if the degree of saturation, calculated with the 2 second rule, would exceed 0.7, see Figure 1. For existing intersections an accident rate of 1.0 accidents/million vehicles indicated a need to change from give way or stop control.

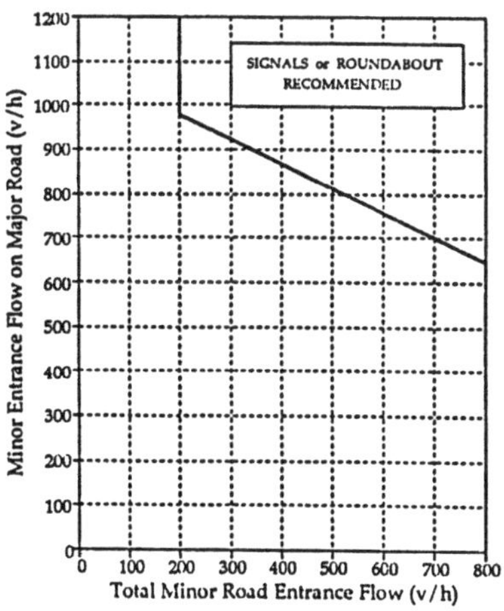

Fig.1. Traffic flow criteria for traffic signals in the Swedish 1973 urban guidelines

Layout

Left turn lanes with kerbed traffic islands were recommended as well as kerbed secondary road islands. Roundabouts were suggested to have inner radius bigger than 40m, implying a weaving mode of operation. Since the mid sixties Swedish roundabouts have operated with circling traffic priority.

Research on Traffic Capacity and Safety

The lack of Swedish empirical data and research on how different intersection types and designs affect capacity, level-of-service, delays, queueing, stops, vehicle operating costs and perhaps most important traffic safety were obvious during the sixties guideline work. Many basic assumptions in the guidelines with great economic impact relied on traffic engineering judgment. Therefore in the seventies the National Swedish Road Administration initiated two long term research programs - Intersection Capacity and Intersection Traffic Safety. The objectives were to fill the lack of empirical data, to produce capacity, vehicle user cost and traffic safety models from these data to support project evaluation studies and if necessary to revise the traffic engineering guidelines from the sixties.

Capacity

The work on traffic capacity in unsignalized intersections was undertaken in cooperation with Arne Hansson. The final results are a manual for capacity calculations (ref 3) and a main-frame computer program CAPCAL (ref 4), later converted to a IBM-PC version. The proposed revisions of these methods reported at the preceding workshop and at the ARRB conference 1986 (ref 5,6) are still incomplete. The empirical data and the theory behind the derived models are described in detail elsewhere (ref 7,8). The models rely on the gap-acceptance theory inspired by the German researchers Jessen, Harders and Siegloch. Roundabouts are treated as a series of give-way controlled T-junctions.

Here follows a straightforward description step by step of the actual manual procedures to calculate capacity, delay, queues and stops for a give-way or stop controlled entrance to an intersection without traffic signals.

STEP 1 - Split the entrance into "lane groups." Rule: All lanes having one traffic movement in common belong to the same "lane group", see Figure 2. All approaches have at least one long lane (>30m). For short lane correction, see step 6. For lane widths, see comment.

Example:

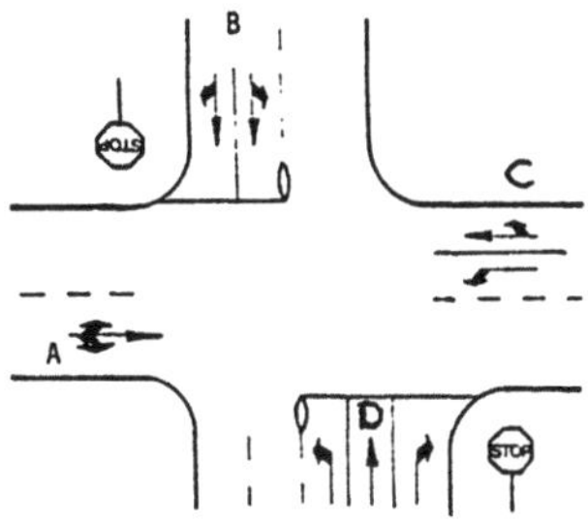

A 1 lane group

B 1 lane group

C 2 lane groups: straight+right/left

D 3 lane groups: right/straight/left

Fig.2. Example of dividing into lane groups.

Comment: In CAPCAL the traffic signal corrections for lane widths and lane markings are applied. This means that lanes wider than 4.9m are to be divided into two unmarked lanes. Resulting capacities are corrected according to Figure 3.

Correction k; Ca'=k*Ca; Ca= lane group capacity (1)

C'a= corr. capacity

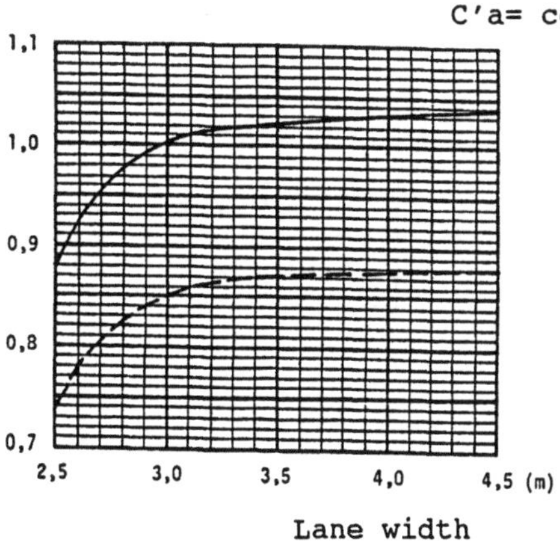

——— marked lane

- - - unmarked lane

Fig.3. Correction of capacity due to lane width and markings

STEP 2 - Calculate higher rank flows q_r for lower rank movements Rules according to following formulae. For notation see Figure 4:

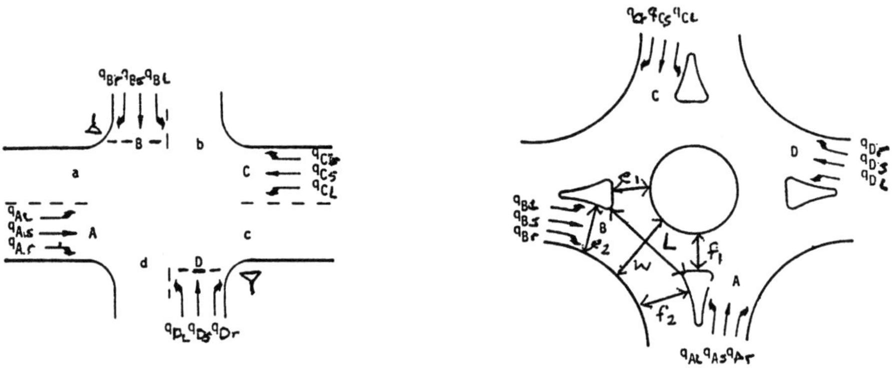

N = number of exit lanes

Fig.4. Notation for calculation of higher rank flows

Minor flow	Four-way intersection Higher rank major flow	Roundabout Higher rank major flow
q_{Al}	$q_{Cs}+q_{Cs}/N_b$ (2)	
q_{Br}	q_{Cs}/N_a (3)	$x_2*q_{Cs}+x_2*q_{Dl}$ (4)
q_{Bs}	$q_{As}+q_{Al}+q_{Ar}/N_d+q_{Cs}+q_{Cl}/N_d$ (5)	$x_1*q_{Cl}+q_{Cs}+q_{Dl}$ (6)
q_{Bl}	$q_{Al}+q_{As}/N_c+q_{Cs}+q_{Cl}+q_{Ds}+q_{Dr}/N_c$ (7)	$q_{Cl}+q_{Cr}+q_{Dl}$ (8)

x_1 and x_2:

$$x_1=1.0-(w-6)*(0.08*f_1-.06) \quad \text{if } (e_1+e_2)<w;\ x_1=e_1+e_2 \qquad (9)$$

$$x_2=1.0-(w-5)*(.07*f_2+.028) \quad \text{if } (e_1+e_2)<w;\ x_2=e_1+e_2 \qquad (10)$$

Comment: The way to derive higher rank flows is partly derived from empirical data for intersections and from literature and traffic engineering judgement for roundabouts.

STEP 3 - Critical gaps a

The gaps for the different movements depend on speed limit and type of control for intersections, see Figure 5, and on the relation width/length for roundabouts, see formulas (11) and (12).

Movement	Primary road speed limit				
	50		70		90
	G	S	G	S	G
Primary left	5.0	-	5.8	-	6.5
Secondary left	5.3	6.0	6.2	6.8	7.5
Secondary straigth	5.2	5.8	6.0	6.5	7.0
Secondary right	4.8	5.5	6.0	6.5	7.2

G = Give-way S = Stop

Fig.5. Basic value for intersection critical gaps a (s)

Roundabouts:

a_r=3.06+1.1*w/L+2.375*(w/L)^2; w and L see figure 4 (11)

a_{sl}=3.06+2.6*w/L+2.75*(w/L)^2; w and L see figure 4 (12)

Corrections for city size according to Figure 6 and percentage of trucks according to formula (13) are made for all movements, for secondary roads for left turners and crossing movements for step-wise crossing according to Figure 6, and for right turners for intersection angle/radius according to formula (14).

a_{corr}=a-.01*(HVm-10); (13)

a_{corr}=a+1-(1+(Rad-12)/18*(1-(Ang-90)/120); (14)

HVm= heavy truck % in movement

Rad= right turn radius (m)

Ang= right turn angle (degrees)

City size correction for all movements		Correction for possibility of two step passing for left turners and crossings from secondary road			
Thousands of inhabitants	Corr	Possibility of two-steps	Lanes primary road		
			2	3-4	>4
< 30	+ 0.2	No	0	+ .3	+ .6
30 - 300	0	Yes/No	+ .2	+ .5	+ .8
>300	- 0.8	Yes	- .5	0	+ .3
rural	+ 0.2				

Fig.6. Gap corrections for city size for all movements and for possibilty of two-steps for left turning and crossing from a secondary road.

Comment: All corrections are additive. The follow-up time is supposed to be 0.6*a. The basic gaps for intersections partly rely on Swedish empirical data. For roundabouts they are estimated from literature. The CAPCAL-program has a correction for grade too, as for traffic signals.

The proposed new method (ref 9) will introduce the following changes based on literature review:

- fairly decreased values for right turners
- gradient and city size corrections are taken away
- 90 km/h is increased .2s for the rural case
- a junction width correction is introduced:

  a_{corr}=a+.006*(W-10); W= primary road width (m) (15)
- a flow correction is introduced:

  a_{corr}=a+1.8*(qr-.139); q_r=higher rank flow (v/s) (16)

STEP 4 - Service-times b

There are four cases A, B, C, and D, describing different major flow headway distributions. The choice depends on movement, number of lanes and distance to traffic signal. Correction is made if it is possible to pass in two steps. For roundabouts there is only one case, D.

CASE A Tanner bunched headways . When 2-lane primary road, for left turners from primary road and for right turners from secondary roads.

CASE B Exponential headways. When 2-lane primary road for crossing and left turning from secondary road. When 4-lane primary road for left turn from primary road and right turn from secondary road.

Case C Pearson type III distribution type. Always when traffic signals within 300m on primary road. When 4-lane primary road for crossing and left turning from secondary road.

CASE D Combination of case b and c. For roundabouts.

CASE A $b=\exp(-1.8*q_r)*(\exp(a*q_r)-\exp(.4*a*q_r))/$
$/(q_r*(1-1.8*q_r))$ (17)

CASE B $b=(\exp(a*q_r)-\exp(.4*a*q_r))/q_r$ (18)

CASE C $b=(5/q_r)*(\exp(.4*a*q_r)-\exp(.24a*q_r))*(1-.395*\exp$
$(-.546a*q_r))\hat{}(1/.73)$ (19)

CASE D $b=(1-x)*CaseB+x*CaseC;\ x=.29*q_r*\exp(.25*a)$ (20)
q_r= higher rank flow (v/s), see step 2
a= critical gap a (s), see step 3

For left turning and crossing from secondary road correction is made depending on the possibility of two step passing according to Figure 7.

Correction b'=b*corr

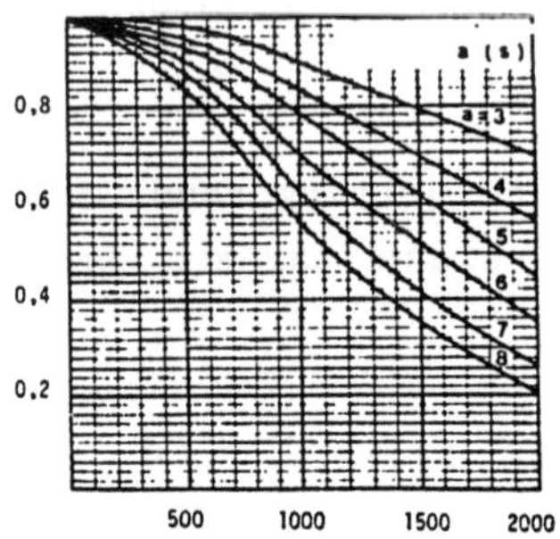

Critical gap a (s)

Higher rank flow (v/h)

Fig.7. Service-time correction for two-step passing for left turners and croosing movements from secondary road

Comment: Two-step passing correction is based on simulation. Analysis of the Swedish empirical data has not been able to find any Pearson III type distributions (variance ratio less than 1). Therefore in the new proposal case C is taken away for intersections. There are also proposals to make bunching steeper in case A and to weight service-times with and without queue at low higher rank flows.

STEP 5 Partial degrees of saturation and secondary conflicts corrections

Firstly partial degrees of saturation Bp are calculated for all movements:

$$Bp=qm*bm; \tag{21}$$

qm= movement flow (v/s), b= service-time, see step 4

Secondly, if give-way or stop control, for movements according to Figure 8 correct the degree of saturation to Bm' because of secondary conflicts:

$$Bpm'=Bpm/(1-Bpm_1/N_{m1})(1-Bpm_2/N_{m2})..; \tag{22}$$

Bpm_1..= partial degrees of saturation for secondary higher rank movements, see Figure 10

$Nm_{1..}$ = number of lanes for these movements

Lower rank movement	Secondary higher rank movement	
	T-intersection	Four-way
q_{Bs}		q_{A1}, q_{C1}
q_{B1}	q_{A1}	q_{A1}, q_{C1}, q_{Ds}
q_{Ds}		q_{A1}, q_{C1}
q_{D1}	q_{C1}	q_{A1}, q_{C1}, q_{Bs}

Figure 8 Secondary higher rank movements

STEP 6 Calculate degree of saturation Ba for the "approaches"

$Ba=(Bpm_1+Bpm_2+..)/Na;$ (23)

Bpm_i=partial degree of saturation, see step 5

Na=number of lanes in the "approach"

For the special case with an approach with two movements with one long and one short lane (<30m) a reduction of the degree of saturation is made, based on probability theory.

Bcorr=Ba-(1-(pBmin)^Nc)*Bmin; (24)

Bmin = lowest of the partial degrees of saturation

p_{Bmin} = percentage of flow for that movement

Nc = possible number of cars in the short lane

Comment: A far more complicated model is proposed for the new Capcal.

STEP 7 Calculate capacity Ca (v/h) for the approach as:

Ca=Qa/Ba (v/h); (25)

Ba=degree of saturation for the approach, see step 7

Qa=approach flow (v/h)

STEP 8 Queues C (no of cars), delays d (s/v) and percentage stops ps for the approach are calculated as follows:

$C=.75*Ba/(1-Ba)$ (26)

$d=C*Na*/q_a$ (27)

$ps=100(1-p_f+p_f*Ba)$ (28)

$pf=exp(a*q_r)$ (29)

Ba=approach degree of saturation

Na=number of lanes in the approach

q_a=approach flow (v/s)

p_f=probability to catch first lag

a=average critical gap (grams/stop swedish car flect 1990) in the approach (s)

q_r=higher rank flow (v/s)

Distributional percentiles P (for instance P=0.01) and corresponding waiting time d_{max} are estimated:

$d_{max}=3*lnP/(ln(.3d-ln(1+.3d))$; d= average delay (s) (30)

<u>Vehicle Operating Costs and Emissions</u>

The traffic engineering evaluation guidelines (ref 10) calculate extra vehicle operating costs, VOC, and emissions, CH, NOx, CO and CO2 for an intersection in the following manner. The methodology is derived from the VTI VETO-computer program (ref 11) made by Ulf Hammarstroem. VETO, requesting very detailed input data on geometry and vehicle parameters, can be used for more detailed calculations.

All turning vehicles and all vehicles stopped or almost stopped by stop or give-way control or by traffic interaction, that is left turning and straight ahead primary road traffic, are expected to have an extra VOC (SEK, 1990) and extra exhaustions of CH, NOx, CO and CO2 according to Figure 9. Emissions are also estimated for 2000 and 2010 forecasting changes in the Swedish car fleet.

In a roundabout with normal size (inner circle 20m) all entering vehicles are assumed to have an extra VOC and extra emissions that is .75* value according to Figure 9.

Speed limit (km/h)	Extra VOC, CH, NOx, CO and CO2 for stopping and "close" stopped/turning for Swedish car fleet 1990									
	Cars					Truck/Bus				
	VOC	CH	NOx	CO	CO2	VOC	CH	NOx	CO	CO2
	(SEK-1990)	g/stop				(SEK-1990)	g/stop			
50	. 11	.15	.3	1.2	37	1.23	0	6	2.5	350
70	. 16	.21	.4	1.2	50	1.76	0	9	6	520
90	. 18	.24	.4	.7	54	1.95				

Fig.9. Extra VOC and emissions in intersections for stopped and "close" stopped/turning vehicles.

<u>Traffic Safety Research</u>

The work on traffic safety was undertaken in cooperation with VTI. As for capacity empirical data and statistical analysis are reported elsewhere. In this paper the objectives are to give an impression of the size of the material and to summarize the main conclusions as they have been put forward to Swedish practitioners. This means conclusions based on historical accident statistics and traffic engineering judgment.

The data collection has been extensive. All state main road at-grade intersections with give-way or stop control have been classified by intersection type, primary and secondary road traffic flows (aadts) and layout and control changes during the analysis period. Accidents in the intersections were analysed for almost a ten year period. Two accident methodologies have been used, "with-without" and non-experimental before-after (refs 12-17). "With-without" means historical comparisons of accident statistics for intersections of different types with approximately the same level of primary and secondary road aadts. Accident rates and injury consequences have also been estimated for grade-separated state intersections on two-lane roads (ref 18). A smaller inventory and "with-without" analysis has also been carried out for urban municipal intersections including give-way, stop control and traffic signals covering two years of accident statistics for 10 of the biggest Swedish cities after Stockholm, Gothenburg and Malmoe (ref 19). A new study on urban intersections is carried out at the time of writing this. Roundabout layout and accident statistics have been reviewed for municipal as well as state roads (ref 20 a,b). The safety feature of LHOVRA traffic signals has also been examined (refs 21,22).

The main results put forward to Swedish traffic engineers are:

Roundabouts perform extremely well. Accident rates are not low but injury consequences are. The average inner circle radius is 20-25m though the guidelines have recommended >40m. State roundabouts, which means semi-urban and sometimes rural environment, have bigger inner circles. Big radius, >40-50m, tend to decrease traffic safety. Accident rates doesn't seem to depend on traffic flow. 3-armed roundabouts seem to operate well.

Traffic signals with LHOVRA-technique behave much better than traditional traffic signals on 70 km/h urban roads. The LHOVRA-strategy decreases amber and red driving. The LHOVRA-experiments on rural high-speed roads have not been succesful, generally speeking. Traffic signals on urban 70 km/h roads have on average high injury consequences and might be questioned from a safety point of view.

4-armed intersections tend to have very high accident rates and injury consequences in rural areas and probably on at least 70 km/h urban roads.

Left turn lanes with kerbed islands, on average, increase accident rates and injury consequences **in 4-armed rural junctions** and probably also on 70 km/h urban roads. Rear-end crashes "caused" by left turners are decreased but more severe left turn and crossing accidents tend to increase. Summarized, the result is a negative impact. These results have been intensively discussed and disputed. Data from at least four undependent studies support the conclusions on kerbed islands. The same indications are reported from Finland.

A theory to explain the results is that left turning lanes, especially if kerbed, widen the intersection and make the straight through path clearer and more distinct. For secondary road users and left turners from the primary road this means longer paths to cross through the intersection, more difficult to overview the intersection and on-coming traffic and perhaps also faster on-coming traffic.

In T-intersections no positive effect can generally be seen.

Left turn lanes with marked islands seem to have a positive impact in rural T-intersections and perhaps on urban 70 km/h roads.

In four-armed rural intersections no positive impact has been found.

Part of the explanation would be that intersections with left turn lanes with marked islands generally are not so wide.

Staggered intersections have superior performance than four-armed intersections. The safety impact might be up to 40% reduction depending on traffic flow distribution according to empirical follow-ups.

Stop control seems to have 20% better injury consequences than give-way.

Right turn lanes are not a safety measure. Rear-end accidents with right turners from primary roads are very rare events. There is an obvious risk that right turning lanes will have the same safety impact as left turning lanes. No statistics are available on right turning lanes.

Traffic Engineering Guidelines for Traffic Safety Evaluation

For rural areas a traffic safety evaluation guideline (ref 23, 10) has been derived that is used in project evaluation schemes and final design evaluations. The procedure is outlined as follows. There is also a similar procedure for urban intersections (ref 10).

STEP 1 Decide the predicted number of severity-cost adjusted traffic accidents per year Af for the proposed intersection types for the estimated aadts according to Figure 11. For give-way/stop control type A, correct for layout according to Figure 10.

Layout	Illumination	Reduction of accidents (%)	
		3 - arms	4 - arms
Secondary road island	No Yes	0 5-10	10 15-20
Left turn lane marked	No Yes	15-20 20-30	10 10-20
Left turn lane kerbed	No Yes	0 10-15	10 10-20

Fig.10. Correction for give-way/stop control type A

Notations:
A=give-way or stop
D=roundabout
E=traffic signal
F=grade separation
(00)=100*Qs/(Qs+Qp)

Average number of severity-cost adjusted accidents per year Af

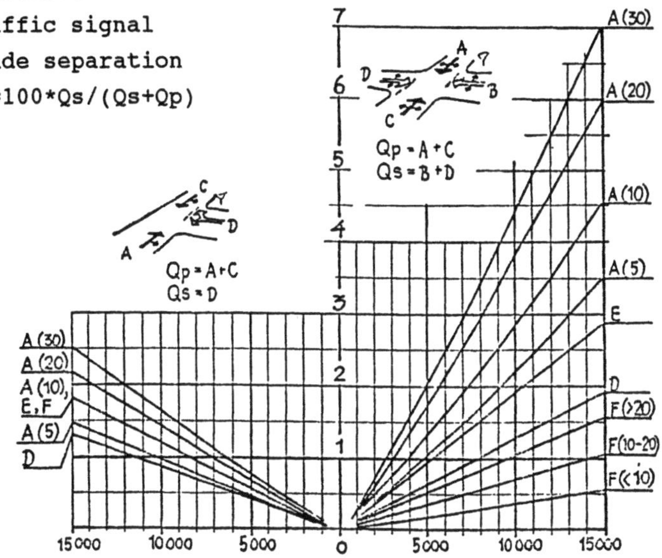

Annual average daily traffic Qs+Qp

Fig. 11 Average number of severity-cost adjusted traffic accidents per year Af for different intersection types.

The research, as mentioned earlier, gives worse results for kerbed islands.

Comment: A staggered intersection is calculated as two isolated T-intersections with a connecting link.

STEP 2 Compare the estimated results for the different proposals.

STEP 3 For an existing intersection you first calculate the observed number of cost-adjusted accidents A by weighting accidents depending on intersection type varying from 0.6 for roundabouts to 2.0 for four-armed intersections with give-way/stop control.

The estimated future number of accidents An is then calculated

$$An= Af+k*(A-Af);\ k=.04*Af/(1+.04*Af) \qquad (31)$$

The impacts of alternative designs are evaluated using information on An, accident type and conflict distribution and standard relations between intersection types according to Figures 10 and 11.

New Guidelines

The research findings, especially in the area of traffic safety, and policy changes in the National Swedish Road Administration, aiming at more decentralized decision making combined with follow-up by means of result analysis, made it necessary to change the content and the character of the intersection guidelines. The new rural guidelines have been operating since 1986 (ref 23) and the urban ones from 1987, slightly changed in 1989 (ref 24).

Choice of intersection type

The new procedure for choice of intersection type is a combination of objective-planning and cost-benefit-analysis. Three general objectives or criteria are raised and could be combined with regional or local objectives depending on road category. If the objectives don't give an obvious choice it is recommended to evaluate capacity and level-of-service effects expressed as delays, vehicle operating costs and traffic safety effects according to the standard procedures described above and compare these with construction and maintenance costs and environmental and other impacts for the intersection types under consideration. The general objectives are as follows:

- if the degree of saturation for the design hour for a standard give-way or stop controlled intersection exceeds 0.5-0.7 other intersection types (staggered, traffic signal, roundabout, grade separation..) should be taken into consideration. Graphs are included to support a rough estimation as illustrated in Figure 12.

Notation: (aadts)
B=degree of saturation
during design hour

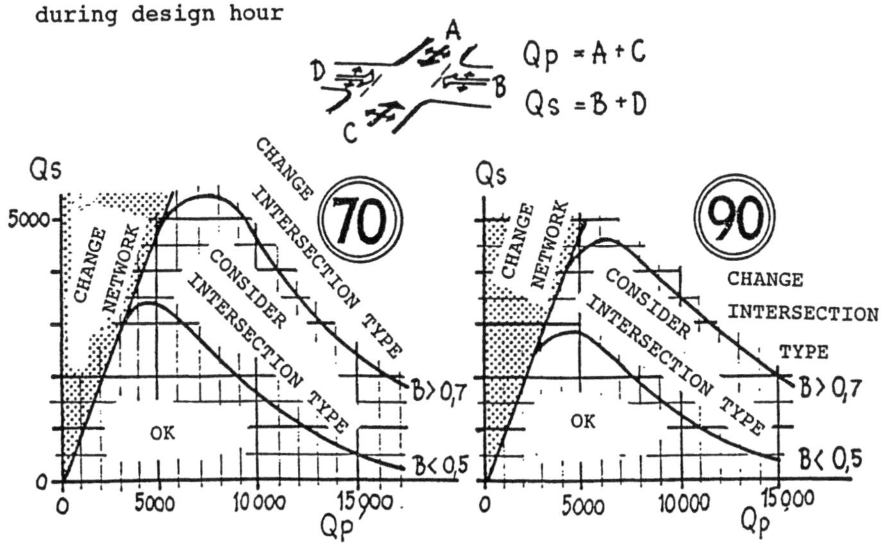

Fig.12. Capacity objectives for four-legged rural intersections

- if estimated number of cost-adjusted accidents exceeds 1 per year for a standard give-way or stop controlled intersection other intersection types (staggered, traffic signal, roundabout, grade separation..) should be taken into consideration. Graphs are included to support the estimations as illustrated in Figure 13.

Notation: (aadts)

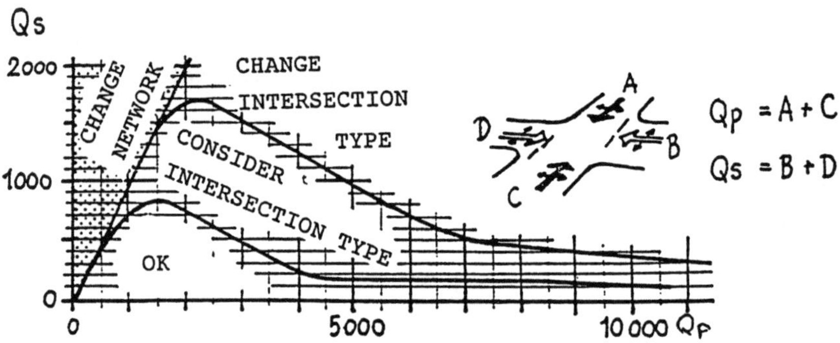

Fig.13. Traffic safety objectives for four-armed rural intersections.

- decide if traffic signals or roundabouts are acceptable depending on road type and desired mean speed in the design hour, if urban, along the major road (varying from 25km/h to 60km/h depending on road class).

Design Recommendations

The design recommendations follow the traffic research results. From a safety point of view four-armed intersections are to be avoided. The same goes for right turn lanes in general and left turn lanes in four-armed intersections. Left turn lanes with kerbed islands should be avoided, if not needed for pedestrians. Staggered intersections both right/left and left/right are recommended. In rural conditions left/right is prefered.

For bicycles and pedestrians two principles are discussed, i.e. with and without setback according to figure 14. Advantages and disadvantages are listed without any accommodation which one to to choose

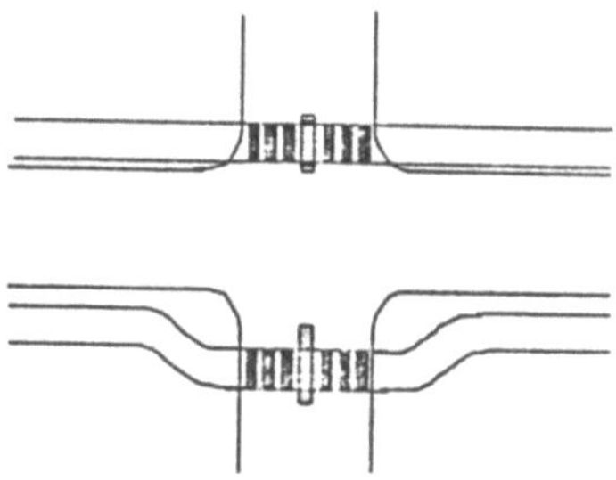

Fig.14. Bicycle and pedestrian crossing with and without set back.

Roundabouts are highlighted for their safety and capacity performance. Problems with pedestrians and bicycles are discussed. The inner circle can be decreased down to 5m on city streets. For principal roads 10-20m is recommended depending on the number of circulating lanes.

The sight distances have been decreased along rural secondary roads and increased along primary roads, see Figure 15. For urban conditions primary road sight distances are increased

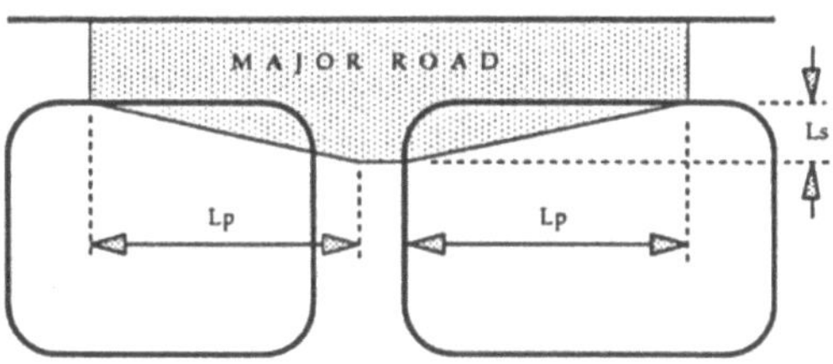

Ls (m)	Stop	Give-way (to improve capacicity)	
	3 - 5	3 - 5	(20)
	Speed limit (km/h)		
	50	70	90
Lp (m)	80 - 110	130 - 170	190 - 240

Fig.15. Recommended sight distances

References

1. National Swedish Road Administration: TV 124 Geometric Guidelines on Rural Junctions (in Swedish). NSRA 1975.
2. National Swedish Road Administration: TV 117 RIGU Geometric Guidelines on Urban Junctions (available in English). NSRA 1973.
3. National Swedish Road Administration. TV 131 Manual on Capacity Calculations (in Swedish). NSRA 1977.
4. National Swedish Road Administration. TU 147 Manual on CAPCAL (in Swedish). NSRA 1981.
5. Anveden J. Intersections without Traffic Signals, p 80-102 Swedish Research on Unsignalized Intersections. Springer Verlag 1988.
6. Hansson A, Bergh T. A New Swedish Capacity Manual - CAPCAL2. Proceedings 14th ARRB/Conference Canberra 1988.
7. Hansson A. Swedish Capacity Manual: Part 2 Capacity of Unsignalised Intersections. TRR 667 1978.
8. Hansson A. Traffic discharge in Unsignalised intersections (in Swedish). LTH Ph.D Thesis 11 1975.
9. National Swedish Road Administration. Calculation procedure - Intersections without signal control. Unpublished NSRA 1987.
10. National Swedish Road Administration. 1989:16 and 16B Effect of Road Investments (in Swedish). NSRA 1989.
11. Hammarstrom U, Karlsson B. VETO - A Computer Program for Calculation of Transport Cost depending on Road Standard (in Swedish). VTI Meddelande 501 1988.
12. Brude U, Larsson J. Rural Junctions on the Trunk Road Network. VTI Rapport 233.
13. Brude U, Larsson J. Effects on Traffic Safety of Road Improvements at Junctions (in Swedish). VTI Rapport 310.
14. Brude U, Larsson J. Staggered Junctions on Rural Roads (in Swedish). VTI Meddelande 544.
15. Brude U, Larsson J. Before-and-after Studies of Accidents at Rural Junctions in the "1983 Junction Inventory" (in Swedish). VTI Meddelande 545.

16. Brude U, Larsson J. Replacement of Mandatory Stop by Mandatory Yield at Junctions in the National Trunk Road System Network (in Swedish). VTI Meddelande 640.
17. Brude U. Traffic Safety at Junctions. Third European Workshop on Recent Development in Road Traffic Safety Research. VTI 1990.
18. Johansson O. Traffic Safety of Grade Separated Junctions on Two-lane Roads (in Swedish). Vagverket TU 153 1983.
19. Cedersund HA, Svensson A. Traffic Safety in Major Urban Junctions (in Swedish). VTI Meddelande 259 1981.
20a. Cedersund HA. Roundabouts (in Swedish). VTI Meddelande 362 1983.
20b. Edmund HA. Intersections without Traffic Signals, page 305-318. Traffic Safety at Roundabouts. Springer Verlag 1988.
21. Brude U, Larsson J. Traffic Safety Evaluation of LHOVRA-technique in Traffic Signals (in Swedish). VTI Meddelande 575.
22. Peterson A, Bergh T, Steen K. LHOVRA - a New Traffic Signal Control Strategy for Isolated Junctions. TEC July/August 1986.
23. Vagverket. TV 124 Geometric Guidelines on Rural funtions NSRA (available in English) 1986 .
24. Vagverket, Kommunforbundet. ARGUS Geometric Guidelines on Urban intersections (in Swedish) Kommunforbundet 1987/89 .

A Strategy to Optimise the Geometric Design of Isolated Priority Junctions

J.P. Silcock and I. Borzym
Transport Studies Group, University College London

Summary
Junction design in Britain is assessed principally in terms of criteria relating to junction capacity, total junction delay, and safety. This paper outlines a strategy for optimising geometric design that includes automatic optimisation of a quantifiable measure of junction performance, namely capacity, within an interactive program.

In Britain, the capacity of a traffic stream at a priority junction can be determined by relationships developed at the Transport and Road Research Laboratory [1]. These relationships can be used to form a set of capacity constraints which allow linear programming techniques to be used to assess the capacity of the priority junction for any given geometry. By the addition to the capacity constraints of other design constraints such as those of safety and cost, a problem of maximising the junction capacity can be completely specified. However, geometric constraints become non-linear, making the optimisation procedure complex. Also, there are a number of discrete choices involved in the design of any junction such as number and location of islands and obstructions to visibility; this further complicates the automatic optimisation process.

The interactive procedure outlined in this paper involves producing an initial design for each discrete choice of layout, assessing capacity and adapting those features that have the greatest effect on capacity. The adaptation of geometry can be done automatically, incorporating rules representing good, safe design, or interactively by the engineer working from a graphical representation of the junction.

1 Introduction

A program to aid the traffic engineer in the geometric design of isolated priority junctions is being developed at the Transport Studies Group, University College London. Previous work on the optimisation of roundabout design [2] and signal design [3] provide a starting point for this paper. It will identify some of the considerations which practising traffic engineers agree determine an improved design and consider the extent to which it is desirable to incorporate these considerations into an automated procedure which may be used to improve the design of any type of priority junction in use in Britain. The design criteria must be assessed and in each case the geometric variables which affect the assessment obtained from the criterion must be identified as must the limits in applicability of the criterion obtained by taking note of tacit assumptions and the effects of implicit approximations.

2 The capacity of a priority junction

To avoid confusion, the terms used in this paper are defined as follows:

A **traffic movement** is the flow entering the junction on a specific approach, leaving by a specific exit.

A **traffic stream** is one or more movements that form the same queue at the junction, occupying one or more adjacent lanes.

The **capacity of a lane or stream** for a non-priority movement at a junction is the maximum flow per unit time that may emerge from the lane or stream in the prevailing traffic conditions.

A stream is **saturated** when the mean arrival rate over an appreciable period of time equals the capacity.

The **reserve capacity of a stream** is the amount by which the flow in a stream may increase before the stream becomes saturated.

The **reserve capacity of the junction** is the least of the reserve capacities under the assumption that all flows of traffic approaching the junction increase by the same proportion.

For comparison with other literature it should be observed that the concepts of lane and stream coincide in the TRRL model and in the exposition in section 2. They are in general distinct because a stream may occupy more than one lane, but the methods in section 2 still apply provided that the capacity of the stream is calculated appropriately from that of the lanes. In doing so, however, linearity of capacity in the geometric variables is lost.

2.1 A model for the capacity of a non-priority stream

The model adopted to assess the capacity of movements through priority junctions was developed initially for T junctions by TRRL in the late 1970s. A regression analysis of a sample of 37 cases of selected typical T junctions provided empirical formulae for the capacities of non-priority streams. These relations were multilinear in the conflicting flows, the average available road widths to each arriving non-priority stream, the major road width, the width of any central reservation and the average visibilities, from arm to arm. The analysis indicated no evidence for any significant effect on capacity due to gradient of approach, angle of intersection of junction arms, turning radii of vehicle paths, acceleration or deceleration lanes, or visibilities other than at the 10m line. It was concluded that the problems of giving way to other traffic dominate service time swamping the effects of all other variables. This should be contrasted with the signalled case where some of these factors do contribute. The sample junctions included sites with low average flows and single lane approaches and also sites with high average flows and dual carriageway major road approaches. The TRRL formulae are a triumph in that even given the severe limitations they have provided a simple versatile formulation of a complicated situation. TRRL have more recently extended the capacity formulae to a semi-empirical model for capacity at crossroads and staggered junctions [4].

For a stream consisting of a single movement i, the general model for the capacity Q_i takes the form

$$Q_i = g_i(a_i - \Sigma_j b_{ij}q_j) \qquad 1$$

where g_i is a parameter dependent on junction geometry, a_i and b_i are constants, and the q_j are the entering flows making movements conflicting with stream i.

If a stream consists of more than one movement, as is often the case for minor road traffic at a priority junction, then the stream capacity is given by the flow-weighted harmonic mean of the capacities of the separate movements. For such a stream i, the capacity Q_i is given by

$$Q_i=\left(\sum_k f_{ik} Q_{ik}^{-1}\right)^{-1} \quad and \sum_k f_{ik}=1 \qquad 2$$

where f_{ik} are the proportions of vehicles making movements k in stream i with capacities Q_{ik}.

2.2 A procedure for assessing the reserve capacity of a priority junction

It is assumed in this paper that there are no movements sharing lanes; this simplifies the presentation and allows the paper to concentrate on describing the techniques used to optimise geometry. The theory is being adapted to include shared lanes as well as streams occupying more than one lane.

For a given junction geometry and pattern of arrivals, from equation 1, capacity constraints for each stream i can be written down as

$$\mu q_i \leq g_i(a_i - \mu(\Sigma_j b_{ij}q_j)) \qquad 3$$

where μ is a flow multiplier. If μ=1, these state that the actual approaching flow in each stream i must not exceed the capacity of that stream. If all constraints are satisfied at μ=1, then the junction is within capacity, and all entering flows q_j are equal to their corresponding approach flows. Let μ increase until, at $\mu=\mu^*$

$$\mu^* q_i = g_i(a_i - \mu^*(\Sigma_j b_{ij}q_j)) \qquad 4$$

for one or more streams i. Let the set of such streams be I^*. Then for $i \in I^*$ stream i is at capacity with a flow $\mu^* q_i$, and the junction reserve capacity for the actual flows is

$$100(\mu^* - 1) \text{ per cent} \qquad 5$$

This method of assessing reserve capacity can be stated as a simple linear optimisation by:

maximise μ, subject to the set of constraints (3).

The linear programming problem is then easily solved using, for example, the simplex algorithm [5].

2.2.1 Sensitivity analysis

A major benefit of using linear programming techniques like this is that it provides sensitivity analysis; this gives the effects on μ (and therefore the reserve capacity) of small changes in the parameters. It provides a value δ_i for each capacity constraint for streams $i \in I^*$ that limit the capacity of the junction. Then for each $i \in I^*$, a small increase in g_i of Δg_i yields an increase in μ^* of

$$\Delta\mu^* = \delta_i \Delta g_i (a_i - \mu^* \Sigma_j b_{ij} q_j) \qquad 6$$

Since the g_i are functions of the junction geometry, these formulae can be used to give the effect on capacity of small geometric changes such as increasing available road width or improving visibility. The effect is evaluated using the chain rule for partial derivatives. Equation (7) shows the effect on μ^* of a small increase, Δw_i, in available width of a lane $i \in I^*$ that limits junction capacity.

$$\frac{d\mu^*}{dw_i} = \frac{\partial\mu^*}{\partial g_i}\frac{\partial g_i}{\partial w_i} \text{ ,and from 6 } \Delta\mu^* = \Delta w_i \delta_i (a_i - \mu^* \Sigma_j b_{ij} q_j)\frac{\partial g_i}{\partial w_i} \qquad 7$$

2.3 An appraisal and interpretation of the capacity model for use in design at specific sites

The sample used to obtain the TRRL model represents a broad variation in junction type, but in the context of all priority T junctions it is a relatively small subclass. High costs preclude more extensive surveys. In the context of future developments of the capacity model and its use as a design tool, the implications of using an empirical model based on a relatively small data set should be borne in mind.

The geometric variables required by the TRRL capacity model are lane half-widths averaged over measurements at 0, 5, 10, 15, 20m from the stopline or turning position, and visibilities from the 10m mark behind the stopline. The visibilities are averaged to give one value for left and one value for right visibility from each arm. An average major road width and an average width for the central reserve are also required. Some of the design variables are illustrated in Figure 1.

2.3.1 Relating the geometry to the existing capacity model

Although it would in principle be possible to express the capacity formulae directly in terms of raw geometric variables there are no advantages and many disadvantages to this approach. The number

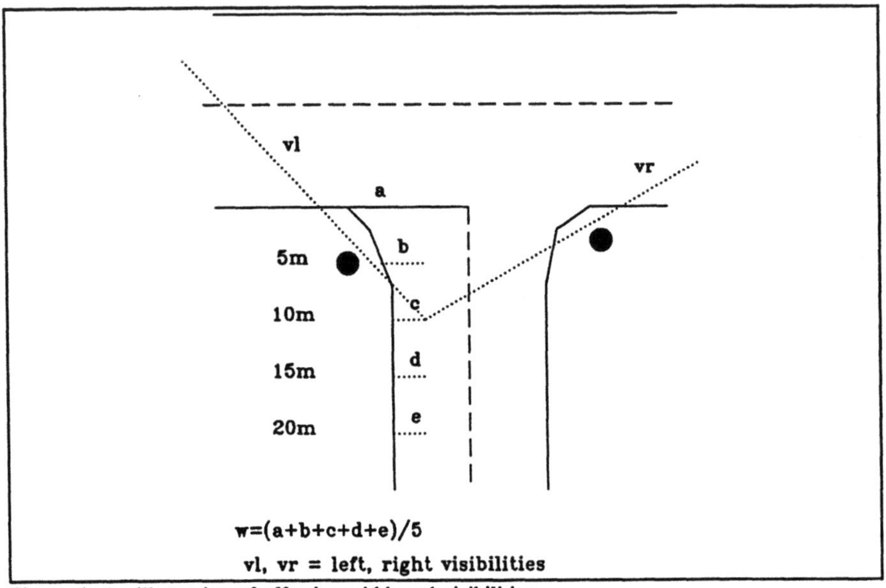

Figure 1. An illustration of effective width and visibilities

of independent variables (lane widths) would be increased. Visibilities would be given as nonlinear functions of these variables.

To facilitate the extension and interpretation of the capacity model a distinction will be made and preserved between three types of variables. The measured raw geometric variables will be distinguished from the model-dependent geometric parameters which are obtained from averaging processes. The latter will in turn be distinguished from the variables as required by the capacity model equations or enhancements of them. This decision has implications for the mathematical formulation of the problem.

The equivalence of many geometries with respect to the capacity model is now expressed by the independence of the model visibilities as variables and the addition of constraints on the geometric variables to capture the relations expressed in the averaging process. The separation of the three components will be required to allow the introduction at a later stage of new features. It is not necessary for the design strategy but would be informative to characterise the equivalence classes of geometries with respect to capacity.

A functional interpretation of the capacity model parameters is required to provide extensions in the absence of survey data. Superseded theoretical models, for example the gap acceptance model, may still provide consistency checks. Providing such an interpretation also highlights features in the original data set which have become tacit assumptions in the model but which may fail in more general situations. The details of the proposed interpretation will be given elsewhere but an important consequence of this interpretation implies that a design strategy using the capacity model may divide a junction into four types of region with many of the geometric variables insensitive to alterations in other regions. The types of region are as follows,

(a) The region before the 10 m mark. Visibility has no effect and the width data is interpreted as defining the capacity available to freely flowing arriving traffic on the road segment. If the road has a constriction then the effective width should be modified to take this into account.

(b) The 10 m mark. This is the point at the lane centre at a distance of 10 m from the stopline or its equivalent. This approximates the location of the first vehicle not in service. The empirical significance of this visibility is interpreted as the effect on capacity due to the second vehicle in the queue preparing to gap accept. This region should be regarded as a thin band and thin obstacles may be ignored.

(c) The region between the 10m mark and the stopline, where the effective width influences the rate at which vehicles are served. The empirical lack of significance of visibility at the stopline should be interpreted as indicating that visibility was usually adequate for the manoeuvre in the sample surveyed. There is a decrease in capacity described by the interaction term with conflicting streams which affect visibility. Obstacles which affect visibility at the stopline and are not thin must be considered on grounds of capacity.

(d) The region of interaction, which the model assumes has sufficient capacity to accommodate all incoming vehicles. This region must be separately assessed to ensure that there are no persistent blockages other than the queue of traffic turning right from the major road. This queue is included in the capacity model, and under the suggested interpretation should be regarded as persistent.

3 Capacity, congestion delay and queue lengths

The selection of a priority junction design at this stage implies the satisfaction of constraints on the total acceptable flow in any given lane. The constraint on right turning flows should be stringent and could be used to select between priority and other designs.

There is a close relationship between capacity and congestion delay at priority junctions. Queues of more than a few vehicles may be undesirable in certain streams. Geometries which generate such queues can be avoided by requiring a capacity for each such stream of at least that which produces the maximum acceptable queue. There should be no new constraints originating from other delay considerations because the undesirable cases should have been rejected earlier on capacity grounds.

4 Design requirements, guidelines and practice

The term road division shall refer to lane, reserved region or footway. A discrete choice of broad junction type identifies the appropriate class of absolute design requirements, namely those set and preserved within the strategy. In the event of an accident the violation of these constraints might be legally regarded as a contributing factor. Provision is made within the strategy to impose and interactively redefine a stronger class of preferred design requirements. These requirements carry no liability implications but are regarded among designers as good design practise. The strategy also permits the interactive addition and redefinition of constraints corresponding to actual design procedure. The latter class are most likely to be relevant to "difficult" designs and special cases.

All classes of constraint define minimal and maximal road division widths and visibility distances. A set of data points describes the division boundaries in terms of piecewise linear segments. Functions of these segments define approximating tangent vectors to the boundary at each data point. Alternatively the description may be given in terms of arc lengths and deflection angles of fitted optimal design curves. Sufficient data points are required by the strategy to ensure that satisfaction of constraints imposed on the divergence and convergence of road division boundaries entails the satisfaction of width and visibility constraints at all points.

Implicit in the above discussion is the satisfaction of constraints on turing radii. There is a minimum turning radius which corresponds to the turning circle of a typical large car. Associated to this is the requirement that with a given kerb radius the road should have sufficient width to accommodate the templates representing turns for large vehicles. For urban junctions this provides access for emergency vehicles.

5 Optimal Design Curves and Geometric Delay.

Euler spiral transition curves, the safest design for constant speed turns on fast roads, provide an optimal road design for joining two fixed tangent directions (see section 7; initial data requirements) along a fast rural major road with a smooth curve. A similar derivation allowing for accelerating or decelerating vehicles may be used to define corresponding optimal design curves. The design strategy includes an option to model the road by a best fit of the optimal design curve through the data points.

The shape of the road has been defined in terms of theoretical design criteria related to road safety through road holding considerations or road space considerations. Inclusion of geometric delay will tend to counter these considerations since the geometries where paths with the highest flows are straightened will correspond to minimal delay. Assessment of geometric delay might provide the best choice of intersection angle for the junction arms, where this can be varied.

6 Discrete Features and Cost Constraints

An interactive procedure allows the inclusion or exclusion of an island. The location and shape are required to conform to guidelines and once selected impose additional constraints on the road space arising from the total width available for the junction.

Cost constraints include the cost of maximum land take which will often define the boundary of the region of design. The cost of removal or addition of obstacles or other discrete objects should be assessed at each discrete decision point. The cost of moving services (drains, cables) can only be fully assessed once the road area has been determined, but some approximate form of running total is required.

Inclusion or exclusion of pedestrian crossings is again an interactive decision. The location is constrained away from the stopline. The inclusion of a crossing reduces capacity but imposes no further constraints.

7 Initial data requirements

The data requirements demand sufficient information to reproduce a design layout diagram with geometric parameters fixed up to a margin of 0.05 m. None of the non geometric criteria approach this level of precision in the determination of the geometric parameters. It follows that for the purposes of evaluating non geometric design criteria the entailed geometric constraints may be greatly simplified by a judicious use of approximations. While this has the advantage of simplifying the potentially complicated assessment and optimisation routines within the program it also implies that the details of the geometric design will be controlled by the specifically geometric design guidelines. This suggests the engineer should have considerable freedom in most circumstances in the choice of final layout. The other implication is that the type and quantity of initial data required is largely independent of the strategy chosen for design modification.

The initial data is ultimately expressed in terms of the positions of specified points with respect to a previously defined grid over the junction. The number of data points depends on the complexity of the junction. An outline of the types of data required is given below.

(a) Data points defining the fixed outside boundary of the junction within which all modifications occur. The tangent directions to arriving and exiting lanes at the boundary are also fixed and must be specified. The boundary will typically correspond to maximum land take either on grounds of cost or building preservation.

(b) Data points defining the extent and position of all immovable features and obstacles. This includes building lines within the junction, street furniture, possibly services and even preserved trees.

(c) Data points defining the extent and position of all movable obstacles.

(d) Data points defining the curves corresponding to all lane markings, reserved regions, footway areas, and any other road divisions imposed on the junction, such as points defining the boundaries of different land take cost contours.

8 Choice of design strategy

The possible strategies vary between two extremes. In a fully interactive strategy the computer is used to produce drawings and carry out calculations estimating the effects of small continuous changes of variables but all decisions and choices are made the engineer. A fully automatic process would recast the problem as a very complicated nonlinear optimisation. At present the geometry is not fully determined by the design criteria and a fully automatic process would require a formal definition of optimal geometry and the imposition of further constraints to fix a unique solution. The criteria used in design are themselves subject to evolution and the models used in the assessment of a design may require replacement or enhancement. It is advantageous to keep the different components of the design process distinct. While the best expert system remains an expert with a system it is reassuring to know that an engineer would be reluctant to accept and implement the output of a black box. It is important that the engineer should have access to data describing the exact way in which the design modification is interacting with the design criteria. The constraint on this is the desire to use the computer to save valuable man-hours. Fortunately at various stages in the design process there is a natural hierarchy in the order in which the variables must be fixed. This allows a substantial degree of automation without loss in flexibility. The exact nature of this hierarchy depends on the type of junction. There are two extreme types of priority junction

(a) Urban Junctions of ambiguous priority. These are commonly junctions found along heavily used side-roads in large cities. Typically T junctions or crossroads they have relatively high flows of traffic on all approaches with no obvious major minor distinction. Priority may be assigned either on grounds of flow or safety. These are the most difficult to design since the dimensions of the roads are all at the same level in the hierarchy. Road space is constrained by poor visibility along some paths and often many restrictions arising from fixed points corresponding to immovable objects.

(b) Rural junctions between minor roads and high-speed A roads: These are relatively easy to modify since the hierarchy fixes the width of the major road and its central reserve at an early stage. Essentially the road cannot be safely widened at a junction without encouraging dangerous overtaking. The shape of the major road is determined by the correct design speed curve. The only remaining freedom is the improvement in the turning refuge provision or in the addition of merging acceleration lanes or slipways. Visibility is likely to be good and the overriding concern is the safety considerations which select the best layout to allow slow moving traffic from the minor road to join the fast moving traffic on the major road.

The majority of junctions fall between these two extreme cases. They fortunately have the advantages of both extremes in that there is a clear division between major and minor, the layout of the major road is fixed at an early stage and there are also some fixed points which help to cut down the possible layouts rapidly.

At each level in the hierarchy, certain variables become fixed. For example the capacity assessment, described in section 2, uses sensitivity analysis to define the relative road space assigned to each stream, according to the distribution of traffic.

The flow chart in Figure 2 shows the general conceptual design strategy but has insufficient detail to adequately describe the algorithm which determines the order in which variables are fixed.

8 Conclusions

The level of interactive design included in the strategy is determined by the high level of flexibility required. The engineer must be able to include the consideration of features which have not been explicitly modelled within the criteria. This implies the ability to add additional constraints into the automatic routines at various points within the strategy, but also to reject a design optimal with respect to one of the criteria in favour of a design which has obvious advantages not considered within the program.

Discrete choices such as the inclusion of islands which define broad junction geometry could not be adequately handled by an automatic process, although given such a discrete choice the constraint structure is greatly simplified and the implications for the criteria under some clearly defined small

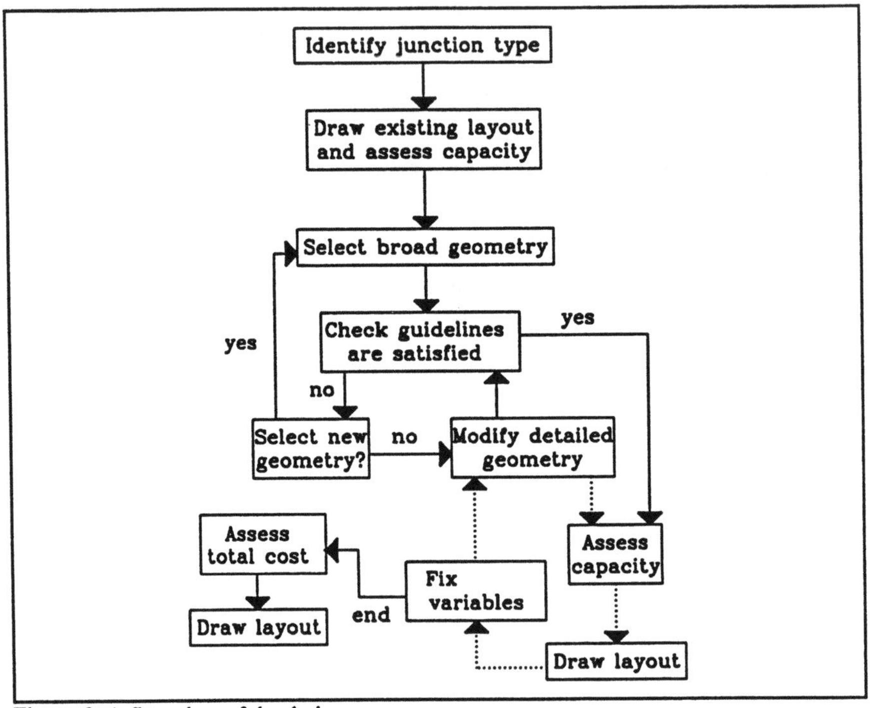

Figure 2. A flow chart of the design strategy

modifications in shape or location could be automatically evaluated. Given the level of under determination of the geometry by the design criteria and the importance of good design for such objects, especially if they are intended to separate traffic or to act as crossing refuges, it is inevitable that this part of the design should contain a large interactive component.

The differences in the order in which variables are fixed arising from differences in junction type could in principle be treated automatically but this might be at the cost of assigning an unacceptably high level of control over the design process to the machine. It follows that menu driven options or interrupted iteration procedures are to be preferred. Since the ultimate design is not optimal in the mathematical sense there is no disadvantage to interrupting the iterations during an automatic optimisation for reassessment.

The usefulness of the automatic processes arises principally in the prediction of the effects of small changes in the variables and the identification of active design constraints. Also in keeping track of the total cost of a design modification at any stage, and the production of a drawn to scale layout.

Acknowledgements

The authors would like to thank Professor Allsop and Dr Heydecker for their help and invaluable comments. This paper describes research funded by the Science and Engineering Research Council.

References

1 Kimber, R.M.; Coombe R.D.: The traffic capacity of major/minor priority junctions. TRRL Report SR 582. Crowthorne 1980.

2 Willumsen L.G.; Kay W.: Computer Assisted Optimisation of Roundabout Design. Proceedings of the International Workshop on Intersections Without Traffic Signals, Bochum, West Germany, March 1988 (Ed. Werner Brilon) 289-304. Springer-Verlag.

3 Allsop, R.E.: Evolving Application of mathematical optimisation in design and operation of individual signal-controlled road junctions. Proceedings of the IMA Conference on Mathematics in Transport Planning and Control, Cardiff, September 1989.

4 Kimber, R.M.: The capacity of some major/minor priority junctions. TRRL Report LR 735. Crowthorne, 1976.

5 Luenberger, D.G.: Linear and nonlinear programming. Addison-Wesley, 1984.

Results of Hungarian Measurements in Connection with Capacity of Unsignalised Intersections Carrying out Capacity Calculation

Dr. András Bényei professor
Department of Road Engineering
Faculty of Civil Engineering
Technical University of Budapest, Hungary

Dr. István Fi associate professor
Department of Road Engineering
Faculty of Civil Engineering
Technical University of Budapest, Hungary

1. Measuring of headways

Hourly traffic volume as result of traditional cross-section counting can be related to the quality of traffic flow only approximately. The analysis in fact show that 5's traffic within an hour for example may be very different and the deviation may be even threefold.

Drivers evaluate traffic conditions on the traffic situation in their own surrounding. This traffic situation is better characterized by the 5's traffic than by the hourly one. At the same time it is expedient to give traffic in hourly traffic. Therefore the concept of fictitious hourly traffic has been introduced, it is 12 time the actual 5's traffic.

Only 3 vehicle categories were to be differentiated for the examinations: passenger car, light and heavy vehicles.

Headway distribution in a cross-section and momentary vehicle density distribution, $Fi = F(x)$ is given best by two curves intersecting each other (Figure 1.) on the basis of the measurement results, its equation is the following:

$$Fi = 1 - e^{\frac{m'}{0,43429}x}$$

$$Fi = 1 - e^{-\frac{B}{0,4329}} \cdot e^{-\frac{M}{0,4329}x}$$

The intersection of the two curves at abscissa

$$x = B/(m'-M)$$

is the validity limit of the two equations.

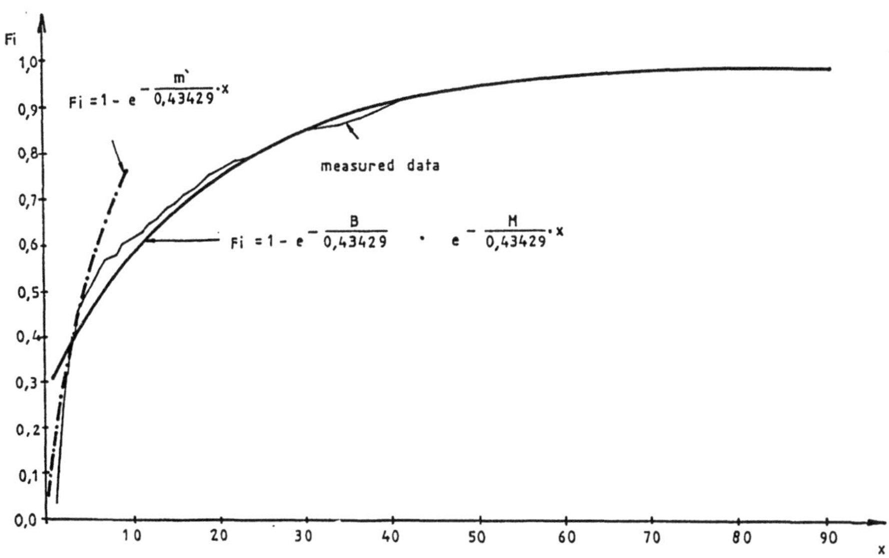

Fig. 1.

Variable x may mean headway t (s) in one-way or two-way traffic or momentary vehicle density S (Vehicle/km).

Distribution function Fi = f(x) is suitable for the determination of the theoretical traffic F (Vehicle/hour) that substituted into the equation results the distribution function of the examined variable x as the distribution function produced from actually measured value in case of traffic J (Vehicle/h).

According to the results of examinations headway distribution does not follow POISSON distribution but the above mentioned two curves fit the best the line Fi = f(t).

Relations describing headway distribution measured in one-way traffic on straight and horizontal section of rural road are:

$$Fi = 1 - e^{-\acute{a}_1 . J_1 . t}$$

that is valid until t = 5 s over J_1 = 350 Vehicle/h. In case of bigger headway relation

$$Fi = 1 - b' . e^{-a' . J_1 . t}$$

is valid.

Constants a_1', a' and b' of the equations can be read from figures 2. 3. and 4.

The theoretical traffic F Vehicle/h that results the same headway distribution as the distribution function produced

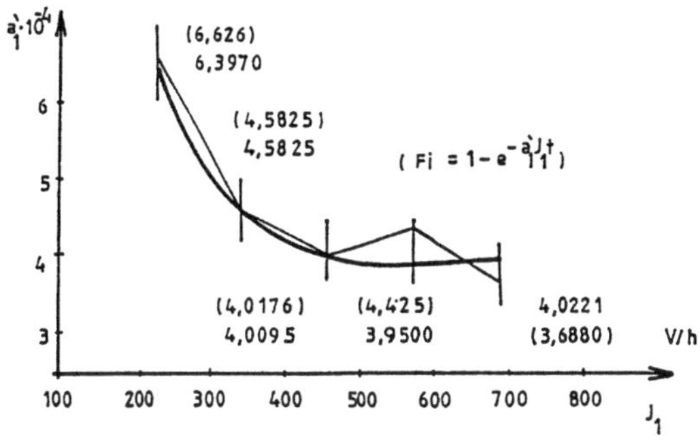

Fig. 2.

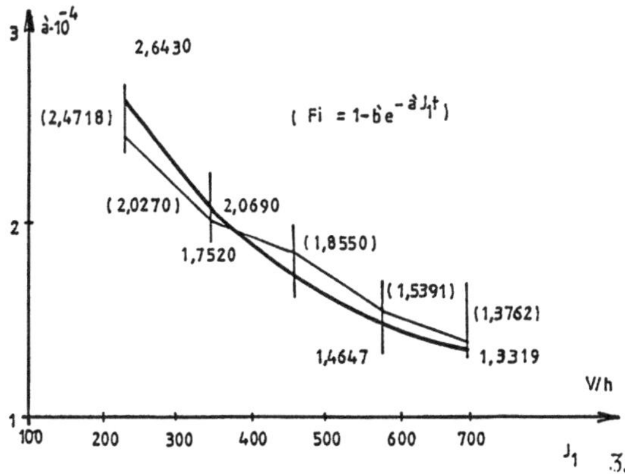

Fig. 3.

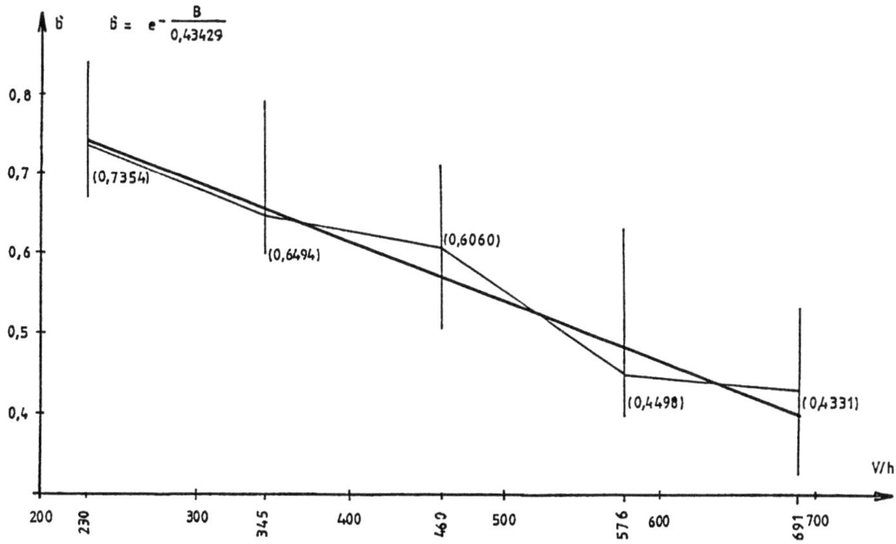

Fig. 4.

from the value measured in case of actual traffic J_1 Vehicle/h may be also road from diagram.

The above relations are valid also in case of two-way traffic. Traffic J (Vehicle/h) here means the traffic volume of the two directions. It has been stated that less full cross-section, theoretical traffic F results the same headway distribution as the actually measured value J.

Comparing the headway distirubtion function of J_1 Vehicle/h measured in one-way traffic and the same value J (Vehicle/h) in case of two-way traffic it has been stated that they are not at all identical. The distribution function of cross-section headway, in accordance with measurement, may be produced from bigger theoretical traffic F than the distribution function identical with the previous one belonging to one-way traffic in accordance with the measurement.

The basis of examinations referring to formation of platoon is the assumption that vehicles go in platoon if at least three vehicles go in one groups and the headway between them is at least 7.2 s.

2. Momentary vehicle density

Mometary vehicle density (S) is the expression

$$S = \frac{1000}{t.v} \qquad \text{(Vehicle/km)}$$

where

t is the headway between two vehicles measured in s and

v is the speed of the second vehicle in m/s.

S expresses the speed of the tested vehicle and the distance from the vehicle goes before it. Road uses appreciate these two paramters the best. Therefore characteristics calculated from distribution function S are suitable for the qualification of the adequacy of traffic conditions.

For example let's regard free flow that traffic where at most 30% of vehicles go is conditions characterized by 10 Vehicle/km or less vehicle density (Vehicle density 10 Vehicle/km already means platoons).

3. Measurements referring to gap time

Gap time measurements at combined intersection where in one traffic direction the traffic increased by test vehicles were analysed the crossing movement of one minor direction.

For comparison with data in references the following were determined

- Gap time of GREENHIELDS (t_G) and
- Critical limit (t_c)

The results obtained are the following:

t_G = 5.7 s

t_c = 6.2 s

That headway distribution in main traffic flow which can be used for crossing vehicle from the other controlled direction was determined from gap time measurements. It has been taken into account that gap time is made up of several part gaps. Distribution function has been defined.

In that case if a gap time is used by several vehicles from the controlled direction the distribution function of headway of vehicle moving in platoon has been defined.

Sign of part gap time is given in Figure 5., value Fi(50%) - if the waiting vehicle is passenger car in Figure 6., and if the waiting vehicle truck is in Figure 7.

We must note that in case of distribution functions of part gap time the unrealistically long measured value was not taken into account (restricted distribution).

4. Number of vehicles from controlled direction

As a basic value to define the number of vehicles from subordinate direction without disturbing the main direction the curve in Figure 8, has been given. The curve has been produced by computer program based on simulation.

The program has defined on the basis of the measurement results of

- the traffic of the main direction and the distribution of headway,
- the traffic of controlled direction and the distribution of headway,
- time gap and its distribution function, taking the restricted distribution into account,
- the number of vehicles from subordinate direction and

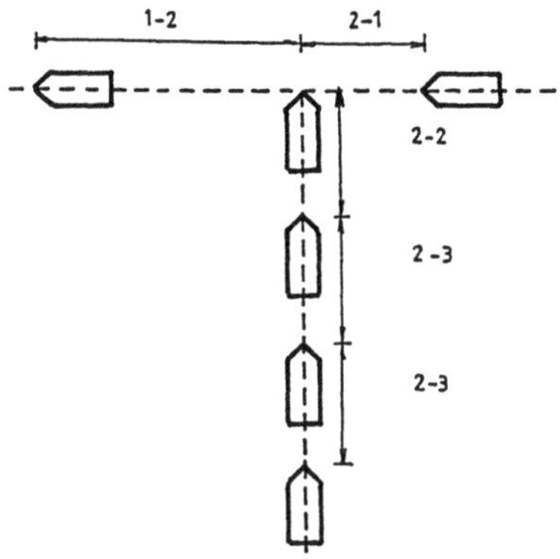

Fig. 5.

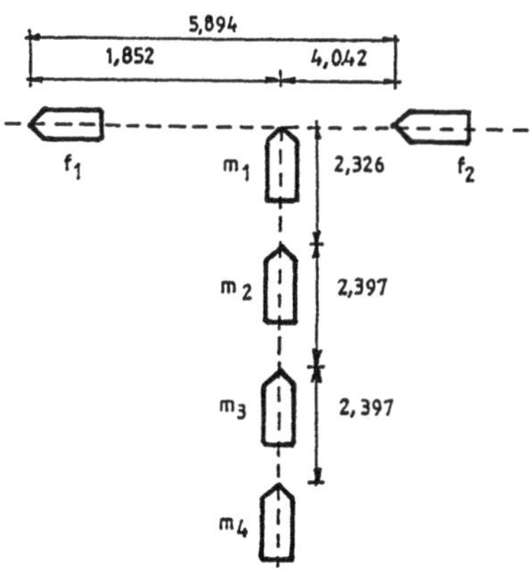

Fig. 6.

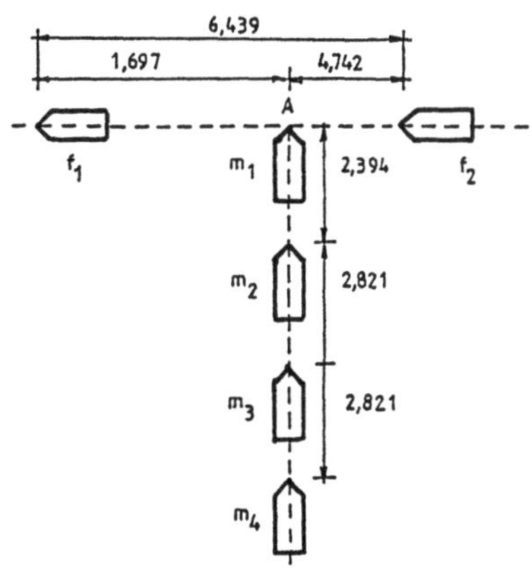

Fig. 7.

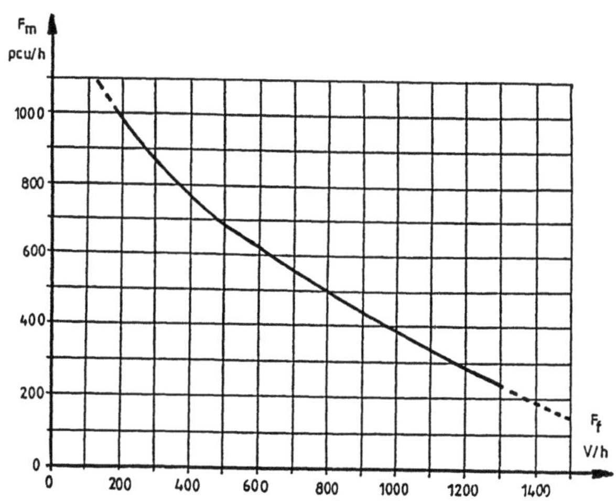

Fig. 8.

- the average lost time of the vehicles of subordinate direction.

Values can be seen in Figure 9. have been given for average lost time t_v(s) as a function of the traffic of main and subordinate directions.

5. Capacity calculation

The essence of the program for capacity calculation is the following:

- site plan and traffic controll arrangement of the examined intersection must be given (among others the method of traffic controll, the number of traffic lanes, usage of lanes, geometrical characteristics),
- value of traffic must be given (traffic flow in main, traffic flow in subordinate directions),
- the program counts the basic value of the number of crossing vehicles from controlled direction in function of main traffic volume and corrects these values taking local conditions into account,
- it counts in the subordinate directions the maximum number of waiting vehicles, the necessary section length for this and average lost time,
- at last it defines the level of service and reserved capacity by subordinate directions.

Table 1. gives level of service as a function of the average lost time of the subordinate direction.

The method is also suitable for the determination of the combined intersection in Figure 10.

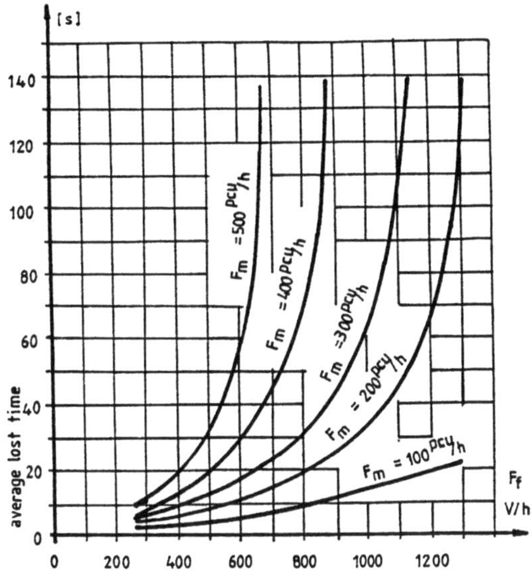
[s]
140
120
100
80
60
40
20
0
average lost time
0
200
400
600
800
1000
1200
Ff
V/h
Fm =500 pcu/h
Fm =400 pcu/h
Fm =300 pcu/h
Fm = 200 pcu/h
Fm = 100 pcu/h

Fig. 9.

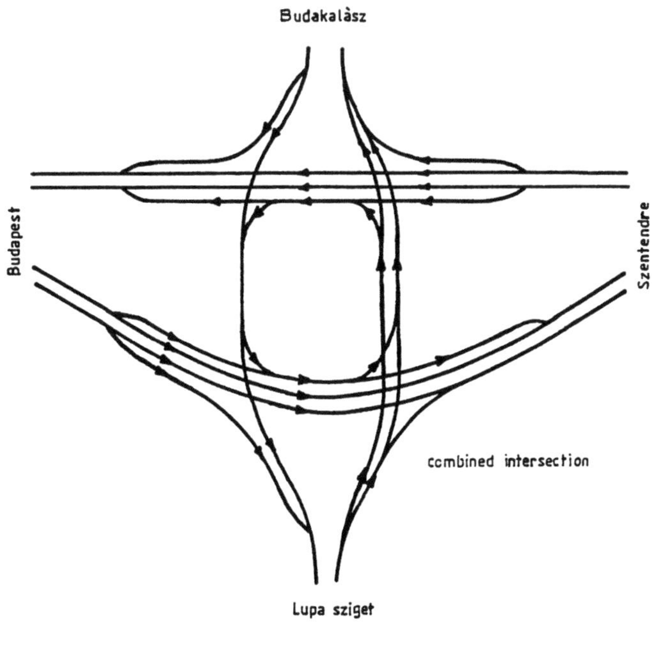
Budakalász
Budapest
Szentendre
combined intersection
Lupa sziget

Fig. 10.

Table 1.4

Level of service as a function of average last time affecting the vehicles of the subordinata direction.

Average lost time by vehicles (s)	Level of service
$t_v < 10$	A
$10 < t_v < 30$	B
$30 < t_v < 50$	C
$50 < t_v < 70$	D
$70 < t_v < 90$	E
$90 < t_v$	F

Calculation is carried out by intersection details. In this case it is checked whether the extent of directions is adequate for waiting of subordinate traffic.

The method is also suitable for the determination of the capacity of roundabouts working on the basis of the priority of vehicles on circular path. In this case the method calculates capacity by connections.

REFERENCES

1. Bényei,A. Ambrus,K. Csorja,Zs.: Recherches four Définir la Congruence des Circonstances de la Circulation des Routes a Deux Voies a Grande Trafic
Université Technique de Budapest
Chaire Construction des Routes, 1984.

2. Bényei, A., Ambrus, K., Csorja, Zs.: The Momentary Vehicle Density as the Measured Value for the Adequacy of Traffic Conditions
News Letter II. 1984. 2.
Technical University of Budapest

Recent Australian Unsignalised Intersection Research and Practices

R. J. TROUTBECK

Department of Civil Engineering,
Queensland University of Technology,
Brisbane, Australia.

Abstract

Australian engineers have been using gap acceptance theory for some time. They understand most of the concepts associated with this theory and are often prepared to modify values in the light of local conditions. The appropriate gap acceptance theory for Australian roundabouts is discussed using the conclusions from observed driver interactions. As a result of these interactions, all circulating streams could be assumed to act as one. The consequences of this assumption are discussed in this paper. The method of estimating the capacity and delays at a roundabout is described here. From the brief discussion of the accident rates at roundabouts it can be confirmed that they are generally safer than other forms of control. A list of current research projects illustrates that unsignalised intersections are still considered to be a major research topic in Australia.

Introduction

During the past few years, Australian research into unsignalised intersections has concentrated on the performance of roundabouts. This paper discusses the recent work on roundabout capacity and safety performance.

Roundabouts are an important traffic control device on urban arterials. Designers need to be able to predict their performance and to be able to estimate when they would be expected to reach the end of their useful life. Gap acceptance techniques were developed to assess the performance of roundabouts.

Basic gap acceptance assumptions

Roundabouts have been traditionally been analysed as a series of T intersections. It is assumed that the behaviour of each entry does not significantly affect the next. For most intersections this is quite true. In other cases the technique can be used in an iterative manner and it will quickly converge to solutions. This technique has been continued in the approach developed in Australia. For a roundabout leg, the circulating streams are considered to have priority and the entering stream to be the minor stream.

The gap acceptance theory has two basic elements. First, the usefulness of gaps of a particular size or opportunities to enter the intersection needs to be quantified. Second, the proportion of gaps of a particular size and their associated arrival patterns are also required. The average queueing delay and capacity can then be estimated using these two basic elements.

For the first of these, the usual technique is to assume that drivers are consistent and homogeneous. No driver will enter a roundabout unless the gap between the priority stream vehicles is at least the "critical acceptance gap", T. It is further assumed that a number of drivers will be able to enter in very long gaps. Usually, headways between the minor stream vehicles entering in a long gap is the "follow-on time", T_0.

Troutbeck [1] indicated that if drivers were heterogeneous, then the entry capacity would be decreased. However, if drivers are inconsistent then the capacity would be increased. If drivers are assumed to be both consistent and homogeneous, rather than more realistically inconsistent and heterogeneous, then the difference in the predictions is only a few per cent. For simplicity, consistent and homogeneous driver behaviour is assumed.

For the second element, Troutbeck [2] has based the capacity and delay predictions on Cowan's M3 model [3]. It was assumed that traffic consists of two types of drivers; those that follow others at a small headway, Δ, and those which have a longer headway. There is no need to model the small gaps accurately as these will all be unacceptable. Consequently, Cowan's headway model was acceptable as it represents the larger gaps accurately [4]. An advantage of this headway model is that it is a general one and simplifies to other more common models.

Driver interactions at roundabouts

A large empirical study of the behaviour of drivers at roundabouts was undertaken in Australia. The data was used to test the assumptions about which streams affect the behaviour of the entering drivers. Troutbeck [5] reports the results of this study which are also summarised below.

Vehicles leaving at the same leg that drivers are waiting to enter the roundabout were first thought likely to affect the behaviour of these entering drivers. Unfortunately, the applicable data sets in Troutbeck [5] were limited, but it would appear that the exiting drivers have only limited influence on the entering drivers. The two sites that did show an increased influence had higher speeds for circulating vehicles given the size of the roundabout. This gave less opportunity for the entering drivers to recognise that the vehicle is exiting at that leg.

Every entering driver needs to give way to the circulating vehicles in the near-side lane. It was thought that they may be prepared to enter the roundabout alongside a vehicle in the circulating lane near the central island. This behaviour would have been expected to increase capacity of the roundabout. Troutbeck [5] observed this simultaneous merging behaviour at only three of the 41 sites studied. These roundabouts were small, and would be expected to have lower circulating vehicle speeds. Speeds were estimated to be about 10 km/h (typical of those in parking areas) and this enabled the entering drivers to be confident when entering alongside a circulating vehicle. Apart from these sites, a subjective assessment of the video tapes confirmed that the entering drivers largely give way to all circulating vehicles. Kimber and Semmens [6] at TRRL agree with this conclusion. They found that the proportion of vehicles leaving at the previous exit, the proportion of entering vehicles turning left, and the proportion of circulating vehicles leaving at the next exit did not have a 'discernible effect on entry capacity'.

Whilst it is not reasonable to extrapolate the measures to driver behaviour, it could be hypothesised that these findings result from the entering drivers being unsure whether the circulating vehicles will change lanes [7]. This would result in them being unwilling to enter the roundabout along side a circulating vehicle. Vehicles in the inside circulating lane often move to the outer lanes on exiting resulting in frequent lane changes in the circulating lanes.

Drivers in different entry lanes are expected to behave differently. The drivers in one lane have to negotiate a tighter radius and enter more slowly. If, during acceptable gaps, drivers in one lane are prepared to follow vehicles in other entry lanes (at the same entry) at much shorter headways than when following a vehicle in the same lane, then the two streams operate relatively independently. Figure 1 illustrates the relative size of headways. This figure but does not indicate significance from a statistical test. Troutbeck [5] found that there was only one approach where the headways between drivers in different entry lanes was significantly greater, at the 5 per cent level, than those between drivers in the same entry lane. That is, there was only one site at which drivers in the entry lanes did not enter independently. This roundabout was very small and has the two entry streams on a narrow approach. Here the drivers tended to wait for each other.

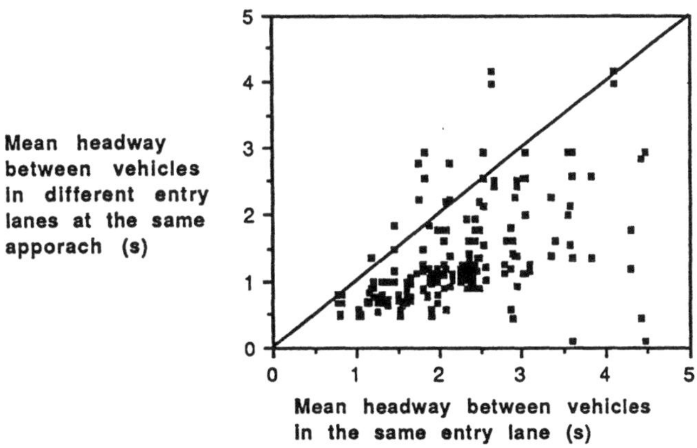

Fig. 1. Entry lane independence

The implications of these findings are to change the current Australian practice only marginally. The effect of the exiting vehicles is not an important one. If the exiting drivers are assumed to affect the behaviour of the entering drivers, then the circulating flow should also include the exiting drivers. This will affect the recorded entering driver behaviour with the entering drivers appearing to accept shorter gaps. Given that there would be more gaps, the estimation of capacity would not be expected to be substantially different. The present Australian Guide [8] recommends that the circulating streams should exclude the exiting traffic. This practice need not change. However, designers may include the exiting drivers in the circulating stream if negotiation speeds are unusually high or if the entering drivers are failing to recognise differences between the paths used by the circulating and exiting drivers.

The conclusion that an entry lane is unaffected by the presence of other entry lanes enables the capacity and delay of each lane to be estimated independently. The results can then be aggregated over all lanes. This is similar to the lane by lane approach used in SIDRA [9,10]. This is an extension to the current assumption used in the NAASRA [8] roundabout guide which assumes that the flow in each lane is the same and consequently the delay and capacity (maximum entry flow) is also the same for each lane. The finding that drivers in different entry lanes have different gap acceptance properties, agrees with the logic proposed by Golias [11] in a simulation model.

For roundabouts with more than one circulating stream, the circulating stream adjacent to the central island has a significant effect on the behaviour of all entering drivers. All circulating streams should be considered to have priority over the entering drivers. This is consistent with the assumptions used in NAASRA [8]. Since it is concluded that all circulating streams can be considered to affect the entering drivers, the headways between vehicles in both streams can be used to estimate both capacity and delay. The next section discusses the use of a single lane headway model to represent multi-lane conditions.

The use of a single Cowan M3 model to represent headways in two streams

Consider a roundabout with more than one circulating lane. Assuming that the headways in circulating lane one have a Cowan's M3 distribution [3] as follows

$$F_1(t) = 1 - \alpha_1 e^{-\lambda_1 (t-\Delta)} \qquad t \geq \Delta \tag{1}$$

and $F_1(t)$ is zero otherwise. Here α_1 is the proportion of free vehicles in the lane. These vehicles are not in bunches and are free of interaction. Δ is the headway between bunched vehicles and λ_1 is a decay constant. The headways in the other circulating stream have a similar distribution and terms:

$$F_2(t) = 1 - \alpha_2 e^{-\lambda_2 (t-\Delta)} \qquad t \geq \Delta \tag{2}$$

Again $F_2(t)$ is zero otherwise. In these equations, the decay constants, λ_1 and λ_2 are given by

$$\lambda = \frac{\alpha q}{(1-\Delta q)} \tag{3}$$

where α, λ, and q are lane dependent. Under these conditions, Troutbeck [2] found that the headways in vehicles in either stream are given by

$$\begin{aligned} F(t) &= 0 && t < 0 \\ F(t) &= 2\beta t && 0 < t < \Delta \\ F(t) &= 1 - \alpha' && t = \Delta \\ F(t) &= 1 - \alpha' e^{-\lambda'(t-\Delta)} && t > \Delta \end{aligned} \tag{4}$$

where

$$\alpha' = \frac{\alpha_1 q_1 (1-\Delta q_2) + \alpha_2 q_2 (1-\Delta q_1)}{q_1 + q_2} \tag{5}$$

$$\beta = \frac{q_1 q_2}{q_1 + q_2} \tag{6}$$

$$\lambda' = \lambda_1 + \lambda_2 \tag{7}$$

This distribution of headways between vehicles in either stream is shown in Figure 2. In this figure, the two major streams are identical. A negative exponential headway distribution is often used to represent the headways on a multi-lane facility. The reader should note the differences in the values predicted by these different distributions.

Both the one lane and the two-lane headway distributions are approximations to the actual distribution of headways. Accordingly, they only fit the measured distribution within the limits of experimental accuracy. Gap acceptance procedures only require that the longer headways or gaps be accurately represented. The shorter gaps need only be noted. Consequently this combined headway distribution can be represented by a single Cowan M3 model with the following properties:

$$F(t) = 1 - \alpha^* e^{-\lambda'(t-\Delta^*)} \qquad t \geq \Delta^* \tag{8}$$

and F(t) is zero otherwise. This modified distribution is illustrated in Figure 3.

Values of α^* and Δ^* must be chosen to ensure the correct proportions and the correct mean headway are obtained. This will ensure that the number of headways greater than t, 1–F(t), is identical from either technique when t is greater than Δ. Hence

$$(1-\Delta^* q_1 - \Delta^* q_2)\, e^{\lambda'\Delta^*} = (1-\Delta q_1)\,(1-\Delta q_2)\, e^{\lambda'\Delta} \tag{9}$$

and

$$\alpha^* e^{\lambda'\Delta^*} = \alpha' e^{\lambda'\Delta} \tag{10}$$

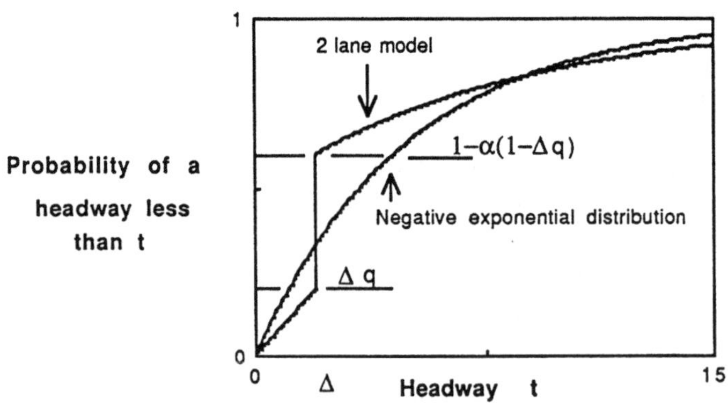

Fig. 2. Distribution of headways between vehicles in two identical streams.

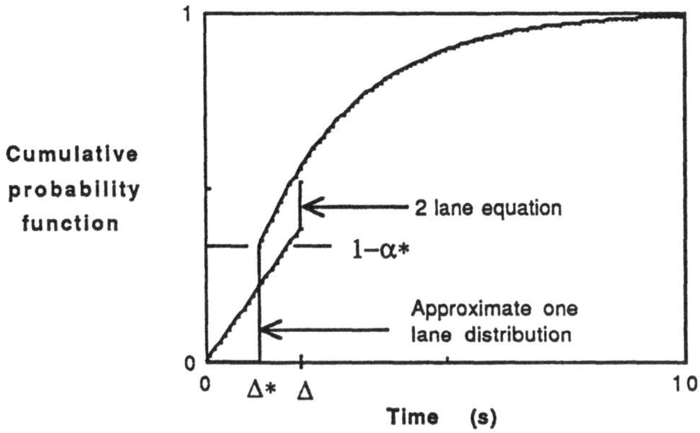

Fig. 3. Modified 'single lane' distribution of headways.

These expressions are solved iteratively assuming Δ_i^* to be equal Δ initially and that a better estimate of Δ_{i+1}^* is found from the previous estimate of Δ_i^* using the equation:

$$\Delta_{i+1}^* = \frac{1 - (1-\Delta q_1)(1-\Delta q_2)\, e^{-\lambda'(\Delta - \Delta_i^*)}}{q_1 + q_2} \tag{11}$$

A value of α^* is given by equation (10).

For example consider the case where stream 1 had a flow of 800 veh/h with 80 per cent of vehicles not in bunches (i.e. free vehicles). The second stream had a flow of 600 veh/h with 60 per cent of the vehicles 'free'. The inter bunch headway, Δ, was 2 s. For this example Δ^* is 1.15 and α^* is 0.662. For headways greater than 2 s, both curves give identical cumulative probabilities and both distributions have the same mean headway. As another example, consider two identical streams with the same flow and proportion of free vehicles. If the minimum headway between vehicles in the one lane is 2 s, then the minimum headway for the approximate one lane distribution is shown in Figure 4. As the lane flow increases so Δ^* decreases, but Δ^* is not as strongly dependent on the proportion of free vehicles. Note that Tanner [12] recommended a value of 1 s when evaluating a priority road with two streams.

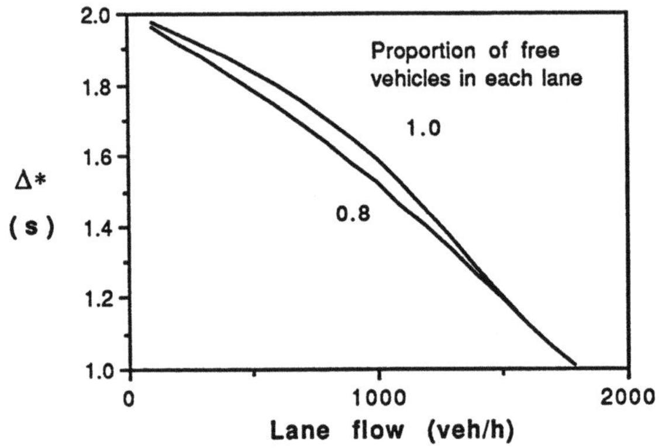

Fig. 4. The minimum headway for the approximate one lane distribution

The single lane capacity equation [2] can then be modified to give:

$$q_{emax} = \frac{\alpha^* (q_1+q_2)\, e^{-\lambda' (T-\Delta^*)}}{1 - e^{-\lambda' T_0}} \tag{12}$$

This is identical to the result for a major stream with two lanes. Similarly, Adams' delay is estimated from a modified version of the single lane equation and is

$$D_{min} = \frac{e^{\lambda'(T-\Delta^*)}}{\alpha' (q_1+q_2)} - T \frac{1}{\lambda'} + \frac{\lambda'\Delta^{*2} - 2\Delta^* + 2\Delta^*\alpha^*}{2(\lambda'\Delta^* + \alpha^*)} \tag{13}$$

Whilst this equation does not give precisely the same results as given by Troutbeck [2], it does give acceptable answers. The percentage error in using this technique is typically less than about three or four per cent. Figure 5 gives the percentage error from equation 13 when

$\alpha_1 = 0.65$

$\alpha_2 = 0.65$

$\Delta = 2.0$ s

$T = 4.0$ s

$T_0 = 2.0$ s

$q_1 = q_2$

The equations for average delay are based on Adams' delay and the degree of saturation (and therefore capacity). Since the modified distribution gives satisfactory estimates of these latter quantities, it will also be expected to give satisfactory estimates of delay.

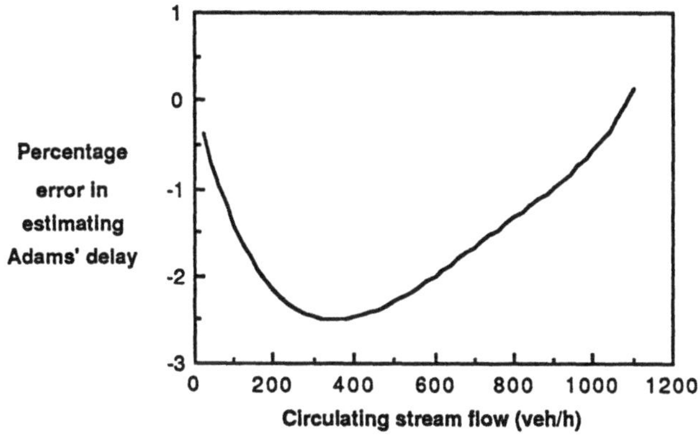

Fig. 5. Percentage error from the use of equation 13 and for the conditions described in the text.

The principal advantage in using this modified single lane equation is that it is simpler to use and from the theory does not introduce significant errors. Since it is concluded that all circulating streams can be considered to affect the entering drivers, the headways between vehicles in both circulating streams can be used to estimate both capacity and delay. It is recommended that equations 12 and 13 be used. The Cowan's M3 type models presented here are all approximations. It is concluded that a modified single lane model can be satisfactorily used to represent the distribution of headways in either one or two lanes.

Drivers were observed at roundabouts and their behaviour was quantified using equations described in Troutbeck [13]. These equations will be presented here but the reader is directed to Troutbeck [13,14] for details about the choice of equations. This section will contain an analysis technique for roundabouts.

As documented above, each approach of the roundabout will be treated as an isolated T intersection. This is consistent with existing Australian practice [8] and the UK practice [15].

Predicting the behaviour of drivers at roundabouts

The behaviour of drivers is affected by the expected entry lane flows. The lane volumes are estimated from turning movement proportions. This is in contrast with other analysis methods for roundabouts where entry traffic is equally distributed to each entry lane [8]. Drivers turning right can be expected to use the right hand lanes, similarly drivers turning left will use the left entry lanes. The designer should proportion the straight through traffic into all entry lanes either equally or by judgement based on other traffic factors. This method is consistent with the signalised intersection guide [10]. The entry lane with the largest volume is termed the dominant entry stream for that particular entry.

The circulating flow is also required when estimating driver gap acceptance characteristics. The circulating flow is calculated from the turning movements at the roundabout and represents the flow of vehicles past the nose of the roundabout. The exiting drivers are excluded from the circulating flow.

Drivers in the dominant stream have shorter gap acceptance qualities. The drivers in the other streams, the sub-dominant streams, will have longer gap acceptance qualities. The critical gap and the follow-on time are estimated for each movement.

The dominant stream follow-on time (T_{0dom}) is a function of the inscribed diameter (Di) the number of entry lanes (n_e) the number of circulating lanes (n_c) and the circulating flow (Q_c). The inscribed diameter of the roundabout is the largest arc that can be drawn inside the kerb line. The diameter will be different for each entry of an oval roundabout. The number of entry and circulating lanes is a step function of their width as shown in Table I. The equation for the dominant stream follow-on time is

$$T_{0dom} = 3.37 - 0.000394Q_c - 0.0208Di + 0.0000889Di^2 - 0.395n_e + 0.388n_c \quad (14)$$

An initial value of the dominant stream follow-on times is estimated using Figure 6 and is adjusted with a factor based on the number of entry and circulating lanes (Table II).

For example, if the roundabout has an inscribed diameter of 30 m and the circulating flow is 1000 veh/h then the dominant stream follow-on headway is 2.4 s. Given that the number of entry lanes equals the number of circulating lanes then the corrected dominant stream follow-on time remains at 2.4 s.

Table I. Entry lane and circulating lane widths

Entry lanes		Circulating lanes	
number of lanes	width (m)	number of lanes	width (m)
1	<6	1	<10
2	6-10	2	10-15
3	10-15	3	15-20

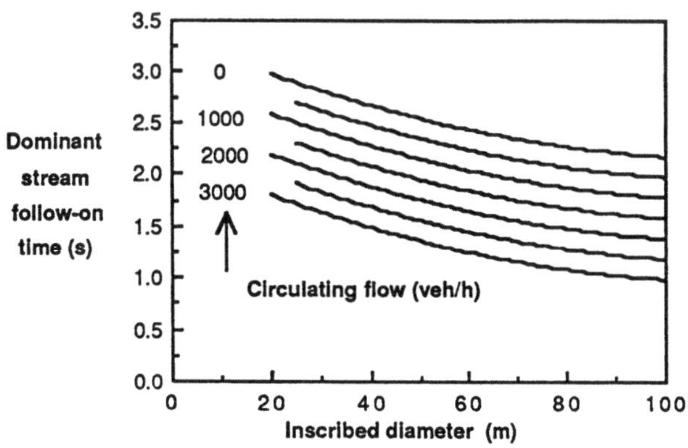

Fig. 6. Effect of circulating flow and diameter on the dominant stream follow-on time

The follow-on time for entry streams other than the dominant stream is calculated from the dominant stream follow-on time (T_{0dom}) and the ratio of the dominant stream flow (Q_{dom}) to the stream flow in question (Q_{sub}). The equation for the sub-dominant stream follow-on time is

$$T_{0sub} = 2.149 + 0.5135\, T_{0dom} \frac{Q_{dom}}{Q_{sub}} - 0.8735 \frac{Q_{dom}}{Q_{sub}} \quad (15)$$

The follow-on time for the sub-dominant stream is constrained to be not less than the dominant stream value. This relationship is illustrated in Figure 7.

Table II Adjustment factors for the dominant stream follow-on time

		Number of entry lanes 1	2	3
Number of circulating lanes	1	0.00	– 0.39	
	2	0.39	0.00	– 0.39
	3		0.39	0.00

Note: Add these factors from the initial values from Figure 6.

If there are three entry streams, one will be the dominant stream and the other two will be sub-dominant. The follow-on time for the dominant stream is evaluated first and then the follow-on times for the two sub-dominant streams. Their characteristics will be related to those from the dominant stream. Since lane flows may be different, so will the follow-on times in all three lanes.

Consider the case where the dominant entry stream flow was 800 veh/h; the sub-dominant stream flow was 400 veh/h; and the dominant stream follow-on time was 2.5 s. The ratio of flows is then 800/400 or 2.0 and the sub-dominant stream follow-on time would be approximately 3.0 s.

The ratio of the critical acceptance gap (T) to the follow-on times (T_0) is given by the circulating flow (Q_c), the number of circulating lanes (n_c) and the average entry lane width ($\bar{e}$) by the equation

$$T = (3.6135 - 0.0003137\, Q_c - 0.3390\, \bar{e} - 0.2775\, n_c)\, T_0 \quad (16)$$

The critical acceptance gap is restrained to be not less than 1.1 times the follow-on time. This is to ensure that the expression for capacity will be a monotonic decreasing function of the circulating flow. This relationship is applied to the follow-on times for all lanes of an entry (Table III).

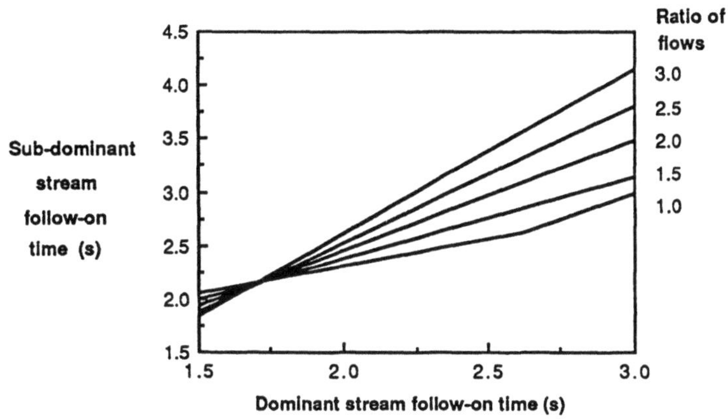

Fig. 7. Sub-dominant stream follow-on time

Table III Ratio of the critical acceptance gap to the follow-on time

Average entry lane width (m)	3	4	5	3	4	5
Number of circulating lanes	1	1	1	2	2	2
Circulating flow (veh/h)						
0	2.32	1.98	1.64	2.04	1.70	1.36
200	2.26	1.92	1.58	1.98	1.64	1.30
400	2.19	1.85	1.52	1.92	1.58	1.24
600	2.13	1.79	1.45	1.85	1.51	1.18
800	2.07	1.73	1.39	1.79	1.45	1.11
1000	2.01	1.67	1.33	1.73	1.39	1.10
1200	1.94	1.60	1.26	1.67	1.33	1.10
1400	1.88	1.54	1.20	1.60	1.26	1.10
1600	1.82	1.48	1.14	1.54	1.20	1.10
1800	1.75	1.42	1.10	1.48	1.14	1.10
2000	1.69	1.35	1.10	1.41	1.10	1.10
2200	1.63	1.29	1.10	1.35	1.10	1.10
2400	1.57	1.23	1.10	1.29	1.10	1.10
2600	1.50	1.16	1.10	1.23	1.10	1.10
2800	1.44	1.10	1.10	1.16	1.10	1.10
3000	1.38	1.10	1.10	1.10	1.10	1.10

Consider the example where the average entry lane width is 4 m, and the circulating flow 1000 veh/h is in one lane. The T/T_0 ratio is then 1.7. Hence the critical acceptance gap would be twice the follow-on time. If the follow-on time was 2.4 s then the critical acceptance gap would be 4.0 s (2.4 x 1.7 s).

Predicting the distribution of headways in the circulating stream

The distribution of headways was quantified using Cowan's M3 model [3]. The proportion of free vehicles and the minimum intra-bunch headway were required. Both of these terms were interrelated and it was decided that the mean bunch size needed to be chosen first. It was recommended that the intra-bunch headway, Δ, be set to 1 s for multi-lane circulating sections and to 2 s for single circulating lane roundabouts.

The proportion of free vehicles, α, is evaluated from the circulating flow, the number of lanes and the proximity of the roundabout to other intersections which increase bunching. The greater the flow the greater the bunching, and the lower is the proportion of free vehicles. Simplified equations produced the values in Table IV. These are average values that should be adjusted according to the proximity of the roundabout to signalised intersections. As an example, if the single circulating lane had a flow of 1000 veh/h then the proportion of free vehicles would be about 0.3. If the roundabout is some distance from a signalised intersection then this value could be increased to 0.4 or 0.5. On the other hand, if the traffic approaching the roundabout was strongly affected by a signalised intersection then this α value should be decreased.

The proportion of free vehicles is expected to range from 1 for random traffic to about 0.3 for heavily platoon traffic. Values as low as 0.2 were recorded in extreme cases. This is equivalent to an average platoon length of 1 to about 3 vehicles in most conditions.

Predicting the capacity and steady state delay at a roundabout

Troutbeck's equation [2] can then be used to estimate the maximum entry capacity.

$$q_{emax} = \frac{\alpha q e^{-\lambda(T-\Delta)}}{1-e^{-\lambda T_0}} \tag{17}$$

where

α is the proportion of free vehicles in the circulating streams

q is the flow of vehicles in the circulating streams

T is the critical acceptance gap

T_0 is the follow-on time

Δ is the minimum headway in the circulating streams, and these are related by equation 3.

Troutbeck [13,14] illustrates values of the predicted capacity. The results are characterised by a distinctive 'S' shape that is unlike the exponential curve usually found from gap acceptance theory. The results from these equations compare favourably with those from Kimber [16]. Figure 8 (from [13]) gives an example of this comparison.

Table IV Proportion of free vehicles, α

No. of circulating lanes	1	>1
Intra-bunch headway (s)	2.0	1.0
Circulating Flow (veh/h)		
0	0.8	0.8
400	0.6	0.7
800	0.4	0.6
1200	0.2	0.5
1600		0.4
2000		0.3
2400		0.2

The minimum delay is given by

$$D_{min} = \frac{e^{\lambda(T-\Delta)}}{\alpha q} - T - \frac{1}{\lambda} + \frac{\lambda\Delta^2 - 2\Delta + 2\Delta\alpha}{2(\lambda\Delta + \alpha)} \tag{18}$$

and for all practical purposes the steady state average delay is given by

$$D = D_{min}\left[1 + \frac{\varepsilon x}{1 - x}\right] \tag{19}$$

where

x is the degree of saturation

ε is a form factor and can be generally set to 1.0 if a better value is not available.

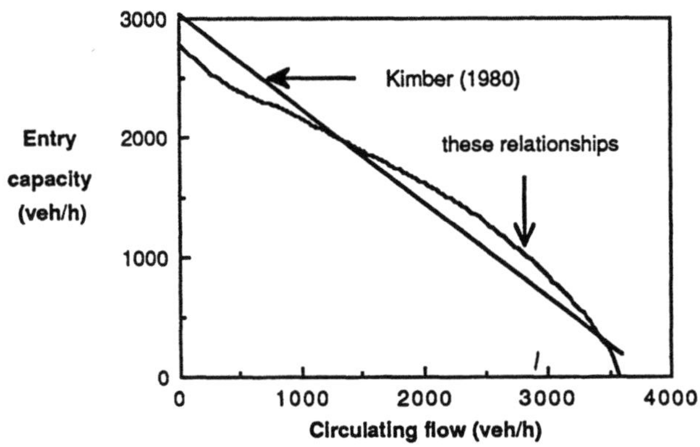

Fig. 8. Comparison between the results from the equations presented here and Kimber's equations [16] for a two-lane roundabout with a diameter of 60 m and a 5 m average entry lane width (from [13])

Accident studies at roundabouts

During the last two years there has been only limited additional research into the accident rates at roundabouts.

Tudge [17] reviewed the accident record for a number of roundabouts on arterial roads and in urban residential areas. These latter roundabouts are to restrict movements rather than control traffic. Some roundabouts constructed at sites where there was not a history of accidents, had accidents after construction. Roundabouts constructed at sites with poorer accident rates demonstrated an improved accident rate after construction. Accident records were useful input data when designing roundabouts. Tudge concluded that accident rates were a function of the geometric design. This is agreement with the work of Maycock and Hall [18] and Arndt [19].

A recent pilot study of accident rates and the geometric design of roundabouts has been undertaken at the Queensland University of Technology. This study by Arndt [19] has identified the entry geometry parameters which affect the accident rates. Arndt concluded that approach roads to roundabouts should have a left hand curve to ensure that drivers are aware of the intersection and to reduce the speed at which drivers will travel through the intersection. The central island must be conspicuous. In general the conclusions were that drivers should be encouraged to slow down before entering the roundabout and that the roundabout should be obvious and conspicuous to approaching drivers.

Future Australian research

A revised Australian guide for designing roundabouts is about to be produced. There will be some changes to the geometric design requirements and guidance for the placement of pedestrian crossings near roundabouts will be given. The capacity and the associated delay calculations described in this paper will be incorporated into the guide.

Research is currently under way to establish a relationship between accidents and roundabout geometry, to develop a simulation model of roundabouts based on the concept of a rotating platter of driver spaces. The concept that roundabouts are significantly affected by bunching in the circulating stream, has identified the need for methods to predict the headway distribution as a function of platoon dispersion. This will quantify the influence of a signalised intersection on a roundabout. This research is currently being undertaken by the author.

Conclusions

In Australia, the gap acceptance approach is considered to be useful. The roundabout capacity and delay techniques described here have gained support and will be incorporated into a national AUSTROADS guide for designing roundabouts.

This paper has described the issues associated with gap acceptance theory and indicated that it is reasonable to simplify the headway distributions if there are more than one circulating lane. The gap acceptance parameters as explained by the equations presented here are significantly affected by the circulating flow and this produces curves which are unlike those from the more traditional gap acceptance theories. Troutbeck [13] has indicated the results from this method are similar to those found by Kimber [16] using an empirical approach. This conclusion has converged the gap acceptance and the empirical theories.

References

1. Troutbeck, R. J.: Current and Future Australian Practices for the Design of Unsignalised Intersections. In "Intersections Without Traffic Signals" - Proceedings of an International Workshop 16-18 March, 1988 in Bochum West Germany. (Ed: W Brilon) Springer-Verlag Berlin.

2. Troutbeck, R. J.: Average Delay at an unsignalised intersection with two major streams each having a dichotomised headway distribution. Transportation Science 20(4) (1986) 272-286.

3. Cowan, R. J.: Useful headway models. Transportation Research 9(6) (1975) 371-375.

4. Troutbeck, R. J.: Does gap acceptance theory adequately predict the capacity of a roundabout? Proceedings 12th Australian Road Research Board Conference 12(4) (1984) 62-75.

5. Troutbeck, R. J.: Traffic interactions at roundabouts. Proceedings of the 15th Australian Road Research Board Conference 15(5) (1990) 17-42.

6. Kimber, R. M.; Semmens, M. C.: A Track experiment on the entry capacities of offside priority roundabouts. Transport and Road Research Lababoratory. Supplementary report SR334. (1977).

7. Kenny J. H.: Driver behaviour and accidents at roundabouts. In "Roundabout and Unsignalised Intersection Workshop" (Ed: R. J. Troutbeck) Queensland University of Technology, November 1990.

8. National Association of Australian State Road Authorities.: Roundabouts- A Design Guide. NAASRA, Sydney. (1986).

9. Akçelik, R.: SIDRA-2 does it lane by lane. Proceedings. 12th Australian Road Research Board Conference 12(4) (1984) 137-149.

10. Akçelik, R.: SIDRA version 2.2 input and output. Australian Road Research Board. Technical Manual, ATM No. 19. (1986).

11. Golias, J.: Gap-acceptance and capacity for vehicles crossing a priority stream. Ph.D thesis. University College, London. (1981).

12. Tanner, J. C.: A theoretical analysis of delays at an uncontrolled intersection. Biometrika 49(1 and 2) (1962) 163-70.

13. Troutbeck, R. J.: Roundabout capacity and associated delay. In "Transportation and Traffic Theory" Proceedings of the Eleventh International Symposium on Transportation and Traffic Theory, in Yokohama, Japan (Ed: M Koshi): Elsevier 1990.

14. Troutbeck, R. J.: Evaluating the performance of a roundabout. Australian Road Research Board Special Report 45. (1989).

15. Kimber, R. M.: The design of unsignalised intersections in the UK traffic capacity of roundabouts. In "Intersections Without Traffic Signals" - Proceedings of an International Workshop 16-18 March, 1988 in Bochum West Germany. (Ed: W Brilon): Springer-Verlag Berlin 1988.

16. Kimber, R. M.: The traffic capacity of roundabouts. Transport and Road Research Laboratory, Laboratory Report LR942. (1980).

17. Tudge, R.: Accidents in roundabouts in New South Wales. Proceedings of the 15th Australian Road Research Board Conference 15(5) (1990) 331-350.

18. Maycock, G. and Hall, R. D.: Accidents at 4 arm Roundabouts. Transport and Road Research Laboratory Report LR 1120 (1984).

19. Arndt, O.: The effect of geometry on accident rate at roundabouts. Queensland University of Technology Undergraduate Thesis. (1990)

Gap Acceptance and Risk Analysis at Unsignalised Intersections

S. LASSARRE ,
Département d'Evaluation et de Recherche en Accidentologie
Institut National de Recherche dans les Transports et leur Sécurité,
Arcueil, France.
P. LEJEUNE et J.C. DECRET
Division Exploitation Sécurité
Centre d'Etudes Techniques de l'Equipement,
Bordeaux, France.

Summary

In order to explore the possibility to use the critical gap acceptance as an intermediate variable related to accident risk, an exploratory study has been designed including eight intersections on rural roads in the south of France.

Traffic data are collected by video, and gaps on the principal roads are calculated by computer. Lag and gap distributions are estimated by a probit model. The median gap acceptance for three manoeuvres: crossing, right and left turn, is a significant discriminant factor between junctions, related to the visibility distance.

Its variability according to different factors such as the age and sex of the drivers, the speed of the approaching vehicle, the time of the day makes difficult a statistical accident analysis.

I - INTRODUCTION

In a road safety context, intersections incorporate specific characteristics that sometimes complicate enormously the search for solutions designed to reduce the real or potential risk they represent. Although accident analysis is currently the most reliable way of analysing and quantifying the risks, it seemed useful to explore other methods, in the spirit of "conflict" type studies [13], that could be developed which would remain compatible with the method of experimentation.

The method is based on the choice of a model for the behaviour of a driver on a secondary road at an intersection with a main road which is not governed by traffic lights. The probabilist model integrates a function describing the decision taken by the driver to accomplish a right turn, left turn or straight across maneouvre depending on the distribution of the time gaps separating two successive vehicles on the main road.

The parameter of interest to be estimated is the median critical gap, which is the time gap accepted by 50% of drivers or refused by 50% of them. A sample of 8 intersections was chosen for observing vehicles during the execution of certain manoeuvres, video recordings of which are subsequently analysed with the help of a computer program that calculates the gaps between vehicles. The median accepted gaps estimated in different experimental conditions will be used to validate the probabilist model and their variations will be analysed in terms of the characteristics of the intersection and traffic conditions.

A concluding discussion will focus on the obstacles to use such a parameter as an intermediate variable in the analysis of the risk of accidents at intersections, and the way to remedy.

II - CHOICE OF A MODEL OF DRIVING BEHAVIOUR AT INTERSECTIONS

Take an intersection between a main road and a secondary road that is not governed by a system of traffic lights. The interest variable is the time gap between two consecutive vehicles on the main road that a driver on the secondary road accepts or refuses in order to carry out a turn-right, left-turn or straight across manoeuvre.

The first hypothesis [3], which is also the simplest and the most classic, considers that each driver makes his own assessment of the critical gap. The critical gap varies from driver to driver, but is invariable for any given driver. A driver will always refuse a gap that is below a critical threshold and will always accept a gap in excess of it.

Consider that $\alpha_1(t)$ is a driver's gap acceptance function, which represents the probability that a driver will accept a given gap: the hypothesis of stable behaviour as regards the critical gap T corresponds to the following stair function:

$$\begin{aligned} \alpha_1(t) &= 0 \quad t \leq T \\ &= 1 \quad t > T \end{aligned}$$

However, it is very possible that a driver may refuse a gap of six seconds but accept one of four seconds. This possibility introduces a second hypothesis [3] and a function $\alpha_2(t)$ which is a better description of the driver's behaviour and is defined as:

$$\begin{aligned} \alpha_2(t) &= 0 \quad t \leq T \\ &= 1 - e^{-\beta(t-T)} \quad t > T \end{aligned}$$

where β and T are driver characteristics that may vary for all drivers taken together. Below the threshold T, a variation in behaviour is introduced by a growing probability of acceptance.

In the field, repetitive measurements of the same driver to estimate β and T are not possible. A given driver passes through the intersection only once. That is why we preferred to use the first critical gap model in which the variation in behaviour arises from differences between drivers, each driver having a coherent behaviour relative to a fixed critical gap, and not from differences in the driver's behaviour integrated in the whole body of drivers.

To go from one driver to all drivers, we introduce the notion of $u(T)$, the density of the critical gap for all drivers. Take for $\alpha_1(t)$, being one driver's acceptance function, a jump function

H(t), then the acceptance function for all drivers $\alpha_P(t)$ is identical to the cumulative distribution function of critical gaps for all drivers:

$$\alpha_P(t) = \int_0^{\infty} H(t-T)\, u(T)\, dT = \int_0^{t} u(T)\, dT$$

At an intersection, the distribution of gaps accepted and refused by the drivers is observed as well as the distribution of gaps offered (i.e. those accepted plus those refused). The theoretical probability [1, 12] that a gap smaller than t is accepted is equal to:

$$P(t) = C \int_0^{\infty} H(t-T)\, f(T)\, u(T)\, dT = C \int_0^{t} f(T)\, u(T)\, dT$$

where $H(t-T)$ is the acceptance function for one driver,

$u(T)$ the density of the critical gap for all drivers,

$f(T)$ the average number of gaps of length T considered by a driver, $f(T) = e^{pT}$ when the gaps are independent and have a negative exponential distribution with a flow p ,

C the normalisation constant.

This modelisation highlights the bias introduced by the distribution of the gaps offered the driver on the distribution of gaps accepted, which as a result is not identical with the distribution of gap acceptance for all drivers.

The estimated acceptance percentages are biased by the fact that a driver with a high critical threshold is going to refuse a larger number of small and medium-sized gaps, than a driver with a low threshold who will accept the first gap that occurs. Hence, a careful driver will be over-represented compared with a high-risk driver. That is seen in the fact that the estimated acceptance probabilities are lower than the real acceptance probabilities.

III - ESTIMATING THE DISTRIBUTION OF ACCEPTANCE AND THE CRITICAL GAP BY PROBIT ANALYSIS

To eliminate that bias, numerous methods have been proposed [12]. Some of them use probit analysis, which was developed for use in trials and experiments in biology [10]. By analogy , Drew [9] considers the driver as the subject of an experiment who responds to a stimulus - here, the time gap offered - by a refusal or an acceptance, which is a qualitative response. In our case, the intensity of the stimulus, the length of the interval, is not controlled, and in addition the same subject responds to several stimuli.

The idea is to describe in linear terms the relationship between the proportion of acceptances and the length of the gap by means of a probit transformation.

$$P = \frac{1}{\sqrt{2\pi}} \int_{-\infty}^{Y-5} e^{-\frac{u^2}{2}}\, du$$

The probit Yi for the gap class i which is below the limit Ti (=1.00, 2.14, 4.19, 6.24, 8.29,10.33 s.), estimated from the proportion $pi = \frac{number\ of\ gaps\ accepted}{number\ of\ gaps\ offered}$ is a linear function of Ti of the type: $Yi = \alpha + \beta\, Ti$.

In place of Ti one can utilize $\log Ti$ if the distribution of the gaps accepted is more akin to log-normal than normal.

By entering on either gausso-arithmetic or gausso-logarithmic paper [4] the low proportions and values for the time gap classes, one necessarily obtains straight lines. Probit analysis starts graphically by entering the probit values and the time gap values (whether transformed into a logarithm or not) in order to obtain the first estimates of α and β, which are initially used in a computer program analysing the probit BMDP3S.

To go from acceptance distribution to the distribution of critical gaps, we use the Ashworth method, which proves to give the best results and be the most economical in calculating terms of nine estimates [12]. Ashworth [1, 2] established the following proposition:

When the flow of traffic follows a negative exponential law for parameter p (veh./s.) and the distribution of critical gaps is normal, with an average μ_c and a standard deviation σ_c, then the acceptance distribution for a given gap is a normal law with an average μ_a and a standard deviation σ_a, such that

$$\mu_a = \mu_c + p\,\sigma_a^2 \qquad \sigma_a^2 = \sigma_c^2$$

When the distribution of critical gaps follows a log-normal law, the distribution of gaps accepted is not identical but close to a log-normal law. The results obtained from the normal law may be a good approximation of the log-normal case.

The parameters μ_c and σ_c, the average or median(with the normal distribution) critical gap and the standard deviation, characterise a type of manoeuvre at an intersection that takes account of the behaviour of a driver on a secondary road and the presence of traffic on the main road. Starting from these parameters, which serve to discriminate the intersections according to their priority regime, the problem is to seek correlations with the frequency and the nature of accidents for studies of safety at intersections.

IV - SELECTION OF A SAMPLE OF INTERSECTIONS

The intersections were selected more for their convenience for observing behaviour than on the basis of a road safety criterion such as the number of accidents. The main condition was that the flow of vehicles had to be sufficient to obtain behavioural data on around 200 drivers per intersection in one or two hours of observation.

Eight intersections were selected around Bordeaux, in the region of Aquitaine, consisting of four crossroads and three T-junctions , plus one Y-junction. Most of them (six) were indicated by a "STOP" sign and two by a "Give Way" sign; there were central islands in half of them but not the other half, and there were varying distances of visibility.

As regards the convergent traffic, the ratio main to secondary trafic volume takes value between 1.3 and 16.2. Annual average daily traffic flow on the main road was distributed normally around 7,000 vehicles.

The manoeuvres were relatively distinctive at each intersection:

- At three of the intersections, there was clearly a similar and strong proportion of R.T and L.T manoeuvres: intersections Nos. 33-01, 07 and 20.
- At three intersections, there were mainly L.T.'s: intersections Nos. 33-03, 16-01, 01, and 10.
- At one intersections, the majority of drivers went straight across: intersections No. 16-02.
- At one intersection, there were mainly right turns: intersection No. 79-01.

V - TECHNIQUES FOR GATHERING AND EXPLOITING VIDEO DATA

Field equipment consists of a video camera with zoom lens positioned in the rear of a van parked at the side of the secondary road at a certain distance from the intersection.

The data are analysed in a laboratory on a screen with a slow advance and freeze frame capability controlled by the VISTA system on Macintosh [17].

As soon as the operator visualises a vehicle crossing a vertical mark on the screen, he uses the mouse to instruct the computer to record the moment at which the vehicle passes on the main road and the times at which a vehicle on the secondary road stops and moves on.

VI - ANALYSIS OF ACCEPTED AND CRITICAL GAPS BY MANOEUVRE

An exploratory analysis of driver behaviour by type of manoeuvre (right turn, left turn and crossing) is useful for formulating certain hypotheses about variations in the median critical gap as a function of the characteristics of the intersections chiefly visibility, but also traffic conditions on the main road and the importance of external influences such as circadian factors.

VI.1. Right Turns

During daylight conditions, the average accepted gap ranges from 4.3 to 7.2 seconds, indicating a diversity of driver behaviour in executing right turns according to the intersection. The standard deviations of between 2.3 and 3.2 second are far less dispersed. The shortest times, 4.3 seconds on INT 20 and 4.5 seconds on INT 01, correspond to intersections where there is good visibility to the left, which is to say over 300 meters (320 meters and 400 meters respectively). The longest times correspond to intersections where visibility to the left is less than 300 meters. The highest average gap, of 7.2 seconds, was recorded at INT 7901, where there is an island on the main road. At none of the intersections is there an acceleration lane to facilitate entry to the main road.

At all the intersections, the percentage of direct acceptance without stopping at the STOP sign is 41%. Nearly *half the drivers of light vehicles (LVs) do not stop at the STOP sign when turning right*. The highest percentages are found at the intersections with the greatest visibility - INT 20 and INT 3301 (Table 1).

Code	Median	Standard deviation	Nb. of vehicles	Direct acceptance	Model validity	Priority	Type	Distance visibility
7	5.4	2.9	75	28 (37%)	yes	Stop	┼	230m
20	4.3	3.2	161	100 (62%)	yes	Give way	┬	320m
7901	7.2	2.3	179	75 (41%)	no	Stop	"	280m
3301	4.5	2.4	55	27 (49%)	yes	Stop	"	400m

Table 1: Median Accepted Gap with Standard Deviation and Intersection Characteristics for Right Turns.

VI.2. Left Turns

As already indicated, the simultaneous traffic flows in both directions are taken into account for the sequence of times between vehicles. We assume that the driver has to choose a gap in a continuous flow of traffic in both directions. This simplified modelisation of traffic flows may produce distortions in the findings, but the joint consideration of the two traffic flows, which is possible theoretically, calls for a greater calculation effort and larger sample size [14].

The median accepted gap varies little between the intersections, ranging from 4.9 seconds for INT 3301 to 7.3 seconds for INT 1601. For five intersections out of seven, in homogeneous conditions it remains relatively stable at between 6 and 7 seconds.

The visibility and the size of the intersection may influence the critical gap, since the minimum of 4.9 seconds was recorded at an intersection where there is the greatest visibility to right and left, while a high median of 7 seconds was recorded at an intersection with a central island.

Code	Median	Standard deviation	Nb. of vehicles	Direct acceptance	Model validity	Priority	Type	Distance left-right
1	6.2	1.9	309	24 (8%)	yes	Give way	┬	170-250
7	6.7	2.4	80	8 (10%)	yes	Stop	┼	230-200
10	6.2	2.6	144	39 (27%)	yes	Stop	"	200-400
10	6.9	2.2	103	9 (9%)	no	Stop	"	200-400
20	6.6	2.3	137	10 (7%)	no	Give way	┬	320-120
1601	7.3	2.2	91	8 (9%)	yes	Stop	┼	400-200
1602	7.0	3.0	31	2 (6%)	yes	Stop	"	330-340
3301	4.9	2.5	44	8 (18%)	yes	Stop	┬	400-300

Table 2: Median Accepted Gap with Standard Deviation and Intersection Characteristics for Left Turns

VI.3. Crossing Over

The dispersion of the median accepted gap is moderate, between 5.6 and 7.7 seconds, for crossing over intersections, and has an identical range to that of the average gap for left turns.

Code	Median	Standard deviation	Nb. of vehicles	Direct acceptance	Model validity	Priority	Type	Distance left-right
7	5.6	3.7	14	1 (7%)	yes	Stop	+	230-200
10	5.8	2.4	41	3 (7%)	yes	Stop	"	200-400
10	6.5	3.3	31	11 (35%)	no	Stop	"	200-400
1602	7.7	2.0	108	16 (15%)	yes	Stop	"	330-340

Table 3: Median Accepted Gap with Standard Deviation and Intersection Characteristics for Crossing Over

VII - INFLUENCE OF EXTERNAL FACTORS ON THE MEDIAN ACCEPTED GAP

VII.1. Influence of the type of vehicle

We are mainly interested in the behaviour of light vehicles (LVs) and heavy vehicles (HVs). There were too few two-wheelers to enable reliable statistics and models to be established.
For all the intersections concerned, the median accepted gaps for HVs were equal to or greater than those for LVs. Drivers of HVs thus accept slightly longer gaps than drivers of LVs, whatever the type of manoeuvre [16].

		L.V.					H.V				
Man.	Code	M.	S.D.	Nb. of veh.	Dir. accep.	Model valid.	M.	S.D.	Nb. veh.	Dir. accep.	Model valid.
L.T.	7	6.7	2.4	80	8	yes	6.7	4.0	14	2	yes
L.T.	10	6.9	2.2	103	9	no	7.7	2.9	24	3	no
R.T.	20	4.3	3.2	161	100	yes	5.8	2.9	36	9	yes
R.T.	7901	7.2	2.3	179	74	no	7.5	3.5	18	5	no
C.	1602	7.7	2.0	108	16	no	8.5	3.7	18	3	no

Table 4: Median Accepted Gap, Standard Deviation and Population by Type of Vehicle

VII.2. Influence of the Time Factor

Traffic conditions affect the behaviour of drivers in the morning. Very dense traffic on the main road results in very short gaps, which are almost all less than one second, and all are refused. On the other hand, the first gap greater than one second is immediately accepted. Hence, there is a very short median accepted gap of 3 seconds, with a dispersion of 1.5 seconds. At 1.30 pm traffic on the main road is more fluid and offers a greater variety of gaps. Drivers no longer seize the first gap that comes along, refusing the short gaps in favour of longer ones. That results in a median accepted gap of 6.2 seconds, with a dispersion of 1.9 seconds. At 7.30 in the evening, traffic is far less dense than at 7.30 in the morning. Hence, drivers choose longer gaps for turning left at night, resulting in a median accepted gap of 7.7 seconds, with a dispersion of 2.4 seconds.

That finding concurs with the conclusions of Darzentas [8], namely that the median accepted gap for turning left is longer at night than in the day, and especially when there is traffic in the vicinity coming from the right. We would point out that the methodology differs as regards the analysis of the time gaps which, in the English study, were analysed for each flow, whereas in our study all flows were analysed en masse.

Time	Median	Standard deviation	Nb. of vehicles	Direct acceptance	Model validity	Priority	Type	Distance visibility left-right
8h	3.0	1.5	119	9 (8%)	yes	Give way	T	170-240
13h	6.2	1.9	309	24 (8%)	yes	"	"	"-"
19h	7.7	2.4	100	37 (37%)	yes	"	"	"-"

Table 5: Median Accepted Gap and Standard Deviation at Intersection 1 at 7.30 am, 1.30 pm and 7.30 pm

VII.3. Influence of Traffic

R. Ashworth [2] suggests correcting the average μ_a for the gap acceptance distribution by a quantity dependent on the traffic (*q in veh/sec*) and its standard deviation σ_a in order to obtain the average critical gap according to the formula:

$$\mu_c = \mu_a - p\,\sigma_a^2$$

The argument is that, in dense traffic conditions, the careful driver will refuse a larger number of short gaps than at times of fluid traffic and hence shifts the acceptance distribution towards greater gaps.

What did we observe? At intersection 1, at the morning peak of 7.30 am when traffic is very dense, drivers adapt their behaviour to the type of traffic and accept shorter gaps than at 1.30 pm, when traffic is fluid.

The Ashworth model is a model assuming rigid behaviour on the part of the driver, which establishes his refusal and acceptance on the basis of a fixed threshold, the critical gap not varying with traffic conditions. However, it appears than in certain traffic conditions the driver no longer has any choice and has to accept shorter time gaps, since he cannot reach a sufficiently long gap, both because it is so rare and because of the pressure from other drivers on the secondary road. Let us look again at intersections 7 and 20, where drivers of all vehicles adapt their behaviour when turning right.

At INT 7 the distribution of gaps offered is greater for short gaps than at INT 20, since the average traffic flow is twice as great at INT 7 as at INT 20.

Starting from different distributions of the gaps offered, acceptance percentages are virtually identical for the two intersections, with the same average gap of 4.9 seconds. According to the Ashworth model, the acceptance percentage of short gaps should be greater at intersection 7 than at intersection 20. However, it appears that the driver regulates his behaviour to adapt it to the volume of traffic on the main road, which invalidates the model and the correction introduced by Ashworth.

Code	Type	Traffic veh/day	% offered gaps <6.2s	% accepted gaps <6.2s	% accepted gaps >6.2s
7	┼	9400	39	19	19
20	┬	4840	15	90	94

Table 6: Annual Average Daily Traffic and Gap Acceptance Proportion for the Two Intersections

VIII RELATIONSHIP BETWEEN ACCIDENTS AND THE CRITICAL GAP

VIII.1. Obstacles to Establishing the Relationship

Attempts to establish the relationship between the number of accidents and the critical gap for the different manoeuvres at the sample of intersections encounter numerous obstacles. To get a sufficient number of accidents, it is necessary to go back five years with the risk that some intersections may have been equipped with islands in the meantime.

At any given intersection and for a given manoeuvre, the median accepted gap fluctuates considerably over time. At intersection 1, it varies by 100% as between the three times of day,

partly because of traffic conditions (flow/speed) on the main road. That variability must be taken into account by a balanced observation spread over the day.
Between different intersections, variations in the median accepted gap are attributable to the physical characteristics of each intersection. In the case of right turns, the accepted gap is reversely proportional to the distance of visibility. On the other hand, in the case of left turns and crossings, the accepted gap is between 6 and 7 seconds, whatever the intersection, and there is not enough of a variance to give any hope of correlating it with the number of accidents.

VIII.2. Recommendations

Even though it has been not possible to establish the relationship between accidents and critical gaps [11], we can draw some lessons for subsequent research from this exploratory study.

VIII.2.a. Planning Experiments

The experiment has to use a sample of intersections that resemble each as far as possible in terms of their geometrical characteristics and traffic flows, but whose accident rates are very contrasted. Controlling the physical characteristics of the intersections - number of lanes on the main road, type of intersection and priority regime, size of intersection, distance of visibility and type of environment - is easier than controlling the characteristics of the traffic that has to be measured.
Five years of accident data are necessary, especially if different categories of accidents are to be distinguished - day/night, LV/HV, and so on.

VIII.2.b. Traffic Data Collection

Video recordings have to be replaced by automatic data collection thanks to a measuring station linked to magnetic loops on the main road and in the stopping area on the secondary road.
An observer can be used to code the characteristics of the vehicles and the lengths of time they stop. The cost of this equipment and its installation is offset by time savings in the extraction of the video data. Moreover, such a system is less easy to see than a van parked at the roadside.
The system makes a continuous recording of the times at which vehicles pass on both roads, on the basis of which the speeds and the gaps between vehicles are calculated and the vehicles categorized.
The planning of the observation periods must ensure a balanced daily and weekly distribution.

VIII.2.c. Exploiting the Times Between Vehicles

Estimating the median critical gap through probit analysis seems to us to be a simple, efficient and robust method that is easy to implement and initialise thanks to Henry's straight line [4,10].
The probit model can be estimated either from one flow or from two flows at the price of a larger sample of vehicles [14].
The Ashworth correction [2] for traffic on the main road is not necessary. On the other hand, the integration in the model of the approach speed of vehicles on the main road is fundamental at the microscopic level. There is a negative correlation between the average speed and the median accepted gap.
Three other parameters are found to have an influence on the accepted gap:
- the sex of the driver. The median accepted gap is longer for women than for men.[16]
- the age of the driver. Older drivers accept longer gaps than younger drivers, and do not adjust their gaps as much in terms of the speed of approaching vehicles [6,16].
- the type of vehicle. Drivers of LVs accept shorter gaps than drivers of HVs [16].
These differences in accepted gaps are reflected in the degree of accident risk, older drivers having higher accident rates at intersections than younger ones.

VIII.2.d. Estimating the Accident/Critical Gap Relationship

For each category of driver, defined in terms of age, sex and type of vehicle, and for each intersection and each manoeuvre, the risk is equal to the ratio between the number of accidents and the volume of traffic on the secondary road.
The risk can be modelised through a Poisson model depending linearly from explicative variables [15]:
- the flow of traffic on the main road;
- the coefficients for the relationship between speed, the median and the standard deviation of the accepted gap.

IX - CONCLUSION

Coming from the field of road safety research, we ran up against difficulties of an experimental rather than an instrumental or methodological nature when it came to treating the application of the critical or accepted gap variable in a context of statistical accident analysis.
In the future, thanks to the acquisition of temporal traffic data using automatic measuring stations, speeds and times between vehicles can be calculated at lower cost and with greater accuracy.

By rigorously monitoring the factors influencing the critical gap by efficient planning of experiments and the use of a model incorporating approach speeds, it seems possible to us to obtain valid results from research into the accident/critical gap relationship at intersections.

Bibliography

1. Ashworth, R.: A note on the selection of gap acceptance criteria for traffic simulation studies. Trans. Res. 2 (1968) 174-175.

2. Ashworth, R.: The analysis and interpretation of gap aceptance data. Trans. Sci. 4 (1970) 270-280.

3. Blumenfeld, D.E.; Weiss, G.H.: The effects of gap acceptance criteria on merging delay and capacity at uncontrolled junction. Traff. Engng. Control 20 (1979) 16-20.

4. Calot, G.:Cours de statistique descriptive..Paris: Dunod 1966.

5. Cooper, D.F.; Storr, P.A.; Wenell, J.C.: Traffic studies at T-junctions, 4. The effects of speed on gap acceptance and conflict rate. Traffic Engng. Control 18 (3) (1977) 110-112.

6. Darzentas, J.; McDowell, M.R.C.; Cooper, D.F.: Minimum acceptable gaps and conflict involvement in a simple crossing manoeuvre. Traffic Engng. Control 21 (1980) 58-61.

7. Darzentas, J.; Cooper, D.F.; Storr, P.A.; McDowell, M.R.C.: Simulation of road traffic conflicts at T-junctions. Simulation 34 (5) (1980) 155-164.

8. Darzentas, J.; McDowell, M.R.C.: Driver behavior at unlit non-urban T-junctions in daylight and darkness. J; Opl. Res. Soc. 32 (1981) 721-727.

9. Drew, D.R.: Traffic flow theory and control. New York/ Mac Graw Hill 1968.

10. Finney, D.J.: Probit analysis. Cambridge: Cambridge University Press 1971.

11. Lassarre, S.; Lejeune, P.; Decret, J.C.: Niveau de risque en carrefour. Bordeaux: Rapport final CETE 1982.

12. Miller, A.J.: Nine estimators of gap acceptance parameters. Proceedings of the fifth international symposium on the theory of the traffic flow. Newel, G. E. Ed. New York: Elsevier 1971.

13. Muhlrad, N.: Technique des conflits de trafic. Manuel de l'utilisateur. Arcueil: Synthèse INRETS 11 1988.

14. Storr, P.A.; Cooper, D.F.; McDowell, M.R.C.: Analysis of gap acceptance in a complex traffic manoeuvre. European J. of Operationnal Research 5 (1980) 94-101.

15. Summersgill, S.: Relations de prévision d'accidents pour quelques types d'intersections en Grande-Bretagne. Arcueil: Actes INRETS 16 1989.

16. Wenell, J.; Cooper, D.F.: Vehicle and driver effects on junction gap acceptance. Traffic Engng. Control 22 (1981) 628-632.

17. Wootton, H.J.; Potter, R.J.: Video recorders, micro computers and new survey techniques. Traffic Engng. Control 22 (4) (1981) 213-215.

Use, Design and Safety of Small Roundabouts in Norway

Kjell Seim
Senior Engineer
Public Roads Administration
Directorate of Public Roads
Norway

Small roundabouts have become very popular in Norway during the last ten years. The total number of roundabouts has increased from about 15 in 1980 to about 350 in 1990. The success of small roundabouts in improving safety and capacity means this trend will probably continue. This paper presents a survey of the development of roundabouts in Norway. Design criteria are also discussed and examples of layouts given. Particular attention has been paid to the problems of **where** to build roundabouts, and **where not** to build them. The paper also deals with roundabouts in tunnels.

In addition to this, the results of a recent study of accident rates are also presented. The study, as earlier studies, shows that small roundabouts normally have the lowest accident rate of all types of one-level junctions. The number of accidents involving pedestrians is remarkably low, but there is a tendency towards problems for two-wheeled vehicles.

Introduction

Prior to 1980 there was little interest in Norway in roundabouts. As in many other countries, it was thought that this type of junction took up too much space. Furthermore, Norway was used to the "priority to the right" (left-hand rule in UK) tradition. Today it is generally accepted that this rule means low junction capacity and major traffic flow problems. Motorists are guaranteed entry into the roundabout, but not exit, which may cause a complete jam in the junction even if the traffic is only moderately heavy.

In the mid-70s the priority rule was changed for some roundabouts by putting up give-way signs at the points of entry. This naturally led to a substantial improvement in traffic flow, and studies also showed a reduction in accidents. This, supported by positive experience from many small roundabouts in Britain, led to the Norwegians starting to build small roundabouts in Norway at the beginning of the 80s.

There was a certain amount of scepticism at first, both among road and traffic engineers and the public at large. It was generally thought that British road users show greater tolerance, flexibility and caution than Norwegian ones, and roundabouts are more suitable there than in Norway. It was assumed that priority rules and traffic signals at junctions were a better idea for Norwegian (continental) drivers.

However, several follow-up studies of the new roundabouts have shown that Norwegian motorists also cope well with this type of junction. There was a remarkably low incidence of accidents and personal injury, and recorded capacity parameters such as critical lags, gaps and follow-up time seem to be just as low as the mean values in British studies. Road users are probably as flexible and considerate as the system allows now.

Much of the scepticism gradually disappeared and the roundabout became increasingly popular the length of the country from north to south. (If you want, you can dive through roundabouts in the midnight sun, far inside the Arctic Circle.) There are plans or wishes for roundabouts nearly everywhere. Figure 1 shows that the total number of roundabouts, a mere 15 or so in 1980, had risen to 50 in 1985 and over 350 by 1990.

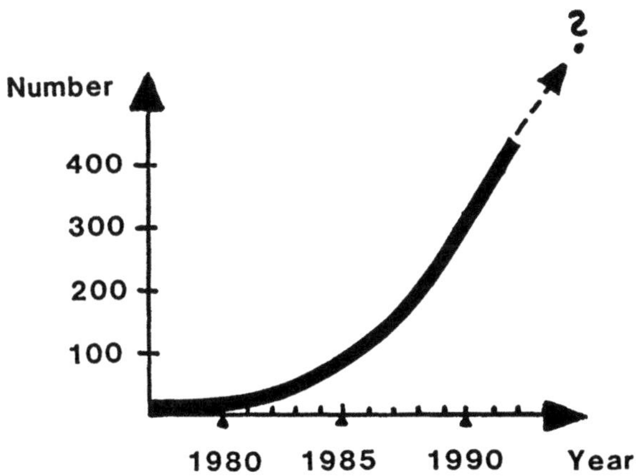

Figure 1. Total number of roundabouts in Norway.

Today (in summer 1991) there are almost 400 roundabouts. This may not be a large number by international standards. But the population of Norway is only about 4 million, which means one roundabout for every 10,000 people - and the number is rising fast.

In the first part of the 1980s it was mainly junctions with priority signs or the normal priority to the right rule that were converted into roundabouts. In recent years a number of junctions with traffic signals have also been converted, and on major roads the building of grade separated junctions is often postponed, and roundabouts are built as a more or less temporary arrangement.

Some old roundabouts still had no give-way signs in the 80s. This naturally lead to some confusion and generated some negative criticism of roundabouts in general. In 1985 the central authorities decided that all roundabouts would be provided with give-way signs at the point of entry.

Typical features of Norwegian roundabouts

The first new roundabouts were constructed partly according to English guidelines. A Norwegian standard has gradually emerged, but the design principles still show UK influence.

Most of the roundabouts are located on two-lane main roads and collector roads through moderately built-up areas. There are also some in built-up areas/town centres. There are few roundabouts in rural areas as yet - an estimated 10 - 20.

Norwegian roundabouts are relatively small. More than 90% have an inscribed circle diameter of 26 to 40 metres and the diameter of the central island varies between 6 and 20 metres. About 15 mini roundabouts have been built, most of which have a central island which is either flat or slightly raised and can be driven over. There are few large roundabouts (less than 10).

The approach to the roundabout is normally widened to give two-lane entry. So far all the central islands have been circular and it is usually possible to drive over their outer part.

There is often a street lightening column in the centre of the island.

Figure 2. Typical of Norwegian roundabout built in 1980s.

The design criteria are discussed later on in this paper.

Safety

Several studies have given clear indications that roundabouts are the safest type of one-level junction. The latest and to date most extensive risk analysis was carried out in 1990, based on accident data from 59 roundabouts and 124 traffic-signal regulated junctions between 1985 and 1989.

Risk is measured by accident rate (AR). This is defined as the number of personal injuries reported to the police per 10^0 (million) vehicles crossing the junction:

$$AR = \frac{PI . 10^0}{AADF . 365 . p}$$

PI = number of personal injuries reported to police
AADF = average annual dayly flow
P = duration of study period
AR = accident rate for junction

The table shows the difference between the accident rate in roundabouts and junctions with traffic signals. Roundabouts with three arms are safest, but the difference between these and those with four arms is not big.

Number of arms	Roundabouts	Traffic lights
3	0.03(0.05)	0.05(0.08)
4	0.05(0.04)	0.10(0.16)

Table 1: Accident rate in roundabouts and junctions with traffic signals (Figure in brackets is from earlier studies).

Junctions with priority signs and junctions with no form of control will normally have an accident rate between 0.10 and 0.30.

The accident rate as a function of roundabout size indicates that the risk is somewhat greater in medium roundabouts than small ones, see Table 2. However, this must be examined more closely before conclusions may be drawn.

Type of roundabout	No. of junctions	Accident rate	Standard deviation
Mini	4	(0.00)	(0.00)
Small	17	0.03	0.02
Small/medium	18	0.03	0.02
Medium	20	0.06	0.02

Table 2: Accident rate related to roundabout size
(Small - inscribed circle = 26 - 30 m)
(Medium " " = 32 - 45 m)

Altogether 33 accidents involving personal injury were recorded in the 59 roundabouts in this study. Only one of these accidents involved a pedestrian (Table 3).

20	car accidents
6	motorcycle/moped accidents
6	bicycle accidents
1	pedestrian accident

Table 3: - Accident distribution according to type of road-user involved

In comparison, pedestrians were involved in 20% of accidents with personal injury in the junctions with traffic signals, (57 accidents out of 287).

Some of the difference in the amount of accidents involving pedestrians in roundabouts and junctions with traffic signals may be explained by the fact that relatively few roundabouts are located in town centres with high pedestrian density. But 40% of the roundabouts have been built on roads or streets with a 50 km/h limit where there would normally be a certain amount of pedestrian traffic.

It can probably be concluded that roundabouts on two-lane roads are safer for pedestrians than other junctions as their geometrical design forces motorists to reduce speed. Moreover, pedestrians often have a clearer view of the traffic situation than in junctions which are controlled by traffic signals.

A substantial percentage of the accidents involve two-wheeled vehicles (36%). Even if the absolute number of accidents is low, this ought to be taken into account in designing roundabouts in the years to come.

It should be noted that no accidents were recorded in 39 of the 59 roundabouts in the study period.

In addition to this, traffic varies between 2,600 to 28,400 vehicles per day with an average of 13,900.

Where are roundabouts suitable?

Continued success with roundabouts will depend on their "correct" location. This is important in a period where interest in this type of junction is so great that it can almost be characterised as a vogue.

New Norwegian guidelines discuss the selection of type of junction. Special attention has been paid to whether a junction should be controlled by traffic signals or built as a roundabout, as this question frequently crops up.

FACTORS IN FAVOUR OF ROUNDABOUTS:

1. Safety
   * Few points of conflict
   * Speed reduction
   * The safest type of junction for vehicles and pedestrians

2. Traffic flow
   * The capacity is usually greater than for similar junctions with traffic signals and other at-grade junctions
   * Fewer delays, especially outside peak-hours
   * Flexible in respect of traffic variations

3. Nature of junction
   * Junction for roads with equal status, e.g. two main roads
   * Even distribution of traffic/substantial turn-off traffic

4. Short junction distances

5. Space
   * No space is available for long separate lanes for left or right turning traffic
   * Available junction space is suitable for a roundabout

6. Possibility of U-turn
   * Advantage to unfamiliar persons
   * Junctions between two roundabouts can be simplified by removing possibility of left turns (clearance)

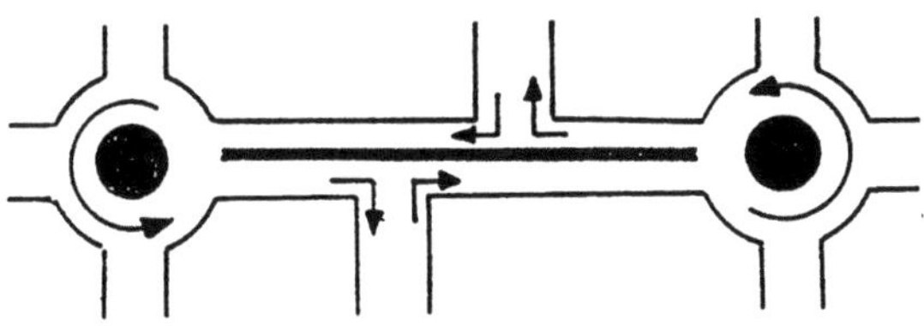

7. Environment
   * Fewer delays reduce energy consumption and pollution
   * The central traffic island can be designed as a positive, aesthetic element
   * Mark the starting point of a street with priority to the environment

8. Maintenance
   * Low maintenance costs
   * Hardly any follow-up required

9. Realiability
   * Functions at all times (does not revert to amber flashing light)

FACTORS AGAINST ROUNDABOUTS:

1. Need for general priority given to primary road traffic

2. Need for absolute priority to public transport

3. Uneven traffic distribution between intersecting roads when there is high-density traffic

4. Speed-reducing geometric design is difficult to achieve

5. Speed reduction is not desirable

6. Substantial cycle traffic (?)

7. Unfortunate location of pedestrian crossings

8. High construction costs for transforming existing at level junctions to roundabouts

If the arguments for having a roundabout are more or less equal to the arguments for traffic signals, then a roundabout should be preferred because of functionality, safety and maintenance aspects.

The responce of a majority of Norwegian motorists to roundabouts has gradually become positive. Motorists, however, are a complex group. Some drivers prefer the strict regulations and safe feeling of a junction with traffic signals. Certain very cautious motorists even make detours to avoid roundabouts.

Some people also complain that there may be too many roundabouts in certain areas. But when confronted with the alternative, which in many cases tunrns out to be several junctions with traffic signals, the criticism generally subsides.

These matters should be taken into consideration when designing and signposting roundabouts. Perhaps the traffic rules ought to be simplified as well.

Design criteria

New Norwegian guidelines for the design of roundabouts will be published this year. This section discusses only some of the design criteria.

As in several other countries, roundabouts are divided into 4 categories according to size

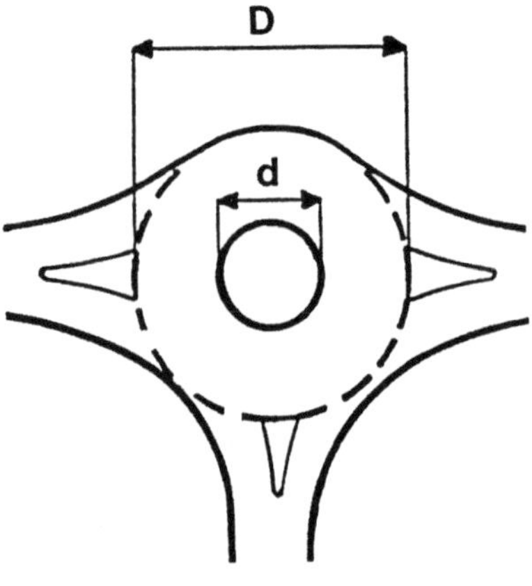

MINI	d = 1,5 - 4 m D < 25 m
SMALL	d > 5 m D = 26 - 30 m
MEDIUM	d > 10 m D = 31 - 45 m
LARGE	d > 25 m

Figure 3. Classification of size

During the first half of the 1980s many roundabouts were constructed with an inscribed circle of 26 - 28 m (small). Due in part to complaints from drivers of large vehicles that the roundabouts were too narrow, the new guidelines state that on primary roads and roads with a large percentage of heavy vehicles "medium roundabouts" should be constructed. An inscribed circle with a diameter (D) of 35 m is recommended for 2-lane roads, and 40 - 45 m for 4-lane roads. To avoid too high speeds, roundabouts should normally not be larger.

A "small roundabout" may be suitable for more local roads with a small number of heavy vehicles and buses.

Caution should be exercised in the use of "mini roundabouts", as they provide little or no deflection of traffic. This type may be suitable in downtown areas where the speed level is low and space is very restricted. "Double mini roundabouts" have been constructed in 4 locations. These are considered successful. The example in figure 4 is the most extreme design that has been built in Norway to date, with a distance of only 8 m between the roundabouts.

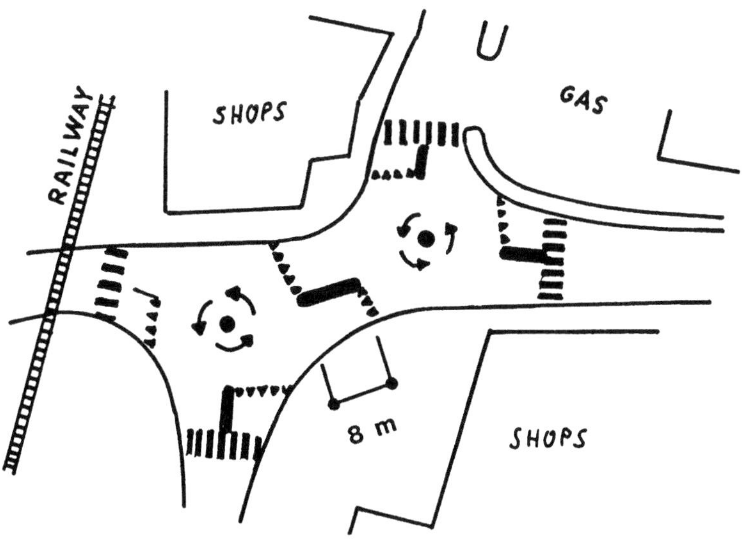

Figure 4. Successful "double mini".

In mini, small and medium roundabouts the centre of the central island should be located in the intersection point between the median lines of the crossing roads. Then the angle to be turned isn´t too small for some traffic directions and unnecessarily great for others. This is also important for the visual leading and the road-user´s understanding of the junction. This important criterion is often deviated from because of lack of space and/or to reduce construction costs. The result may in part be poor driving comfort and less observance of priority, and hence, a greater chance of accidents occurring.

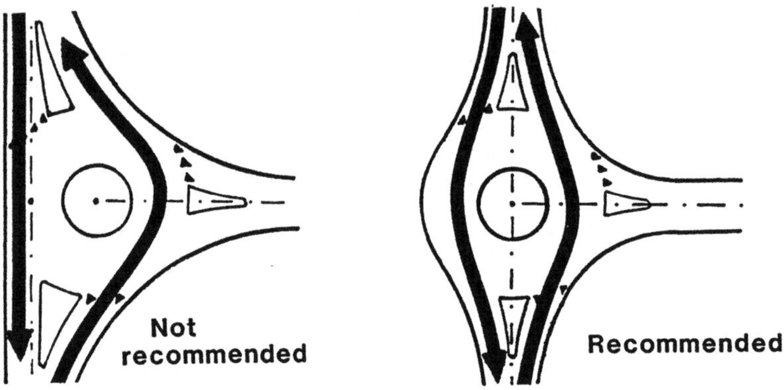

Figure 5. Roundabouts should be located symmetrically around the median lines.

In addition to the size and location of the central island, the design of the entry road is crucial to safety. When there are two or more entry lanes in the approach road, the routing of the dividing/deflection island should be such that the extension of the right-hand border in the form of a straight line will be a tangent to the central island, see figure 6. Experience shows that if this line hits further to the left on the central island, conflicts will arise between cars driving parallel into the roundabout. This also leads to fewer motorists using the left entry lane.

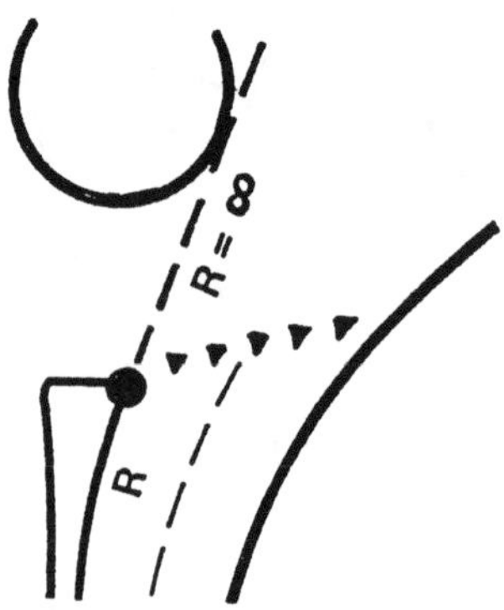

Figure 6. Routing of dividing/deflection island in small and medium roundabouts with two or more entry lanes.

Experience shows that roundabouts have such great capacity that it is often unneccessary to widen the approach to two entry lanes. With only one lane a tighter geometrical formation can be used, providing greater speed reduction. This is perhaps the way to reduce accidents involving cyclists and mopedists.

If a motorist intends to go straight ahead in a roundabout, and there are several entry lanes, in Norway the motorist can select the lane that is most appropriate for the subsequent driving. This rule is favourable to the capacity, but also causes some conflicts and may be a reason why some cautious drivers dislike roundabouts. If only one lane is constructed in the approach, this problem is eliminated.

It has been usual to build the dividing deflection island with an even width (1.5 or 2.0 m). In order to avoid having an unused area in the exit where sand and snow can accumulate, it is now recommended that the dividing island should be widened towards the roundabout. This also makes it more difficult to take a short cut on the wrong side of the central island (figure 7).

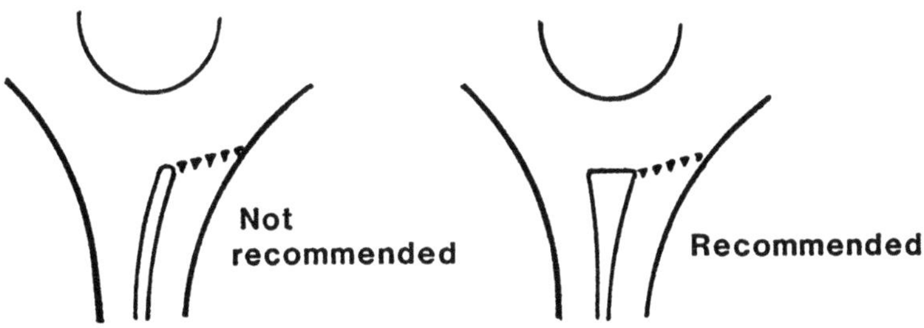

Figure 7. The dividing deflection island should be widened towards the roundabout where there is enough space

Pedestrian crossings should be located at a distance of 10 - 12 m from the roundabout and lead across a physical dividing island. The purpose is to give motorists the opportunity to concentrate on the pedestrians before and after the roundabout itself. Pedestrians on their part can concentrate on vehicles from one direction at a time. These principles are probably very important in order to achieve a low accident risk for pedestrians.

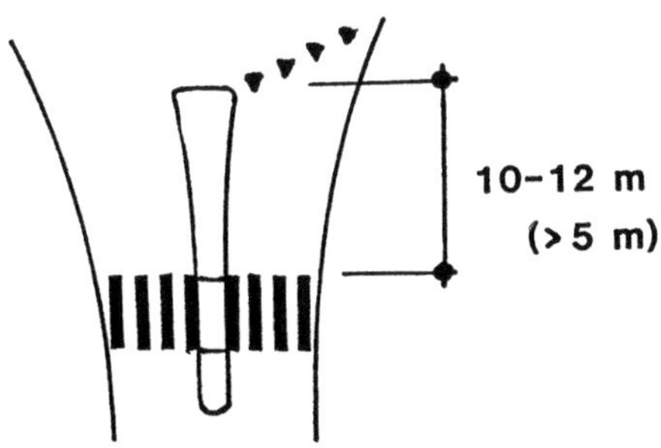

Figure 8. Location of pedestrian crossings.

One should consider placing pedestrian crossings closer to the roundabout in city areas if the speed limit is low, but no closer than 5 m. Raised pedestrian crossings may be a good solution on arms where there is no room for a dividing island.

Markings in the circulation area is generally not recommended in Norway. However, it has proved necessary to use leading lines in roundabouts where there are (due to the capacity) three entry lanes.

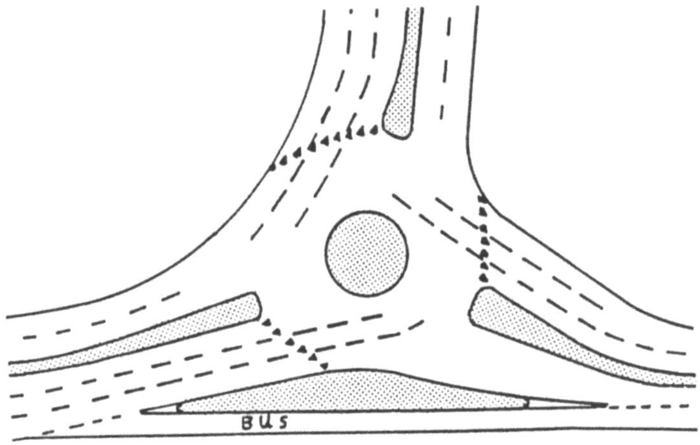

Figure 9. Example of use of leading lines as a "makeshift" in connection with 3 entry lanes.

Roundabout in tunnel

Because of the steep landscape, there are lots of tunnels in rural areas in Norway. In recent years tunnels have become more common in citys as well. The traffic congestion problems are solved by going underground.

In some circumstances there is need for a junction in the tunnel. Probably it is better to have a well constructed junction inside the tunnel than a bad solution just outside with regard to safety, functional and environmental aspects.

Normally junctions in heavy loaded tunnels are grade-separated. But, in accordance with the Norwegian guidelines, roundabouts may be used when the total traffic through the junction does not exceed 20 000 veh/day (AADF).

The demand flow should not be higher than 75% of the capacity on any arms.

Entering a tunnel, the eyes need some time to adapt to the lower luminance. Therefore a roundabout can not be placed closer to the entrance than treshold zone + about 75% of the transition zone. The luminance in the roundabout ought to be over 3 cd/m^2.

Requirements for standard layout in typical Norwegian rocks are shown in figur 10. A diameter of 40 to 50 m seems to be optimal with regard to rock-mechanical aspects and visibility distances both to the left and around the rock "pillar" in the middle of the roundabout. If the angle between the arms is about 120 degrees and there are only 3 arms, the visibility is sufficiently good for a mini-roundabout to be considered.

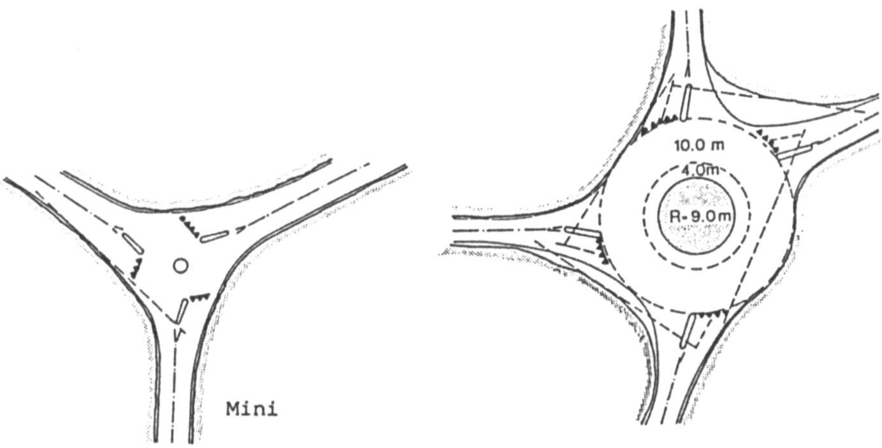

Figur 10. Roundabout in tunnel

In the city of Tromsoe in the north of Norway three roundabouts in tunnels are soon to be opened. One of them is already partly in function with traffic. In all these roundabouts the central island consists of a rock "pillar". Similar solutions are planned in other citys.

Future

The poularity of roundabouts will certainly continue in Norway.

During the years to come more care should be taken of two-wheelers. Consequently it may be more common to have a tight geometry with only one entry lane on the approach to reduce speed. There is need for more research about the problems for the cyclists.

Quite a few of the existing roundabouts are looking too grey and unsavoury. The present trend is to build more aestetic solutions. Especially the centre island can, by an appropriate design, provide sculptural elements of high aestetic quality. The authorities, therefore, regard roundabouts as a means to improve the visual environment regarding roads and streets.

Accident Prediction Models and Applications for Unsignalized and Signalized Intersections

Yuanlin Huang Adolf D. May

Institute of Transportation Studies
University of California at Berkeley, Berkeley

Summary

A 3-level modelling approach is employed for the accident prediction models of unsignalized and signalized intersections. Developments and applications of the accident models for both unsignalized and signalized intersections are shown. The models are developed on intersection characteristics of design, control, traffic flow, and environments. Furthermore, based on the results of the models, the relative safety of unsignalized and signalized intersections is discussed, and comments are provided. All data used in this study are drawn from Traffic Accident Surveillance and Analysis System in California.

I. Introduction

Research[2,3,4] about the accident prediction model development for unsignalized and signalized intersections as well as for highway segments has been undertaken in this institute. This paper is based on the results of the two previous projects[2,3] and recent research. The purpose of this paper is to apply the prediction models and to understand the cause-and-effect relationship of accidents in unsignalized and signalized intersections. All the data have been drawn from the Traffic Accident Surveillance and Analysis System (TASAS) in California.

First, this paper introduces the methodology of a 3-level modelling approach. Then the characteristics of unsignalized and signalized intersections are presented. Furthermore, developments and applications of the accident models for unsignalized and signalized intersections are shown. Finally,the relative safety of unsignalized and signalized intersections is compared and comments are provided.

II. Methodology

A 3-level prediction approach has been employed. The Level I model is a regression model. A grouping and classification technique called Classification And Regression Tree (CART)[1] has been used as a building block for developing Level II model. CART is also a computer program. The Level III model has been based on information including individual accident history. The differences between the predicted and observed values have been improved from Level I to Level III.

1. Level I: Base Model

As a first approximation, the concept of a base model is proposed. A base model here refers to a model which contains the most basic predictor variable. In many studies, traffic intensity variables have been found to be very important in predicting accident occurrences. They are generally used to normalize the number of accidents, and to form a new response variable (e.g. no. of accidents per vehicle) in other studies. In this study, instead of putting some forms of the traffic intensity variables in the denominator of the number of accidents, a base model was built with the number of injury or PDO(Property-Damage Only) accidents per year as the response variable, and traffic intensity, expressed in millions of vehicles entering the intersection per year from all approaches, as a predictor variable. One of the advantages of these approaches is to allow researchers to see the relationship between the two variables in an undistorted manner. Estimates of the parameters can be obtained by techniques such as least square, maximum likelihood, etc. The base model so obtained is referred to as Level I prediction in this study.

2. Level II: Grouping Intersections by CART

Further information, such as design, control, proportion of cross street traffic, and environmental features of the intersections, can also be considered as other major factors affecting the safety of intersections. The importance of these factors can sometimes be reflected in the large variations found in most scatter plots between the number of accidents and traffic intensity. One of the approaches that can be used is to analyze the residuals of the base model based on other characteristics of the intersections. In other words, one could group together those intersections with similar characteristics and which have higher or lower accident records than other intersections in general. The residual is defined as the difference between the observed values and the predicted values by the base model. The CART program is used to analyze the residuals of the base model. The refinement of estimates by grouping intersections based on other information characteristics is referred to as Level II prediction in this paper.

3. Level III: Adjustment by Accident History

The above discussions refer to estimates of accident prediction for groups of intersections. In other words, all intersections within a certain group will have an equal estimates. One can argue that the grouping made was based only on information that was available in the list of predictor variables, and they may not contain all of the factors that affect accident occurrences. The importance of the missing factors might partly be reflected in the accident history of individual intersections. As a result, the accident history of individual intersections becomes a very valuable source of information reflecting the actual safety performance level of individual intersections, in addition to their group characteristics.

Hauer, et al[5] have derived an estimator, Z, to predict the safety of an individual intersection by the following equation:

$$Z = aE\{m\} + (1-a)X$$

where: $a = 1/(1+var\{m\}/E\{m\})$
m = expected accident statistic
X = accident account.

III. Characteristics of Intersections

There were 11,786 unsignalized intersections and 2,488 signalized intersections under study. The Intersection characteristics and variable identifications are shown in Table 1.

Table 1
Intersection Characteristics and Variable Identifications

Factors	%*	%**	Codes and Descriptions
Terrain	20.4	0.9	1 - Mountainous
	30.0	17.7	2 - Rolling
	49.6	81.4	3 - Flat
Design Speed (SPEED)	2.4	2.4	1 - less than 30 MPH
	2.2	4.6	2 - 30-34
	7.1	9.2	3 - 35-39
	8.4	8.2	4 - 40-44
	13.6	17.8	5 - 45-49
	16.0	21.5	6 - 50-54
	7.4	4.8	7 - 55-59
	21.5	18.0	8 - 60-64
	21.4	13.3	9 - Greater than 65
Rural/Urban (RORU)	60.2	4.5	1 - Rural
	39.8	95.5	2 - Urban
Inside/Outside City(IORO)	35.2	87.0	1 - Inside
	64.8	13.0	2 - Outside
Intersection Type(ITYPE)	25.6	72.5	1 - Four-legged
	0.8	2.8	2 - Multi-legged
	3.8	3.9	3 - offset
	61.9	18.2	4 - Tee
	6.7	1.0	5 - Wye
	1.1	1.7	6 - Other
Control Type (CTYPE)	11.1	--	1 - No control
	86.9	--	2 - Stop signs on cross street only
	0.8	--	3 - Stop signs on main street only
Unsignalized Intersection	0.6	--	4 - Four-way stop signs
	(	--	5 - Four-way flashers(red on cross)
	(	--	6 - Four-way flashers(red on main)
	0.7(	--	7 - Four-way flashers(red on all)
	(	--	8 - Yield signs on cross street only
	(	--	9 - Yield signs on main street only

(continued)

Control Type (CTYPE) Signalized Intersection	--	22.7	1 - Signal pretimed 2-phase
	--	4.2	2 - Sinal pretimed multiphase
	--	10.2	3 - Signal semi-traffic actuated 2-phase
	--	7.2	4 - Signal semi-traffic actuated multiphase
	--	11.1	5 - Signal full traffic actuated 2-phase
	--	44.6	6 - Signal full traffic actuated multiphase
Lighting Type (LIGHT)	--	0.8	1 - No lighting
	--	98.8	2 - Intersection is lighted
	--	0.4	3 - Invalid data
Mainline Signal Mastarm(MSM)	--	8.3	1 - No mastarm
	--	91.2	2 - Yes, signal on mastarm
	--	0.5	3 - Missing
Mainline Left Turn Channelization (MLT)	6.0	42.3	1 - Curbed median left turn channelization
	71.3	20.2	2 - No left turn channelization
	22.5	36.5	3 - Painted left turn channelization
	0.2	0.6	4 - Raised bars
	0.0	0.4	5 - missing
Mainline Right Turn Channeli -zation(MRT)	94.3	80.1	1 - No free right turns
	5.7	19.7	2 - Provision for free right turns
	0.0	0.2	3 - missing
Mainline Traffic Flow (MTF)	4.6	4.0	1 - Two-way traffic no left turns permitted
	93.1	89.0	2 - Two-way traffic left turn permitted
	0.1	0.8	3 - Two-way traffic left turns restricted-peak hours
	2.2	5.2	4 - One-way traffic
	0.0	1.0	5 - Other
Mainline Number of Lanes(MNL)	68.3	9.2	2 - Two lanes
	1.8	3.9	3 - Three lanes
	26.1	66.9	4 - Four lanes
	0.3	2.2	5 - Five lanes
	3.5	16.5	6 - Six lanes
	0.0	0.2	7 - Seven lanes
	0.0	0.7	8 - Eight lanes
Mainline ADT	--	--	Packed Numeric (999999)
X-street Signal Mastarm(XSM)	--	46.3	1 - No mastarm
	--	53.2	2 - Yes, signal is on mastarm
	--	0.5	3 - Missing

(continued)

X-street Left Turn Channelization (XLT)	0.3	13.8	1 - Curbed Median left turn channelization
	98.7	58.7	2 - No left turn channelization
	1.0	26.7	3 - Painted left turn channelization
	0.0	0.2	4 - Raised bars
	0.0	0.6	5 - missing
X-street Right Turn Channeli-zation(XRT)	96.4	78.0	1 - No free right turns
	3.6	21.6	2 - Provision for free right turns
	0.0	0.4	3 - missing
X-street Traffic Flow(XTF)	4.5	2.2	1 - Two-way traffic no left turn permitted
	93.4	90.5	2 - Two-way traffic left turn permitted
	0.0	0.2	3 - Two-way traffic left turns restricted-peak hours
	1.9	6.4	4 - One-way traffic
	0.2	0.7	5 - other
X-street Number of Lanes(XNL)	95.1	57.4	2 - Two lanes
	0.1	6.0	3 - Three lanes
	0.6	32.1	4 - Four lanes
	--	0.7	5 - Five lanes
	--	2.3	6 - Six lanes
	--	0.0	7 - Seven lanes
	--	0.1	8 - Eight lanes
X-street ADT	--	--	Packed Numeric (999999)
Median Indicator (MI)	29.0	23.4	1 - Undivided
	71.0	76.4	2 - Divided or independent alignment
	0.0	0.2	3 - Missing
IADT	--	--	= XADT/(XADT+MADT)
MVYR	--	--	= (XADT + MADT)*365/1000000

* : Unsignalized intersections;
** : Signalized intersections.

The defined variables listed in Table 1 will be used throughout this paper.

IV. Accident Prediction for Unsignalized Intersections

As a typical application of the 3-level prediction approach, injury accident prediction is presented in this section. Since the purpose of this paper is to discuss the safety of each intersection group with similar characteristics, Level III model , which is for each individual intersection, will not be presented here.

1. Level I: Model Based on Traffic Intensity

It has been demonstrated in our data set that traffic intensity was indeed the most important single factor in predicting injury accidents. Thus, a base model was built on this fact.

Different functional forms, including power and logarithm transformations, have been used to fit the data , and a straight line relationship has finally been selected because it is as good as any other form. The following equation has been selected as our base model:

$$FIACCYR = o.1567 + 0.0612*MVYR \qquad (1)$$

$$R = 0.495$$

where: FIACCYR = forecasted value of average number of injury accidents per year;

MVYR = million vehicles per year.

2. Level II: Model Based On Intersection Characteristics

The residuals of the base model, as estimated by Equation (1), were analyzed by CART, and the factors and levels that were available and used in the analyses were those listed in Table 1 of Section III.

A 5-fold cross-validation tree was shown in Figure 1. The resulting tree with 12 terminal nodes and a relative error of 0.93 was selected by the 0.3 standard error rule. The 0.3 standard error is one of the features in CART for selecting the right sized tree.

The tree starts with a parent node of 11,786 intersections, and a mean residual of 0 and a standard deviation of 0.55, and the residual is derived by subtracting the Level I predicted values from the observed values. For prediction purposes, this shows that one can divide the 11,786 intersections into 12 groups with similar accident characteristics based on the tree structure as shown in

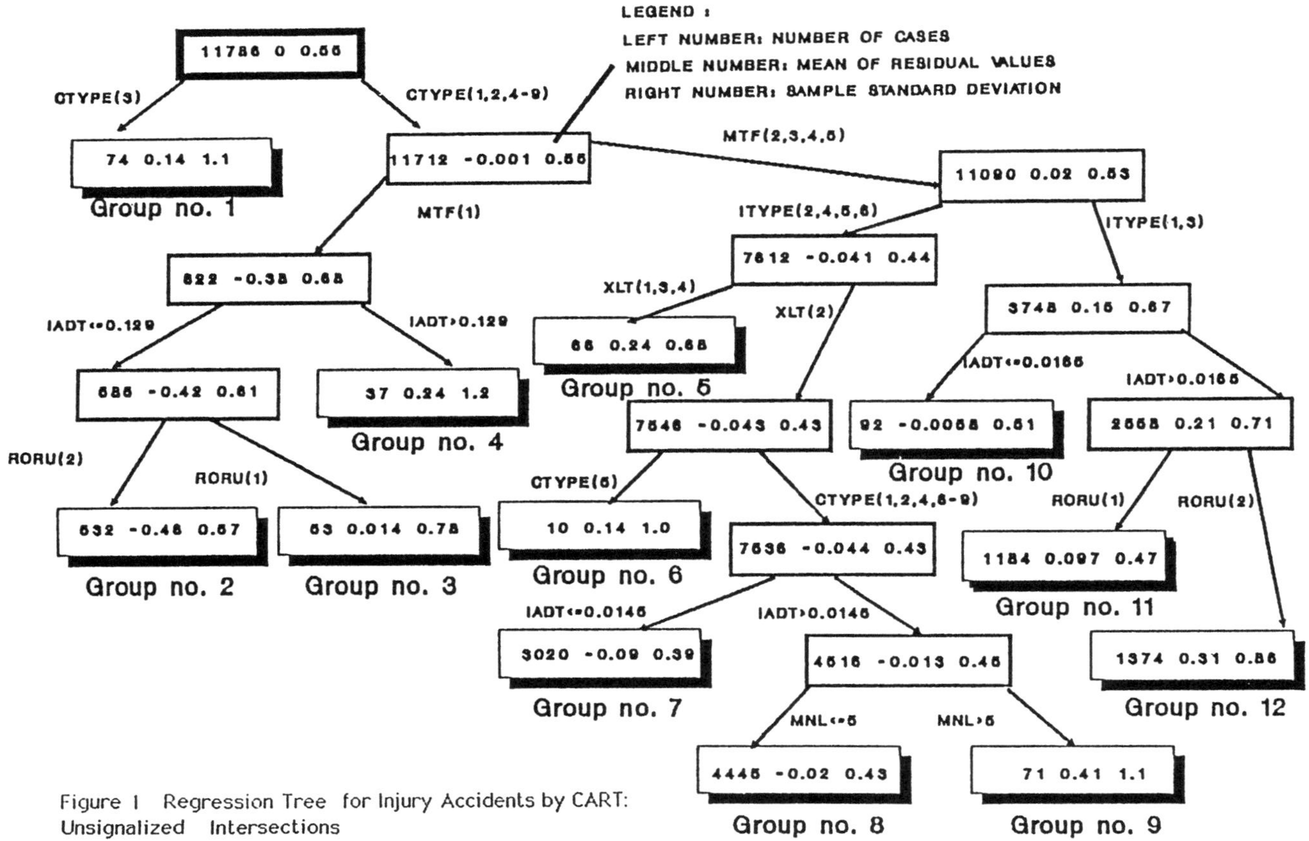

Figure 1 Regression Tree for Injury Accidents by CART: Unsignalized Intersections

Figure 1. Number of intersections, mean residual,standard deviation, and other features are also shown in Figure 1 for each group.

Group 1 intersections are characterized by intersections with stop signs on main streets only. Intersections in this category have a higher risk in general since stop signs on main streets might not generally be expected by motorists.

Group 4 has intersections with control features other than that in group 1, they have two-way traffic with no left turns permitted on main streets, and 12.9% or less of the entering traffic comes from the cross street. This group has 37 intersections with mean residual 0.24 and standard deviation 1.2. The remaining groups (5 through 12) have left turn capabilities or one way traffic on main streets. Observation of the split on MTF, and its associated intermediate nodes with mean residual -0.38 and 0.02, indicates that intersections with two-way traffic and no left turn permitted on main streets(group 2,3 and 4) have a slightly lower risk than others(group 5 through 12).

Group 5 is characterized by intersections with control features other than that in group 1. They are simple intersections with some forms of left turn channelization on side streets and left turns permitted on main streets. This group contains 66 intersections, a mean residual of 0.24 and standard deviation of 0.68. The split on XLT separating group 5 and 6 through 9 indicates that intersections with no left turn channelization on side streets have a lower risk than that with some forms of left turn channelization (0.24 versus -0.043). Two possible alternative explanations are possible: Perhaps there were a large number of left turn movements at intersections with left turn channelization. If not, perhaps some forms of left turn channelization were put in because of a large number of accidents occurring there.

Group 10,11, and 12 have intersections with control features other than that of group 1. They are mostly four-legged or offset(more complicated) intersections with left turns permitted on main streets. Group 10 has 1.65% or less entering traffic from the cross street. Group 11 has greater than that and its intersections are in rural areas. Group 12 resembles group 11 except that its intersections are in urban areas. There are 920, 1184, and 1373 intersections in group 10,11 and 12, with mean residuals -0.006, 0.097, and 0.31, and standard deviation 0.51, 0.47, and 0.86, respectively. Comparing the splits on RURO, including group 2 and 3 against group 11 and 12, it appears that the relative dangers of rural and urban areas depends decisively on preceding nodes. In the case of group 2 and 3, it could indicate that rural intersections (group 3) have a higher risk since they usually have a higher speed. In the case of group 11 and 12, left turns are permitted, it might indicate that left turn conflicts are more important than speed. So urban intersection (group 12) have a higher risk since they have a much more left turn movements.

V. Accident Prediction for Signalized Intersections

For comparison, Level I and Level II models of injury accidents for signalized intersections are presented in this section.

1. Level I: Model Based on Traffic Intensity

The following equation was selected as our base model for signalized intersections for the forecasted number of injury accidents per year:

$$FIACCYR = 0.6186 + 0.1691*MVYR \qquad (2)$$

$$R = 0.475$$

where: FIACCRR = The forecasted value of the average number of injury accidents per year.

MVYR = million vehicles per year.

This model shows, in our data set, the injury accidents in signalized intersections is more than that in unsignalized intersections.

2. Level II: Model Based on Intersection Characteristics

A 9-fold cross-validation tree was obtained by CART program using the 2488 signalized intersection data set. The tree is shown in Figure 2. The resulting tree with 9 terminal nodes and a relative error of 0.90 was selected by the 0.3 standard error rule. For prediction purposes, this shows that one can divide the 2488 intersections into 9 groups with similar accident characteristics based on the tree structure as shown in Figure 2.

Group 1 is characterized by intersections which have an index of conflict of less than or equal to 0.065, i.e., 6.5% of all entering traffic is on the cross street. A simple classification rule in this situation seems to indicate that when the level of conflicts is at a low level no other factors, except traffic intensity, are significant in affecting the safety of intersections. There is a total of 492 intersections in this group with an average residual of about -0.73 and a standard deviation of about 1.5.

Group 2 is characterized by intersections that have an index of conflicts of larger than 0.065 and are either 'Tee', 'Wye', or multi-legged intersections. Since less than 3 percent of the intersections in the data set are multi-legged, one could say that this group represents those intersections that have fewer legs. This might be the reason why they have lower accident statistics as compared to intersections with more legs such as four legged and offset intersections as shown in the intermediate node with an average residual of 0.36 as compared to -0.62 in this group.

Group 3 has an average residual of 0.05. Intersections in group 3 have narrow cross streets (less than or equal to a total of 3 lanes), and they tend to have smaller residuals than the remaining groups (group 4 through 9).

As compared to the intermediate node corresponding to intersections with pretimed controllers and two-phase fully actuated controllers, group 4 has a smaller average residual than the other groups. One

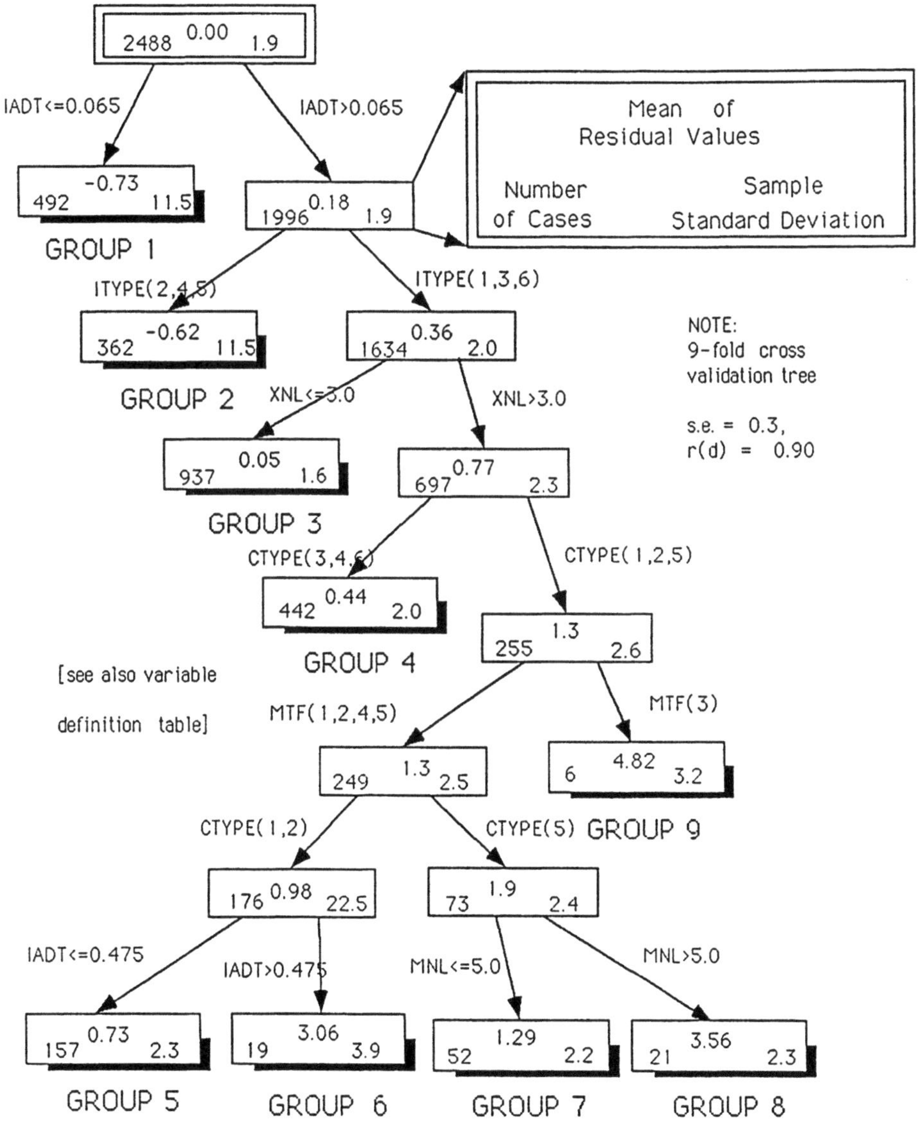

Figure 2 Regression Tree for Injury Accidents by CART : Signalized Intersections

could argue that it might be worthwhile to spend additional money on installing actuated controllers from a safety point of view.

Group 6 is similar to group 5 with the exception that the index of conflicts is larger than 0.475. Both groups have pretimed signal controllers. It is obvious that group 6 has a much higher accident potential as compared to group 5 due to higher level of conflict.

Group 7 and 8 are similar to group 5 and 6 with the exception that all of them have two-phase fully actuated controllers. Whereas group 7 includes intersections that have a total number of lanes on main streets of 5 lanes or less, group 8 includes only those intersections that have a total number of lanes on main streets of 6 or more lanes. The higher residual of 3.56 in group 8 as compared to group 7 seems to indicate that very wide intersections might require more than two phases in order to accommodate the turning movements required in wide intersections.

Group 9 has an average residual of 4.82. There might be a large number of left turn movements in this group of intersections, since left turns are prohibited in peak hours.

VI. Summary and Comments

From section IV, we see that unsignalized intersections with stop signs on main streets have a higher risk, while unsignalized intersections with two-way traffic and no left turn permitted on main streets have a lower risk. Speed is another factor affecting safety, but it is secondary to left turn conflicts, which more seriously affect the unsignalized intersection safety.

From section V, we see that, for signalized intersections, actuated controllers are better from a safety point of view. For very wide intersections, multiphase actuated controllers are necessary in order to accommodate the turning movements required in wide intersections.

In general, we found that there are two categories of important factors affecting accident occurrences in both unsignalized intersections and signalized intersections. One is the level of conflicts. Wherever there is a higher level of conflicts, such as a higher ratio of cross street traffic, more left turn movements, etc, there is a higher risk. The other is the severity of conflicts, such as higher speed, more direct conflicts, etc.

Specifically, the following factors were found to be significant: traffic intensity, proportion of cross street traffic, intersection type, control or signal type, number of lanes, and left-turn arrangements.

Sometimes, safety and efficiency in the situations of unsignalization and signalization are conflict, we have to make a trade-off between safety and efficiency or capacity. It is suggested that, except delay, safety should also be considered in assessing level of service of intersections.

In order to understand the cause-and-effect relationship of accidents in unsignalized and signalized intersections and to trade off properly between safety and efficiency, further studies are needed. Checking model prediction results with the results of field observations, analyzing performance differences of unsignalized and signalized intersections, and detail studies of each design feature, control feature and traffic movements are expected to be fruitful research fields.

References

1. Breiman, L., Friedman, J. H., Olshen, R. A., & Stone, C. J.: Classification and Regression Trees. Wdsworth & Brooks 1984.

2. Lau, M. Y., & May, A. D. : Accident Prediction Model Development: Signalized Intersection. Research Report UCB-ITS-RR-88-7, 1988.

3. Lau, M. Y., & May, A. D.: Accident Prediction Model Development for Unsignalized Intersections: Final Report. Research Report UCB-ITS-RR-89-12, 1989.

4. Shoumaker, W. , Cayford, R., & May, A. D.: Accident Prediction Models for Signalized and Unsignalized Intersections: Update on CART Models and the APSI Program. Working Paper UCB-ITS-WP-90-1, 1990.

5. Hauer, E. et al: Estimation of safety at Signalized Intersections. Transportation Research Board, Washington D. C., Jan., 1988.

Gap Acceptance and Wait Time at Unsignalized Intersections

JON D. FRICKER, MARLENE GUTIERREZ, AND DAVID MOFFETT

School of Civil Engineering
Purdue University
West Lafayette, IN 47907

Summary

The primary purpose of this study is to determine if driver gap acceptance behavior changes as the time spent waiting for an acceptable gap increases. We propose three hypotheses dealing with this issue and present a sample of results pertaining to the first two hypotheses. Special attention is given to the methods used to collect the data used in the study and to define the key events involved in driver behavior at unsignalized intersections.

Previous Studies

While this study was still in the planning stage, Adebisi and Sama [1] published results of a study with many similarities to ours. However, there are some important differences between the Adebisi and Sama study and ours. They observed low-volume conditions to accommodate their capacity-oriented analysis. While our study should have applications to capacity analysis, signal warrants, and intersection geometrics, our study's emphasis is on to what extent driver risk behavior associated with gap acceptance changes as the time spent in a queue increases. This means that we are interested in observing peak period conditions, which are more likely to generate long queues.

During the course of our study, several important issues have arisen. These issues involved methods of data collection, definitions of terms, and interpretation of events at the intersections observed. None of these issues were mentioned by Adebisi and Sama. We address a few of these issues in this short paper. When the final report for our study is written, we will discuss the impact that these issues have on the analysis, results, and applications of such a study.

While we were analyzing the data we had collected during 1990 at the sites described below, a paper by Kittleson and Vandehey [2] addressing some issues similar to ours was presented at the 1991 Annual Meeting of the Transportation Research Board. In some cases, our paper will provide a basis for corroborating or revising the results of Kittleson and Vandehey's work. In addition, this paper introduces different ways of studying and presenting issues related to those examined by Kittleson and Vandehey.

Event Definitions

To avoid the complications of conflicts from opposing minor street traffic, and to focus our study on the behavior of minor street drivers with respect to gaps in *cross* traffic, we made our observations at T intersections. These observations were actually accomplished by making videotapes of the site and superimposing both clock and stopwatch times on the taped scene using a character generator. The times at which specific events occurred were recorded to the nearest 0.1 second by viewing the videotape at a later time. Figure 1 illustrates a typical intersection in our study and depicts the events that we recorded for inclusion in our data base.

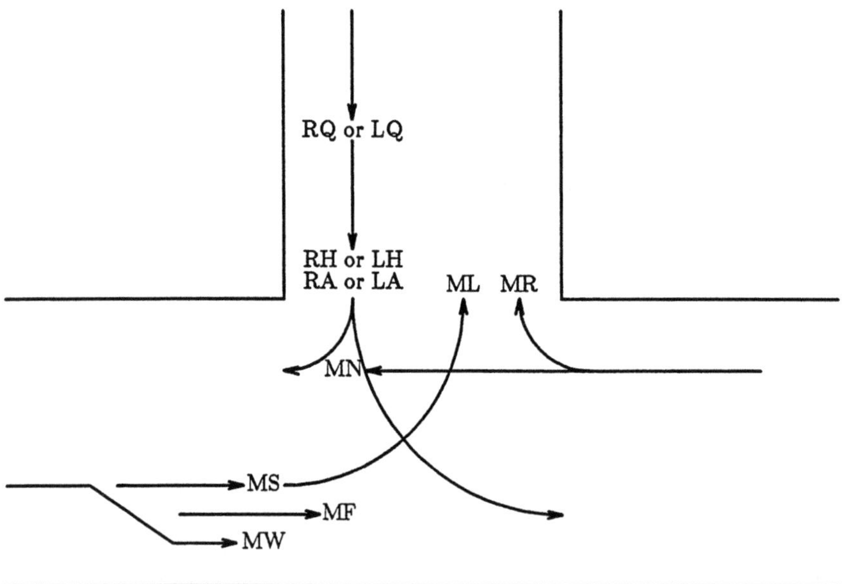

Fig.1. Events diagram for gap acceptance video data

For each vehicle approaching the intersection on the *minor* street, three of the following events will apply:

- LQ = A vehicle that will eventually turn Left onto the major street enters the end of the minor street Queue.

• RQ = A vehicle that will eventually turn **R**ight onto the major street enters the end of the minor street **Q**ueue.

• LH = A vehicle that is about to turn **L**eft onto the major street reaches the **H**ead of the queue, that is, it reaches the intersection in a position to enter the major street traffic in the desired direction.

• RH = A vehicle that is about to turn **R**ight onto the major street reaches the **H**ead of the queue, that is, it reaches the intersection in a position to enter the major street traffic in the desired direction.

• (Note: If a vehicle on the minor street approach never has to wait behind another vehicle in a queue, events LQ and LH, or RQ and RH, are both given the same time of occurrence.)

• LA = A **L**eft-turning vehicle on the minor street **A**ccepts a gap on the major street and completes the turn onto the major street.

• RA = A **R**ight-turning vehicle on the minor street **A**ccepts a gap on the major street and completes the turn onto the major street.

• (Note: If a vehicle reaches the head of the minor street queue and never has to wait for vehicles on the major street, use the same time for events LH and LA or events RH and RA, as applicable.)

For the *major* street, the following events were noted:

• MN = A vehicle on the Major street in the lane(s) Nearest to the minor street passes in front of the vehicle at the head of the minor street queue. From the point of view of the minor street driver, this vehicle is moving from left to right.

• MF = A vehicle on the Major street in the lane(s) Farthest from the minor street passes in front of the vehicle at the head of the minor street queue.

• MR = A vehicle on the Major street makes a **R**ight turn onto the minor street. Although this movement does not cross the path of the head vehicle on the minor street, the approach of any vehicle from the left -- even with turn signal on -- has been observed to usually cause a minor street driver to hesitate until the major street vehicle's right turn is assured. (It is felt that the degree of caution reflected in such behavior may diminish as the wait time at the head of the minor street queue increases.) The time of this event should be recorded as having taken place just after the minor street vehicle's left or right turn has begun. If a minor street LA or RA event results in conjunction with this MR maneuver, but does not actually happen before the time you record for this event MR, assign the LA or RA event associated with this

MR event a time 0.1 seconds before the MR event. Otherwise, the LA or RA event would appear to take place in a gap after the "MR" vehicle.

• MS = A vehicle on the Major street Stops or Slows in preparation for a left turn onto the minor street. If an event LA is not deterred by this event, make sure the time of event MS is recorded as taking place after the event LA. If an event LA is delayed by this event MS, the time of MS occurs at the point the viewer feels that the major street vehicle has begun to block the vehicle seeking to turn left out of the minor street.

• ML = The Major street vehicle completes its Left turn onto the minor street. The time of this event must be recorded as being before any associated event LA. Otherwise, the LA event would appear to take place in the gap before the "ML" vehicle.

• MW = A vehicle on the Major street behind a vehicle causing event MS goes Wide around the right side of the "MS" vehicle. Although this vehicle has little or no impact on vehicles on the minor street, its presence adds to the flow rate on the major street and must be recorded. The time of event MW is the time at which the "MW" vehicle passes the "MS" vehicle.

These events were converted from videotape images to input files in the format shown in Table 1. Computer code was written to process these data.

Table 1. Input File Format for Gap Events

Zionsville Road			<---Major Street
West 96th Street			<---Minor Street
15 June 1990			<---Date Data collected
LQ	04:35.2	(1) truck	
LH	04:35.2	(1)	
RQ	04:35.2	(2)	
LQ	04:35.4	(3)	
MN	04:35.6		Column 1 = Event Code
MF	04:36.1		
RQ	04:44.2	(4)	Column 2 = Time event occurred
MS	04:46.0		
MW	04:46.1		Column 3 = Vehicle sequence
ML	04:49.3		number and notes
LQ	04:49.5	(5)	(to assist data reduction only)
MN	04:52.4		
LA	04:53.2	(1)	
RH	04:57.0	(2)	
RA	04:57.0	(2)	
LH	05:02.8	(3)	
MS	05:03.7		

Description of Sites

The sites included in this study were selected both for what they had in common and for the variety of conditions they represented. The desirable features the sites had in common were:

a. The minor street was stop-controlled.

b. The development of one or more distinct peaks in minor approach volumes during the period of observation, creating queues on the minor approach

c. No sight distance problems or other geometric problems

d. Similarities in traffic volume and approach speeds on the major roadway

e. No through movements on the part of the minor roadway traffic

Whereas Kittleson and Vandehey combined the data from their three sites, we preferred to examine each site's data separately. We felt that, by observing sites with a *variety* of commonly-seen situations, we could accumulate data that would make possible a better investigation of driver behavior. These situations are described below, as we introduce the sites where gap acceptance observations were made.

PGNW at N. Grant Street. The Northwestern Avenue Parking Garage (PGNW) is located on the northeast edge of the Purdue University campus. This is the only site that does not have a single, shared approach lane on the minor street. The exit from PGNW onto North Grant Street has side-by-side left-turn and right-turn lanes. There is no minor approach opposite the parking garage exit, making the site act as a T intersection. Being an exit only, the site had no problem with MR or ML events to complicate the analysis. Video taping took place between 4:54 PM and 5:34 PM on Thursday 5 April 1990. There were peaks in exiting flow rates from the parking garage just after 5 PM and 5:20 PM.

Harrison Bridge Off-Ramp at North River Road. This site (HBOR) offers the opportunity to explicitly separate gap acceptance behavior by direction of the major street traffic. A vehicle reaching the end of the WB ramp from the Harrison Bridge over the Wabash River and wanting to turn left must first cross two lanes of NB traffic on North River Road, then (if necessary) seek refuge in a median cut with capacity one vehicle before entering the SB lanes of North River Road. At this site, we chose to analyze only the first gap acceptance decision, although the videotape showing the MF events is still available for study, if desired. The videotape of this site was taken on Wednesday 16 May 1990 from 4:51 to 5:31 PM.

Eller Road and East 116th Street. Eller Road (the minor street) runs north into East 116th Street in an area north of Indianapolis/Marion County, between the fast-developing towns of Carmel and Fishers. Most of the traffic on the minor street eventually turns left. This site (eller116) produced the longest queues. It also experiences a large number of MR events, but few ML events. (See Table 2.) Video observations were made from 4:54 to 5:36 PM on Monday, 30 July 1990.

West 96th Street and Zionsville Road. This site (zion96) is located at the north boundary of Indianapolis (Marion County), Indiana. This T-intersection, soon to become signalized, offers a site with a minor approach flow rate that is more uniform during the period of observation than the other sites. The minor street is 116th Street. There are numerous occurrences of MR and ML events to complicate the analysis. This site also had a better balance of LA vs. RA events and MR vs. ML events than did "eller116". Videotaping at "zion96" took place on Friday 15 June 1990, between 4:53 and 5:25 PM.

Table 2. Summary of volumes at study sites, by event type

Site	Duration	LA	RA	MN	MR	MF	ML
PGNW	38:47	126	161	237	--	398	--
HBOR	40:07	97	--	426	--	412	--
zion96	32:14	73	93	347	87	100	63
eller116	42:33	145	12	578	236	411	4

The First Hypothesis

As this study evolved, so did the hypotheses associated with the basic issue of driver gap acceptance behavior. Three hypotheses have been formulated and are being studied because of their potential for shedding light on whether driver gap acceptance behavior is influenced by wait times. The first two are well-known and are covered in this paper. The third hypothesis is one we have not seen in the literature, but to be properly tested, it requires more data than we have processed to date. It will be covered in a future report.

The First Hypothesis can be stated (in "null hypothesis" form) as follows:

> There is no difference between the critical gap for left-turning (LT) vehicles and right-turning (RT) vehicles.

In four previous studies comparing critical gaps for left and *through* movements [3,4,5,6], two of them found the left-turning critical gap to be larger, and the other two found the through movement's critical gap to be larger. Likewise, the LT versus RT critical gap issue is far from settled, and there may be reasonable arguments for two alternatives to The First Hypothesis in that context. Let us call the alternative hypotheses I-A and I-B.

> I-A:"[H]igher critical gaps are to be expected when drivers execute left-turning maneuvers". [1,7]

"This could ... be due to the additional time allowed by minor road drivers when they have to cut one traffic stream and merge with another". [1] The implication seems to be that LT drivers are more conservative than RT drivers, because of the more complicated turning maneuver.

> I-B: Critical gaps for vehicles turning left will be smaller than for right-turning vehicles.

If the influence of wait time on the risk-averse behavior of drivers has validity, this alternative hypothesis might be substantiated. The philosophy underlying this hypothesis is "a gap is a gap", regardless of whether it is formed by consecutive "LH" events or a combination involving "RH" events. This philosophy discounts the influence of the more complicated turning maneuver for LT vehicles, but instead recognizes that LT vehicles will always have at least as much wait time as RT vehicles. If LT vehicles have to wait significantly longer than RT vehicles, and if longer wait times lead to less risk-averse behavior, then the critical gap for LT vehicles will be smaller than for RT vehicles.

Testing The First Hypothesis

For the three sites that have both left and right turns made from the minor street, the critical gaps summarized in Table 3 were found.

Table 3. Critical gaps at three sites

Site	Left Turn	Right Turn
PGNW	5.55	5.2
zion96	5.8	4.7
eller116	4.8	5.2

Although these results are not conclusive with respect to either alternative hypothesis, the interpretation of these critical gaps leads naturally to the Second and Third Hypotheses. The only site at which the critical gap is shorter for LTs than RTs -- "eller116" -- is an intersection at which RTs are a rarity and LTs have become a "specialty" of the drivers who use Eller Road regularly.

The Second Hypothesis

The Second Hypothesis can be stated in this way:

> The longer a driver has to wait for an acceptable gap -- either at the head of the queue or including time spent in the queue -- the shorter the eventual critical gap becomes.

Adebisi and Sama [1] and Kittelson and Vandehey [2] offer rare quantitative analyses to support this common sense notion. Our test of this second hypothesis is based on precise data taken from videotapes, a variety of site characteristics, and some new ways of analyzing the data.

Testing The Second Hypothesis

Figure 2 is typical of the plots we have generated to study the relationship between waiting time in queues and gap acceptance behavior. This figure shows the effects of *total* wait time in the minor street queue at site "HBOR". Similar plots have been made with wait time at the *head* of the minor street queue on the horizontal axis for HBOR and the other sites. In Figure 2, the "X" marks indicate the length of an *accepted* gap and the vertical dotted lines span the range between the minimum and maximum rejected gap durations for each driver. The larger dots, or "bullets" are used when only one gap was rejected. As total wait time increases, the probability that a driver's accepted gap will be smaller than one or more previously rejected gaps increases. This behavior is shown in Figure 2 whenever an "X" lies *within* its corresponding dotted line, instead of above it. Note that two of the six drivers with the longest total wait times (the six rightmost vertical dotted lines or bullets) behaved in this way. While statistical summaries and analyses are being conducted, we have found this kind of graphic display to be useful.

Future Research

While we continue to expand our data base and refine our analyses to test the first two hypotheses, we will also be investigating a Third Hypothesis:

> The longer the expected wait time for minor street traffic before an acceptable gap is found, the smaller the critical gap becomes.

The second hypothesis considered drivers waiting on a minor street at a given intersection, and considered whether those who had to wait longer were inclined to accept smaller gaps than those who had to wait less time. Not considered in the second hypothesis is whether most of the drivers we observed traveling through the intersections use that intersection on a regular basis. They probably have accumulated their own set of experiences about the intersection they are approaching. This information most likely includes what the "waiting penalty" is for rejecting a gap that is, under lower flow conditions, not clearly acceptable.

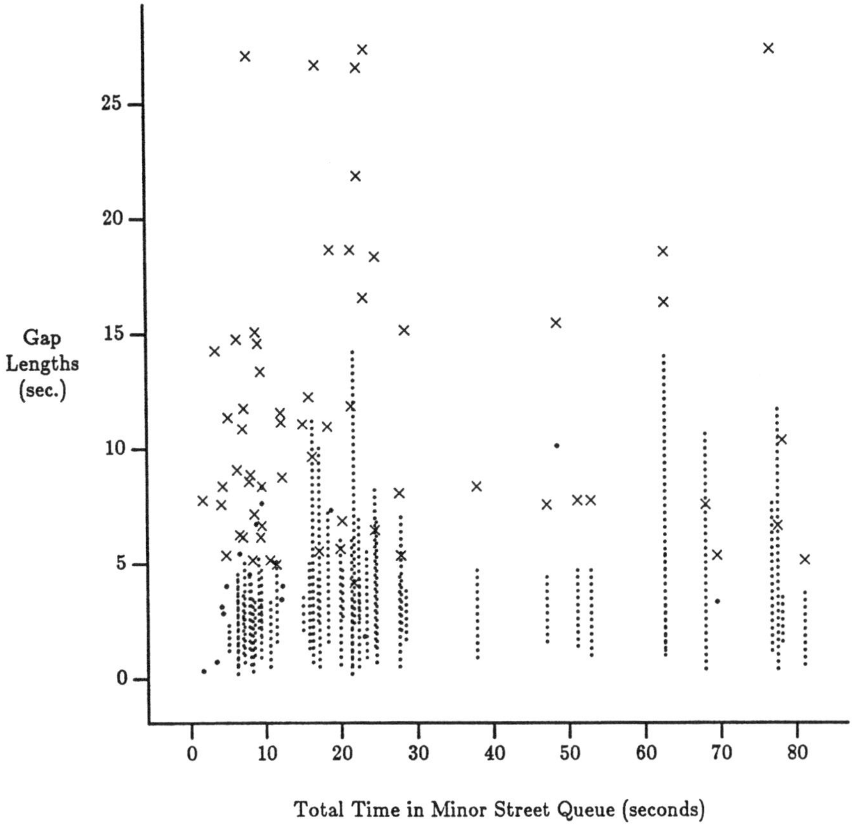

Fig.2. Gaps accepted and rejected vs. wait time in queue

Under higher flow conditions, but it may be worth the slight added perceived risk to accept such a gap, so as to avoid the subsequent added wait time. Thus, the second hypothesis considers the difference between individual drivers at a given intersection, based on the wait time(s) experiences by each individual driver. The third hypothesis considers the difference in behavior between intersections, i.e., how the "typical" driver behavior is influenced by the "typical" wait time at the intersection the driver is approaching.

We will also be looking into alternate definitions and implementations of the events listed earlier in this paper. For example, we will no longer use "MR" to define the end of a gap. We are also interested in separating the analyses of LT and RT gap acceptance behavior.

Results and Conclusions

While we have not settled the matter of whether the LT critical gap tends to be larger or smaller than that of the RT, plots such as Figure 2 lend support to the growing evidence that gap acceptance behavior becomes more aggressive as wait time increases. However, a large part of this paper has been devoted to the explanation of our methodology. This has been done because our results depend upon the event definitions we have adopted and the way in which we have collected our data.

A major element of our study has been the use of videotapes. While previous studies have used time lapse photography, road tubes with three-pen recorders or lap-top computers to record data, we have accepted the tedium of manually transcribing the video images into data files in exchange for certain major advantages. Each minute of video requires about 12-15 minutes per to transcribe, but the benefits in our research are many.

- The use of a character generator allows accuracy to 0.1 second, if necessary.
- The researcher can use careful and unhurried judgment (or refer to others for guidance) when unusual, complicated, or rapid events occur.
- The observer, supervisor or a co-worker can check the data file against the database for errors and make corrections. There is no such chance for verification or correction with other methods. In fact, some errors may go undetected.
- If we change our definitions or methods of analysis, we can return to the same "raw data" on the videotape and repeat the study.

Our ultimate objective is to determine if there is a useful rule of thumb to relate wait or delay time and driver gap acceptance behavior to the provision of adequate capacity at unsignalized intersections. Does a certain level of wait time call for separate LT or RT lanes, based on gap acceptance behavior? Perhaps a different capacity threshold for minor street design than exists in the Highway Capacity Manual can be developed.

References

1. Adebisi, Olusegun; Sama, George N.: Influence of Stopped Delay on Driver Gap Acceptance Behavior. *Journal of Transportation Engineering,* American Society of Civil Engineers, vol. 115, No. 3, p. 305-315, May 1989.
2. Kittelson, Wayne K.; Vandehey, Mark A.: Delay Effects on Driver Gap Acceptance Characteristics at Two-Way Stop-Controlled Intersections. Presented at the 70th Annual Meeting of the Transportation Research Board, Session 179, 16 January 1991.
3. Ramirez, E.A.: Gap Acceptance. Project, CE617 Traffic Operations, Texas A&M University, 12 November 1983.
4. Radwan, E.A.; Sinha, Kumares C.: Gap Acceptance and Delay at Stop-Controlled Intersections on Multi-Lane Divided Highways. *ITE Journal,* p. 38-44, March 1980.

5. Bissell, H.H.: Traffic Gap Acceptance for a Stop Sign. University of California, Berkeley, 1960.
6. TRB Special Report 209, *Highway Capacity Manual*. Transportation Research Board, Table 10-2, p. 10-7, 1985.
7. Polus, A.: Gap Acceptance Characteristics at Unsignalized Urban Intersections. *Traffic Engineering and Control,* 24(5), p. 255-258, 1983.

Simulation of Multilane Stop-Controlled T-Intersections by Knosimo in Canada

Paul L. Chan and Stan Teply
University of Alberta,
Edmonton, Alberta, Canada

1. SUMMARY

This paper reports on an application of the KNOSIMO traffic simulation program for priority-ruled intersections in Edmonton, Alberta, Canada. This University of Alberta study was restricted to urban multilane stop-controlled T- intersections and its main objective was to investigate the relationship between delays and safety. As a result, it required a reliable representation of delays. A relatively large sample of delay data was collected, but it was necessary to supplement it by simulation. The KNOSIMO program was validated and used to that end.

The paper describes the experience with the program and comments on its applicability to various conditions at the studied T- intersections.

2. DATA

Delay data, together with volume and headway information, were collected by using a video camera in January and February 1990, at 26 T-intersections in Edmonton. The surveys were limited to off- peak periods on weekdays with normal driving conditions. In total, over 15 hours of real time events with 21,069 vehicle arrivals were analyzed and transferred to a computer data base with the assistance of the TDIP program [1].

The collected delay data had a high degree of variability. In order to improve the statistical significance of the samples, more surveys would have been necessary. This approach, however, was not considered practical.

To generate additional delay information, it was decided to employ the KNOSIMO simulation model [2]. The data measured in the field were considered sufficient to test the suitability of the program for the needs of the overall project.

3. THE KNOSIMO PROGRAM

KNOSIMO is a traffic model for intersections without traffic signals developed at Ruhr University in Germany for the Federal Minister of Transport [2,3]. It is a microscopic, event-oriented simulation program and is applicable to 3- or 4-legged intersections. Shared and exclusive turning lane situations, as well as time-varying traffic demand and different vehicle types can be considered.

The vehicles are generated in the program using hyperlang headway distributions, calibrated to German conditions to realistically represent single lane traffic streams. The operation of non-priority traffic maneuvers is determined from gap acceptance principles using erlang distributions. The program is interactive regarding data input and file handling operations.

4. DIFFERENCES IN APPLICATION CONDITIONS

In order to simulate traffic conditions for multilane intersections, it was necessary to address the differences in the conditions considered in the program and at the studied locations.

The investigation started from the premise that the program itself would not need modifications and that the input values could be appropriately adjusted.

4a. Geometric features, critical gaps and move-up times

The differences in geometric features between the KNOSIMO and the studied intersections are illustrated in Figure 1. They imply the need for a different application of the critical gap and move-up time criteria [4]. The critical gaps and move-up times used in KNOSIMO are for 2-lane conditions but longer time periods are required to cross a multilane roadway. Moreover, the critical gaps and move-up times used in KNOSIMO were calibrated for driver behaviour in Germany. Whether and how these critical gaps could be transferred to North American conditions was unknown.

Unfortunately, the scope of the overall project did not allow a detailed investigation of these problems. To accommodate them, higher main road speeds were used to raise the critical gaps used by the program. For all approaches, test results showed that a 10 km/h higher main road speed gave better delay predictions.

4b. Headway distribution

KNOSIMO "feeds" traffic from one- lane approaches only. This was a serious problem because the headway distributions for the two- to three- lane intersection approaches included in the study were different. While on single lanes zero headways are not possible, and the number of very short headways is small, they frequently occur on multilane roadways.

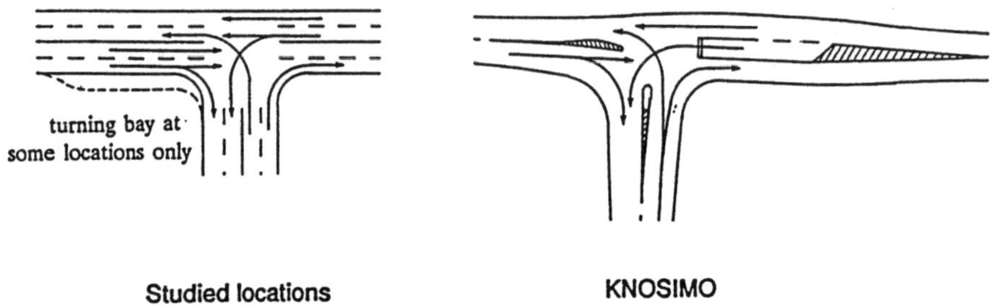

Figure 1: Comparison of intersection conditions in Edmonton and in the KNOSIMO program

The most important parameters are the values of critical gaps which are normally in the range from 4 to 10 seconds. The availability of headways above the critical gap value presents the "practical headway" for the non-priority traffic, and the number of these headways in the traffic stream on the main road determines capacity and delays on the minor road.

In comparing measured multilane headways to KNOSIMO headways in the range greater than 4 seconds, it was found that the difference was relatively consistent. Figure 2 shows that, for headways greater than 4 seconds, an equivalent "one-lane volume" of 400 vph in the KNOSIMO program can be used to approximate the actual headway distribution at a multilane approach with a volume of 690 vph. Figure 3, which includes thirteen intersection approaches, exhibits a linear relation between the single lane and multilane distributions.This linear transformation was employed for input volume conversions.

4c. Traffic interactions

The way individual traffic movements influence each other at a multilane intersection is different than at a 2-lane roadway. For this reason, conflicting volumes for certain non-priority maneuvers needed further adjustments.

It was found, for example, that when the full "eastbound" through volume used for the "northbound" right-turn maneuver from the minor road, delays was overestimated. This maneuver was only slightly affected by the inside lane of the main road traffic. As a result,

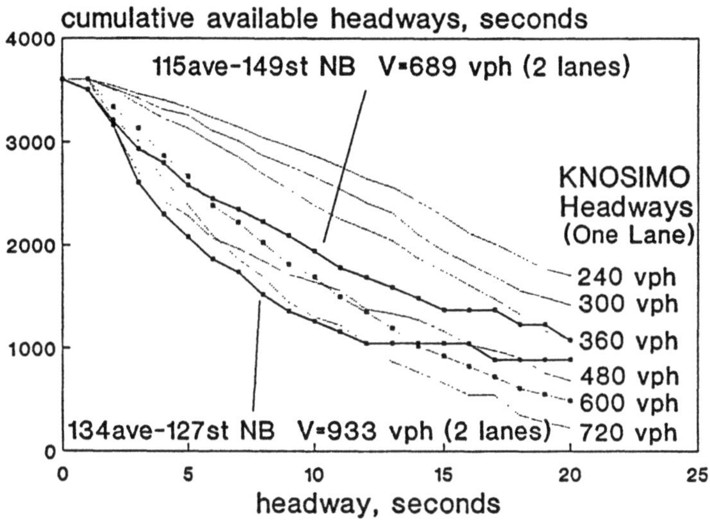

Figure 2 Comparison of measured cumulative available headways (2-lane) to various profiles assumed in KNOSIMO for different 1-lane volumes

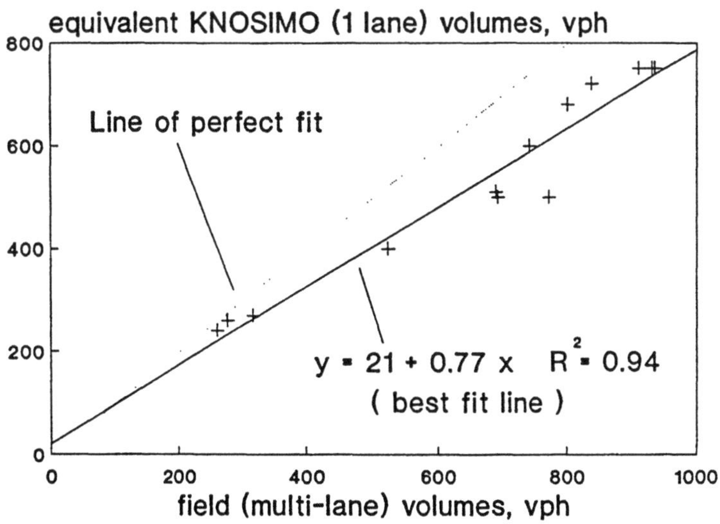

Figure 3 Correlation between actual field (2-lane) volumes and the equivalent 1-lane KNOSIMO volumes

only the curb-lane volume was applied, but since it already represented a one-lane situation, it was not adjusted further to an equivalent one-lane volume.

Nevertheless, for the "northbound" left-turn traffic from the minor road, it was necessary to keep the full volume of the "eastbound" traffic. To represent the actual conflicting "westbound" through traffic volume from the right, only the inside lane volume was used.

As a result, it was necessary to run the KNOSIMO program twice with somewhat different volumes in order to obtain proper delay generating situations.

4d. Effect of platooning

In an urban setting, priority intersections are frequently close to signalized intersections. The operation of these signals which creates platoons in the traffic stream can have a considerable effect on non- priority movements at intersections without signals. The platooning effect diminishes gradually with increasing distance downstream because of platoon dispersion.

No adjustments to the KNOSIMO input were made to accommodate these effects. Nevertheless, the results obtained from the KNOSIMO runs were compared to the measured delays, and are discussed in the next section of this paper.

4e. Type of delay

KNOSIMO calculates the average *overall* delay experienced by the drivers which includes time needed to decelerate and accelerate. The field delay data, however, were recorded as *stopped* delay. To modify the field data, the "zero- volume delays" (11 seconds for minor road left turns and 10 seconds for other movements) were added to make them compatible with delays generated by KNOSIMO.

5. RESULTS AND EVALUATION

The measured and the KNOSIMO generated delays for all of the three non- priority maneuvers at the studied T-intersections are compared in Figure 4. Considering the multiplicity of conditions and the differences among the three movements, the graph exhibits a reasonable degree of trend consistency. On average, however, the KNOSIMO generated delays are about 35% shorter than those measured. The underestimation of the actual delay is especially severe for higher values.

The fit, however, was quite different for individual movements. This was not surprising, since each non-priority maneuver depended on different exposure to other traffic streams, degree of priority and gap acceptance conditions which are discussed below.

5a. Comparison by the type of movement

The simulation results for the left- turn maneuver from the minor road were only about 11% lower than the measured delays in the range below 40 seconds. Above that value, however, KNOSIMO underestimated actual delay by about 60% (Figure 4).

For the right-turn movement for the minor road, KNOSIMO underestimated the actual delay quite significantly. The average was about 40% for shorter delays to over 100% for delays longer than 20 seconds. The discrepancy, however, was caused in part by the previously discussed volume adjustment, which was probably somewhat exaggerated, and partly by the impact of different sight distances and other local conditions. For instance, the two locations with the longest measured delays for minor road right-turners had sight distance problems.

The simulation results for left-turn movement from the main road across the path of the opposing main road traffic stream were very satisfactory for delays under 20 seconds. On average, they were only 8% lower than the measured values. The underestimation beyond 20 seconds, however, was higher.

5b. Comparison by volumes on the main road

Figures 5 illustrates the trend of increasing delay to minor road left-turners with increasing conflicting priority volumes. The conflicting traffic was taken as the sum of the main road volumes in both directions.

The underestimation, again, is noticeable and, as would be expected, the simulated values are more consistent than the measured data. The relatively constant randomness of the differences along all volume ranges suggests that the discrepancies cannot be explained by different volume categories. Similar trends were observed for minor road right- turners and major road left- turners.

Figure 6 shows that the measured probability of short headways is higher for approaches with higher traffic volumes. As a result, there are less of the available headways, resulting in longer delays. The mode of the distribution shifts to the left with decreasing volumes which means that there were less followers in the traffic stream.

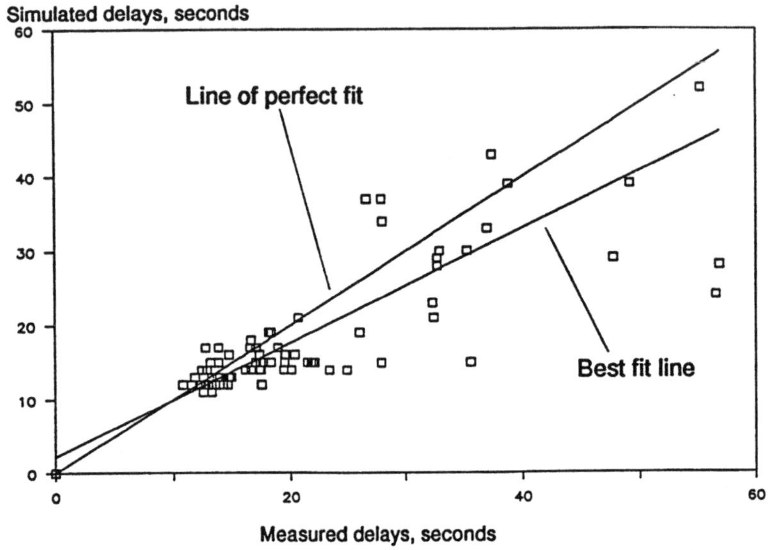

Figure 4: Comparison of measured and simulated delays for all three non-priority maneuvers

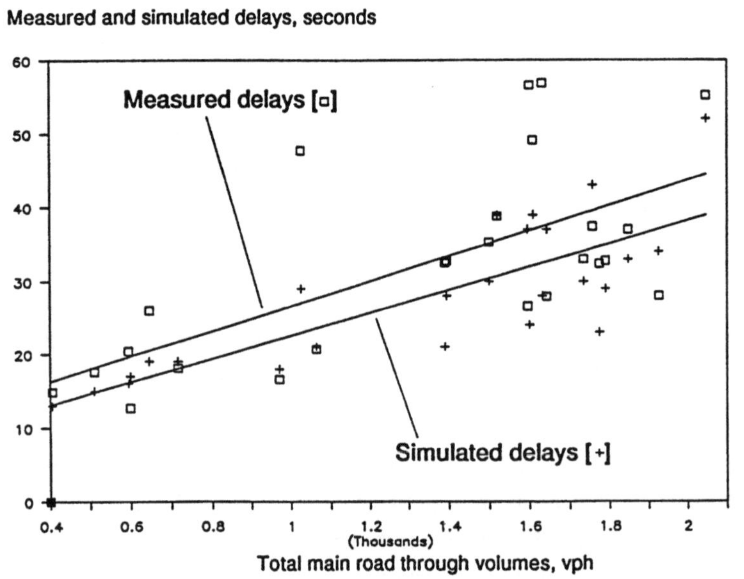

Figure 5: Variations in measured and simulated delays of minor road left turns with respect to major road through volumes

5c. Comparison by distance from upstream traffic signals

It was observed that headway distributions at intersections with low volumes were almost unaffected by the distance to the nearest traffic signal. The relative uniformity of the distributions was interesting.

The greatest impact of distance on headway distributions was for intersections with medium volumes between 400 to 700 vph, as illustrated in Figure 7. Similar to Figure 6, the mode of the distributions is shifted to the left with increasing distance from traffic signals. This shift indicates a substantial reduction of the proportion of short headways resulting from platoon dispersion. Examination of headway distribution data at other locations with high volumes exhibited a similar, but less pronounced, mode shift.

The time shift between the arrivals of the main platoons in the two opposing traffic streams on the main road also influenced non-priority maneuvers. Delays for minor road left- turners were much longer if the two platoons in the opposite directions on the main road arrived in a staggered fashion.

Figure 8 shows the effect of the volume in both main road directions on available headways for left- turn traffic from the minor road. The low points on the graph indicate that at these locations there were less available gaps than at the high point locations. The intersections in these two groups are significantly different. For example, two of the four low points represent a street (99th Street) with short blocks, close spacing of traffic signals, high level of business activity and more trucks, compared to the street with five of the seven high point locations (149th Street).

6. COMPARISON WITH HIGHWAY CAPACITY MANUAL

All of the 26 locations were also analyzed using the 1985 Highway Capacity Manual (HCM) [5] method. The outcome was compared with the results of the simulation and surveys.

HCM gives reserve capacity as the output for each non-priority maneuver and relates its value to a particular Level of Service. There is no direct link to the corresponding delay. Brilon [4] estimated the upper delay limits for different reserve capacity levels and Tracz [6] defined a set of functions representing the relationship between reserve capacity and delay. These suggested transformations, together with the correlation of the measured delay to the calculated reserve capacity of each non-priority maneuver are shown in Figure 9.

In order to convert the HCM reserve capacity levels to delay, the suggested limits were applied, but the results did not agree well with the measured delay. The minor road left turns, with relatively low reserve capacities, were an exception. For the other two non-priority maneuvers, with reserve capacities often over 400 vph, the measured delays are considerably higher than the suggested delay levels. It appears, from this limited evidence, that the HCM method provides only a general indication of the degree of difficulty in making a non-priority maneuver. Figure 9 shows that similar measured delays do not necessarily coincide with similar reserve capacity. Moreover, definitive trends for all three movements are noticeble, different from the relationships proposed by Brilon and Tracz. It seems that a relationship based on a single average delay for all maneuvers may not be appropriate.

Another comparison showed that KNOSIMO generally gives good estimates when compared to the measured delay, except in cases when the reserve capacity is below 100 vph, or for the right turns from the minor road, possibly because the problems of our liberal input modification.

7. CONCLUSIONS

Traffic volumes, sight distance, degree of platooning generated by upstream signals, and other conditions at the 26 studied T-intersections varied considerably. Although many of these variables could not be considered in the KNOSIMO program and its application, the results of the verification were considered very satisfactory. As a consequence, KNOSIMO was used to generate additional information for the main part of the project, i.e. an investigation of the relationship between delay and safety.

Adjustment of arrival volumes and speeds on the main road proved to be a good instrument to account for the effect of local geometric conditions and, possibly, for the different driver behaviour. These adjustments successfully affected main road arrival headway distribution and minor road gap acceptance in the critical ranges.

Naturally, this study had a number of limitations. The major constraint was the fact that the verification of the KNOSIMO program formed only a minor objective of the overall project. Nevertheless, the study demonstrated that KNOSIMO is a sufficiently robust investigation tool and can be forced by relatively simple input adjustments to perform well in a variety of conditions. Additional output information on queues provided by the program was found useful, although it was not employed in this project.

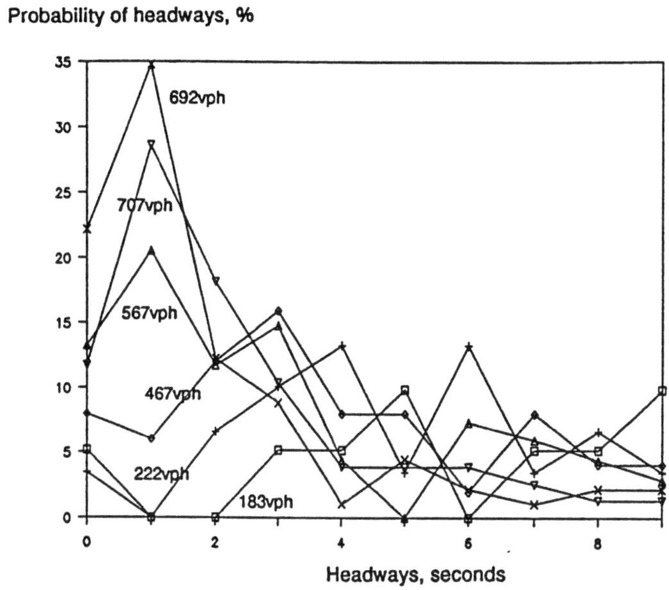

Figure 6 Comparison of short headway distributions for various approach volumes.

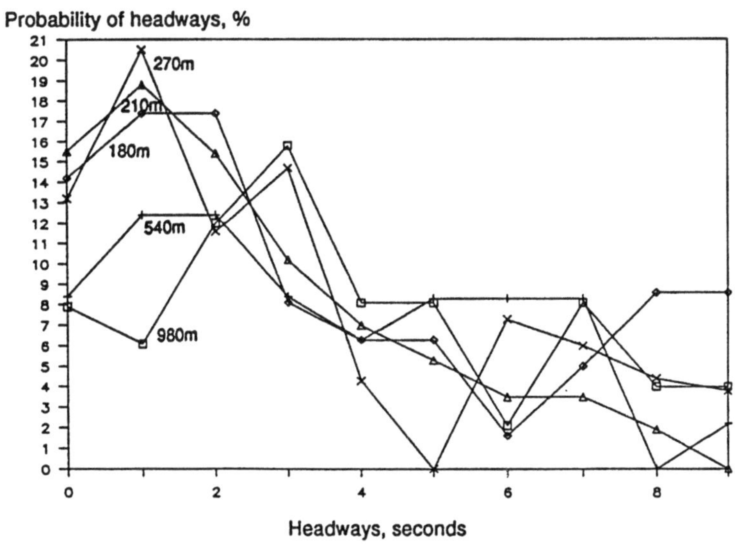

Figure 7: Comparison of short headway distributions for various distances from traffic signals (for 450–650vph)

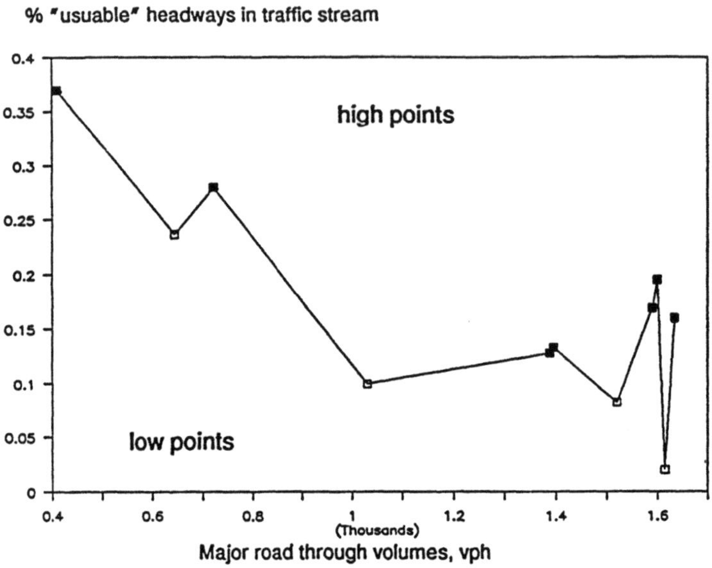

Figure 8: Effects of main road traffic interaction on availability of "usuable" headways for minor road left turn traffic

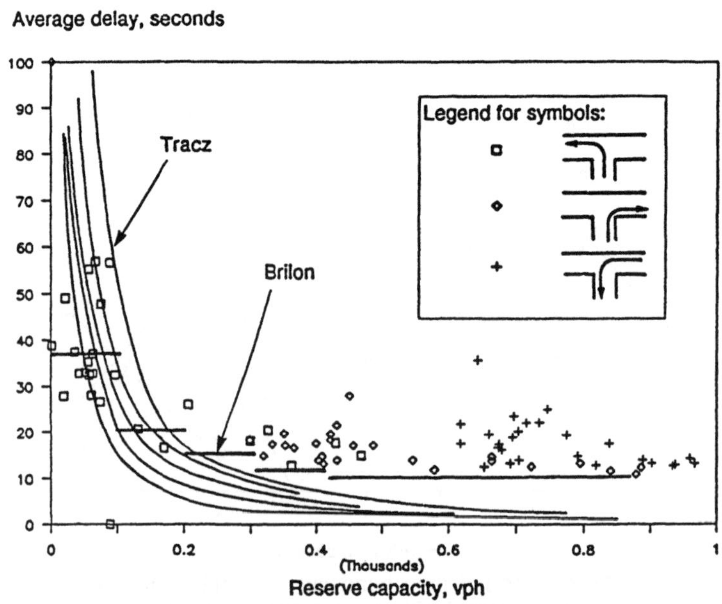

Figure 9: Comparison of reserve capacity [5] and measured delays. Relationships suggested by Brilon [4] and Tracz [6] are included.

8. ACKNOWLEDGMENTS

The support for this project has been provided by the Natural Sciences and Engineering Research Council of Canada as a part of an operational grant, and by the City of Edmonton in the form of traffic and other data. The authors also wish to acknowledge gratefully the help of Professor Brilon in providing the access to KNOSIMO and for additional comments, Professor Kyte for the TDIP program, and Ms. Butt for the assistance in the translation of the KNOSIMO User Guide.

9. REFERENCES

1. Kyte,M., Boesen, A., Traffic Data Input Program", Program documentation and Iser's Manual, Version 2.0, Department of Civil Engineering, University of Idaho, Moscow, Idaho, April 1989

2. Grossmann, M.," KNOSIMO - A Practicable Simulation Model for Unsignalized Intersections", Proceedings of an International Workshop on Intersections Without Traffic Signals, W. Brilon, editor, Springer Verlag, 1988, pp. 263-273

3. "KNOSIMO Version 3.4, Benutzeranleitung", (KNOSIMO User Guide, English translation by Chan, P. and Butt, K.), Lehrstuhl fuer Verkehrswesen, Ruhr-Universitaet Bochum, September 1988

4. Brilon, W., "Recent Developments in Calculation Methods for Unsignalized Intersections in West Germany", Proceedings of an International Workshop on Intersections Without Traffic Signals, W.Brilon, editor, Springer Verlag, 1988, pp. 111-153

5. "Highway Capacity Manual", Special Report 209, Transportation Research Board, Washington, D.C., 1985, Chapter 10.

6. Tracz,M., Chodur,J., Gondek,S., "The Use of Reserve capacity and Delay as the Complementary Measures of Junction Effectiveness", Proceedings of the Eleventh International Symposium on Transportation and Traffic Theory, Elsevier Science Publishing Co., 1990.

Estimating High and Low Flow Rates at Unsignalized Intersections

MARK R. VIRKLER, Department of Civil Engineering
RICHARD W. MADSEN, Department of Statistics
JOANNE H. SUTTON, Department of Statistics
University of Missouri-Columbia

Summary

Unsignalized intersections are often modeled with an assumption of random vehicle arrivals. Vehicle arrivals are not random where nearby upstream signalized intersections cause platoons to form. A method is presented to identify periods of high and low flow. A statistical procedure is developed to estimate the high and low traffic flow rates caused by an upstream signal. The procedure is demonstrated with data from unsignalized intersections affected by platoon flow. Ramifications of using the procedure in the analysis of unsignalized intersections are then discussed.

Introduction

Vehicle arrivals at unsignalized intersections are often modeled as random. Any upstream flow interruption, such as that caused by a traffic signal, would make the assumption of random vehicle arrivals invalid. This paper describes a technique for modifying the assumption of random arrivals at unsignalized intersections that are affected by upstream signalized intersections.

A brief review of past work is given below. This is followed by the derivation of the statistical procedure. The procedure is then demonstrated with five data sets from unsignalized intersections. A discussion of potential applications for various traffic analyses and the conclusions are then presented.

Review of Literature

Non-randomness of vehicle arrivals, or platooning, has been studied for a variety of applications (1). Platoons caused by traffic signals have received particular emphasis because of the effects upon delay at downstream signalized intersections.

Rouphail (2) described three types of platoon dispersion models: no dispersion, TRANSYT model dispersion and Passer II model dispersion. In each case traffic from an upstream traffic signal is dispersed in a particular manner by the time the traffic reaches a point downstream.

In the no dispersion model, vehicles pass through the signalized intersection at a certain rate of flow and maintain that rate of flow indefinitely (i.e., headways remain constant). In the TRANSYT dispersion model (3) a platoon departs from a signalized intersection during the green phase at a particular flow rate, but the platoon is dispersed by the time it passes a point downstream. The platoon flow rate is modified by a platoon dispersion function and arrival histograms are constructed for a downstream point. Non-platoon flows are also represented in the TRANSYT model. In the Passer II model (4), a platoon departs from a signalized intersection at a particular flow rate during the green phase but, downstream from the signal, the platoon flow rate is reduced and the time required for the platoon to pass a point is increased.

Rouphail also presents a platoon dispersion model described as a middle ground between the TRANSYT and Passer II models. The model assumes that traffic downstream from a signalized intersection arrives at two distinct flow rates: a high flow rate within the progression band and a low flow rate outside the progression band. A platoon dispersion formulation is used to relate the high flow rate during the green phase of the upstream signal to the reduced high flow rate downstream from the signal. Delay expressions for a downstream signalized intersection are then developed. An assumption of two distinct flow rates with time periods similar to that of the Rouphail model is used in the method presented in this paper.

Unsignalized Intersection Analysis

The Highway Capacity Manual (5) includes a procedure for analyzing the operation of two-way stop or yield controlled intersections. The procedure depends upon the assumption of random vehicle arrivals. Traffic flows which have a higher priority over a subject movement are assumed to be randomly distributed. The Highway Capacity mentions that the method will not be as accurate when the major street traffic is strongly platooned.

The capacity manual describes how a time-space diagram of the major street traffic could be used to divide parts of the hour

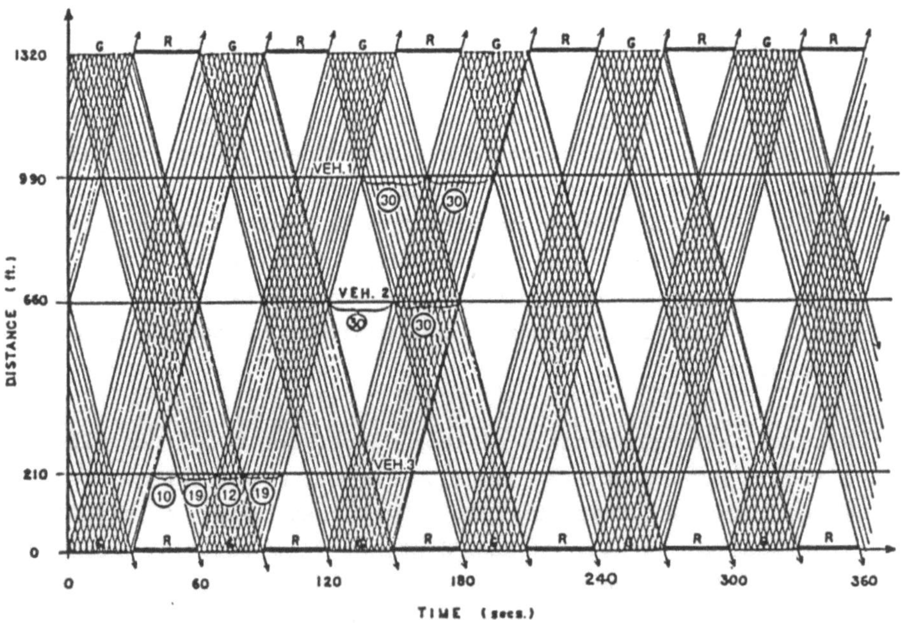

Fig. 1. Time-space diagram with platoons (5)

under study into differing flow periods. Consider Vehicle #2 of Figure 1. The vehicle is waiting to cross the major street. Beginning at time = 30 sec, traffic on the major street is heavy in both directions for one-half of a minute. Few gaps would exist for Vehicle #2's maneuver. During the next one-half minute the major street traffic is light. The through traffic streams have been held back by the traffic signals. Many gaps would exist for the crossing movement. If the percentage of traffic in the major street high and low flow periods could be determined, the unsignalized intersection analysis procedure could be applied separately to the high and low flow time periods. Vehicle #1 and Vehicle #3 would face much different flow conditions because of their locations. A means of identifying these high and low flow periods is described below.

Statistical Procedure

A question related to determining high and low flow rates approaching an unsignalized intersection is: "How long must one count vehicles to estimate rates of flow with a specific accuracy?" The following assumptions were made:

1. Traffic flows at two rates, one high (μ) and the other low (β). The high rate lasts for s seconds and the low rate lasts for t seconds. The cycle length, C, is the sum of s and t.
2. The times s and t are known and the pattern repeats in this same way during the course of data collection.
3. When flows are heavy there is less variability in the counts per time interval than when the flow is light.
4. The low flow rate vehicles arrive according to the Poisson distribution.

The statistical question then becomes: "For how many cycles should traffic be counted in order to estimate μ and β to within a given percentage of their true value?" If we let:

X_i = number of vehicles counted during the low rate period of the ith cycle, and

Y_i = number of vehicles counted during the high rate period of the ith cycle,

from the above assumptions we conclude that the variance of X will be larger than the variance of Y. If n is the number of cycles large enough to satisfactorily estimate ß, it will also be large enough to estimate μ. It will therefore suffice to try to estimate ß accurately since accurate estimates of μ should follow.

If the low flow rate vehicles arrive according to the Poisson distribution, then X_i will be a Poisson variable with an expected value of ß·t. If $\overline{X}$ is based on n cycles, then:

$$E\left[\frac{\overline{X}}{t}\right] = \frac{E(\overline{X})}{t} = \frac{\beta \cdot t}{t} = \beta \text{ and} \tag{1}$$

$$V\left[\frac{\overline{X}}{t}\right] = \frac{V(\overline{X})}{t^2} = \frac{\beta \cdot t}{n \cdot t^2} = \frac{\beta}{n \cdot t} \tag{2}$$

where E denotes the expected value and V denotes the variance of a random variable.

For example, assume we wish to estimate ß to within 10% of its value with 95% probability. If Z denotes a standard normal random variable and if Z_p denotes the $100 \cdot p^{th}$ percentile of the standard normal distribution, then using the Central Limit Theorem we can write:

$$P\left[\left|\frac{\overline{X}}{t} - \beta\right| \leq 0.1 \cdot \beta\right] = P\left[\,|z| \leq \frac{0.1 \cdot \beta}{\sqrt{\frac{\beta}{n \cdot t}}}\right] = 0.95 \tag{3}$$

This equality holds if and only if

$$\frac{0.1 \cdot \beta}{\sqrt{\frac{\beta}{n \cdot t}}} = Z_{.975} \tag{4}$$

This last expression can be solved for n to find the number of cycles that need to be counted.

More generally if the estimation error e is to be no more than a proportion r of the true value of ß with probability $1-\alpha$, we write:

$$P\left[\left|\frac{\overline{X}}{t} - \beta\right| \leq r\cdot\beta\right] = P\left[\,|z| \leq \frac{r\cdot\beta}{\sqrt{\frac{\beta}{n\cdot t}}}\,\right] = 1 - \alpha \tag{5}$$

This inequality holds if and only if:

$$\frac{r\cdot\beta}{\sqrt{\frac{\beta}{n\cdot t}}} = Z_{1-\alpha/2} \tag{6a}$$

or equivalently

$$n = \frac{Z^2{}_{1-\alpha/2}}{r^2\cdot\beta\cdot t} \tag{6b}$$

Here n represents the number of time intervals of length t that must be observed to meet the required estimation accuracy.

Now let f = t/(s+t) = t/C denote the proportion of the cycle in which the traffic flow is light. Also let H denote the total time that one must count vehicles. For notational convenience, write $z_{1-\alpha/2}$ as z. Then we have:

$$H = n\cdot C = \frac{Z_2}{B\cdot t\cdot r^2}\,.C = \frac{Z^2}{\beta\cdot t\cdot r^2}\cdot\frac{t}{f} = \frac{Z^2}{\beta\cdot f\cdot r^2} \tag{7}$$

As an example, suppose that we wish to estimate flows within 15% of their true values with 90% probability and the flow rate is low for one-half of the cycle (i.e., r = 0.15 and f = 0.50). The value of z would be 1.645 and:

$$H = \frac{1.645^2}{\beta\cdot(0.5)\cdot(0.15)^2} = \frac{240.5}{\beta} \tag{8}$$

If β were 1000 veh/hr, then we would count for 0.24 hours. The estimate would be within 150 of the true value. If β were 500 veh/hr, then we would count for 0.48 hours. The estimate would then be within 75 of the true β (with 90% probability).

Note that the question of how many vehicles (L) to count during the low flow period (to ensure a particular proportion of allowable error) would be answered by:

$$L = n\cdot\beta\cdot t = \beta\cdot t\cdot \frac{Z^2}{\beta\cdot t\cdot r^2} = \frac{Z^2}{r^2} \tag{9}$$

In the above example: $L = \dfrac{1.645^2}{0.15^2} = 120$ vehicles.

That is, 120 vehicles should be counted in the low flow period to determine the low flow rate to within 15%. Estimating the time period required for data collection to achieve a specific accuracy requires an educated guess of the low flow rate.

The total number of vehicles counted, on the average, would be:

$$E[\Sigma(X_1 + Y_2)] = n\cdot\beta\cdot t + n\cdot\mu\cdot s \tag{10}$$

If $k = \mu/\beta$ and $b = s/t$ then this total would be:

$$n\cdot\beta\cdot t + n\cdot k\cdot\beta\cdot b\cdot t = (1+b\cdot k)\cdot n\cdot t\cdot\beta = (1+b\cdot k)\cdot\frac{Z^2}{r^2} \tag{11}$$

One should note that flow rates can vary substantially within a given time period, including the peak hour of the day.

There is a practical problem that one faces when trying to collect data on a street with high and low flow rates. The problem is in determining which segments of the collected stream of data correspond to high rates and which correspond to low flow rates. If the stream of data is recorded every m seconds, there will be a pattern of s/m time increments corresponding to a high rate followed by t/m time increments corresponding to a low rate. This pattern is expected to repeat continuously (see Figure 2). The problem then is to determine where in the observed stream of data the high rate starts. An intuitively reasonable method for deciding how the data should be divided into high and low rate periods is to place (by trial and error) the start of the first complete cycle so that the sum of Y_1, the high rate counts, is a maximum. Equivalently one could choose the starting point so that the sum of X_1 is minimized.

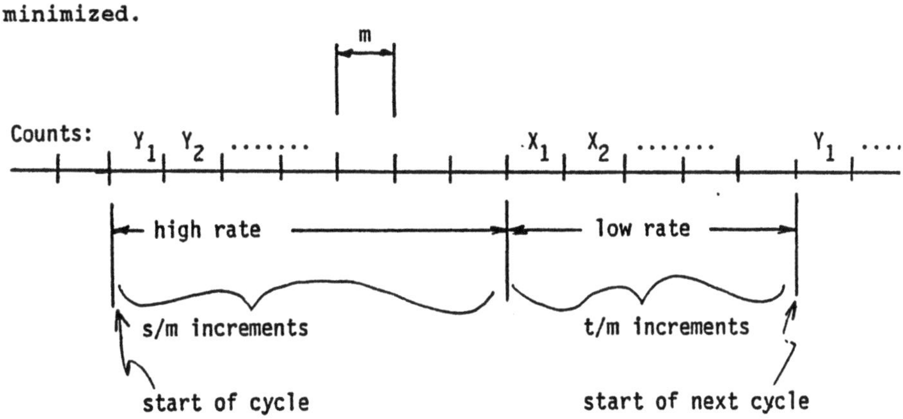

Figure 2. Data stream divided into high and low flow rates

After the starting point is determined, β is estimated by the total vehicles observed in the light period divided by the total time for the light period. A confidence interval would be given by:

$$\beta \pm Z_{1-\alpha/2} \quad \frac{\beta}{\sqrt{n \cdot t}} \tag{12}$$

Similarly the mean rate for the high traffic flow is estimated by:

$$\mu = \frac{\Sigma Y_1}{n \cdot s} \tag{13}$$

The variance σ^2 is estimated by $S_y^2 = \Sigma[(Y_i - \overline{Y})^2]/(n-1)$.

A 1-α confidence interval for μ is $\mu \pm z_{1-\alpha/2}[Sy/(m\cdot s^2)]$ for large n.

<u>Illustration with Data from Field Study</u>

To demonstrate the types of estimates which can be derived form the statistical procedure, traffic counts were collected in five second increments at two sites. Each site was just upstream from an unsignalized intersection and a significant distance downstream from a pre-timed signalized intersection. The first site (data sets A1 and A2) involved an unsignalized intersection approximately 0.24 km (0.15 miles) from a signalized intersection on a two lane road with speeds of about 48 km/hr (30 mph). The second site (data sets B1, B2, and B3) involved an unsignalized intersection approximately 1.6 km (1.0 miles) from a signalized intersection on a two-lane highway with speeds of about 72 km/hr (45 mph).

A computer program was written to find the starting point of a cycle and the high flow and low flow periods, given the cycle length and expected length of the high flow time period. The high flow time period was assumed to equal the upstream through movement green time. Table 1 shows a tabular example of this process for data set A1. The resulting point estimates of high and low flow rates along with their 95% confidence intervals are shown in Table 2. In the first four data sets the mean high and low flow rates were significantly different (p-value < .01) In the fifth case the difference in rates was close to significant (p-value = 0.07).

Given that two distinct flow rates seemed to exist within each cycle, the next question to be answered dealt with whether or not the arrival rates within each of these periods was random. The results of a Chi-square test on the data (i.e., number of 5 second increments in which 0,1,2, etc. vehicles were observed) are in Table 3. In all cases but one, the assumption of Poisson or random data was accepted for the low volume

TABLE 1

TABULAR EXAMPLE OF FINDING HIGH AND LOW FLOW RATES
(Data Set A1)

80 second cycle length

Vehicles counted every 5 seconds (16 observations/cycle)

Expected green band = 30 seconds (6 observations/cycle)

CYCLE \ OBS	1	2	3	4	5	6	7	8	9	10	11	12	13	14	15	16
1	2	2	1	2	1	1	1	0	0	1	0	0	2	0	1	1
2	3	2	2	2	1	2	2	1	0	2	1	0	0	2	1	0
3	1	2	2	2	0	0	1	0	0	2	1	1	0	0	0	2
4	2	0	3	3	1	0	0	0	0	1	2	0	0	0	0	1
5	2	2	3	2	1	1	1	2	2	2	2	1	0	0	1	0
6	2	2	2	2	2	2	1	2	1	1	0	0	0	0	0	0
	12	10	13	13	6	6	6	5	3	9	6	2	2	2	3	4
	← HIGH FLOW					→	← LOW FLOW									→

$$\text{High Flow Rate} = \left(\frac{60\ \text{veh}}{36\ \text{obs}}\right) \Big/ 5\ \frac{\text{sec}}{\text{obs}} \quad X \quad \frac{60\ \text{sec.}}{\text{min}} = 20.0\ \frac{\text{veh}}{\text{min}}$$

$$\text{Low Flow Rate} = \left(\frac{42\ \text{veh}}{60\ \text{obs}}\right) \Big/ 5\ \frac{\text{sec}}{\text{obs}} \quad X \quad \frac{60\ \text{sec.}}{\text{min}} = 8.4\ \frac{\text{veh}}{\text{min}}$$

TABLE 2: HIGH AND LOW FLOW RATES FROM FIELD STUDY

Data Set	Initial Time	Number of Cycles	High Flow Rate $\left(\frac{veh}{min}\right)$		Low Flow Rate $\left(\frac{veh}{min}\right)$	
			Point Est.	95% C.I.	Point Est.	95% C.I.
A1	5:02 p.m.	6	20.0	16.7 to 23.3	8.4	6.0 to 10.8
A2	5:27 p.m.	7	12.9	9.4 to 16.3	8.4	6.3 to 10.5
B1	4:48 p.m.	12	9.8	7.0 to 12.6	4.6	2.7 to 6.5
B2	7:52 a.m.	15	9.6	6.9 to 12.3	5.0	2.8 to 7.2
B3	1:22 p.m.	30	22.0	19.9 to 24.1	19.8	17.8 to 21.8

TABLE 3

Probability of Random Arrivals Having Equal or Greater Chi-Square (for number of five second increments with 0, 1, 2, etc. vehicles)

Data Set	Initial Time	Low Flow Period	High Flow Period
A1	5:02 p.m.	0.5	0.001
A2	5:27 p.m.	0.5 - 0.25	0.75
B1	4:48 p.m.	0.01	0.005
B2	7:52 a.m.	0.95	0.95
B3	1:22 p.m.	0.5	0.995

period. Therefore the capacity manual's assumption of random arrivals is probably valid for the low flow period. In the periods of high traffic flow there is no clear overall indication of fit to the Poisson distribution. In three cases the data fit well. In two cases the data did not fit at all. The assumption of random arrivals during the high flow period appears questionable.

It should be pointed out that on both of these two-lane roadways virtually no opportunities exist for passing a slower moving vehicle. One would expect that vehicles would be less regimented, and therefore more closely follow the Poisson distribution, on a four lane roadway where passing opportunities are readily available.

Potential Applications

Assuming random flow at unsignalized intersections when varying flow rates exist can lead to significant errors. Consider as an example the flow rates determined in Table 1 for data set A1. The high flow rate of 20.0 vehicles per minute (1200 vph) for

30 seconds of each 80 second cycle and the low flow rate of 8.4 vpm (504 vph) for 50 seconds of each cycle yield an average hourly volume of 765 vph. If a movement against this opposing flow required an eight second gap, the hourly average method would yield a potential capacity of 220 vph (see Figure 3). With two discrete flow rates, 34 gaps per hour would exist during the high opposing flow period and 219 gaps per hour would exist in the low opposing flow period, for a total of 253 gaps in the hour (a difference of 15%). The result of using the average approach could be that a poorer level of service would be indicated than actually exists. At the extreme, over capacity conditions could be indicated when, in fact, the intersection is under capacity. This difference is due to the concavity of the curves in Figure 3.

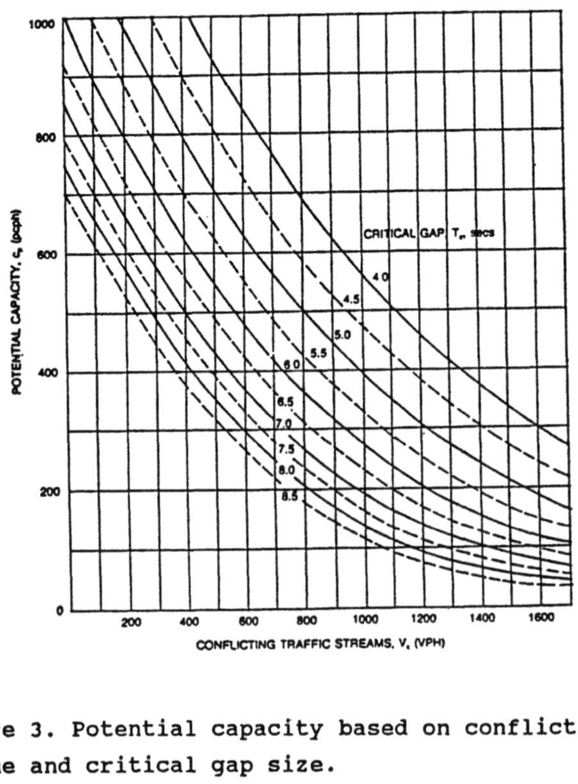

Figure 3. Potential capacity based on conflicting traffic volume and critical gap size.

Determining high and low flows from a single upstream signalized intersection is rather straightforward. If major street signalized intersections exist on both sides of an unsignalized intersection the problem becomes more complex. One option would be to determine arrival conditions for each approach separately. If the signals are coordinated, a time-space diagram could then be used to divide the cycle length into two or more discrete time periods. Another option would be to conduct traffic counts concurrently with synchronized watches.

The above discussion implicitly assumed that the upstream traffic signal has a fixed cycle length and constant green time for the movements in question. With a variable cycle length the procedure would not be directly applicable.

The advantages of identifying platoon flow rates might be achievable without the specific field data collection called for in this procedure. This might be accomplished by relating signalized intersection turning movement counts and signal timing to downstream flow rates. For instance, knowledge of platoon dispersion characteristics might be used to provide characteristic high and low flow rates at downstream locations. Simulation models might be used to develop tables or equations directly applicable to a wide variety of traffic flow, signalization, and intersection location combinations. These models could be tested and verified against the procedure described in this paper.

Conclusions

The method described above provides a viable means to identify different flow rates caused by upstream flow interruptions. With an appropriate data sample, high and low flow rates can be reasonably estimated. The results can be used to modify the assumption of random arrivals to yield more accurate estimates of capacity and level of service at unsignalized intersections. One would expect that this procedure would most often be used for an unsignalized intersection when it is known that the intersection is, or will soon be, near capacity. The use of platoon flow rates will likely yield a higher capacity than would otherwise be apparent. The extra work involved for this procedure would yield the benefit of revealing extra capacity that would otherwise be overlooked.

References

1. Denney, Richard W. Jr.: Traffic Platoon Dispersion Modeling. Journal of Transportation Engineering, Vol. 115 (No.2), American Society of Civil Engineers (1989) 193-207.

2. Rouphail, N.M.: Delay Models for Mixed Platoon and Secondary Flows. Journal of Transportation Engineering, Vol. 114 (No.2), American Society of Civil Engineers (1988) 131-152.

3. Robertson, D.I.: TRANSYT: A traffic network study tool. Report No. LR 253, Road Research Laboratory, Crowthorne, Berkshire, England, 1969.

4. Messer, C.J.; Fambro, D.; Anderson, D.: A Study of the Effects of Design and Operational Performance of Signal Systems. Texas Transportation Institute Research Report 203-2F, College Station, Texas, 1975.

5. Highway Capacity Manual: Special Report 209, Transportation Research Board, Washington, D.C., Chapters 10, 1985.

Empirical Models For Estimating Capacity and Delay at Stop-Controlled Intersections in the United States

MICHAEL KYTE

Assistant Professor of Civil Engineering
University of Idaho
Moscow, Idaho, United States

JOHN ZEGEER

Principal Associate
Barton-Aschman Associates, Inc.
Ft. Lauderdale, Florida, United States

B. KENT LALL

Professor of Civil Engineering
Portland State University
Portland, Oregon, United States

Summary

This paper reports on a series of research studies that have been conducted in the United States on stop-controlled intersections between 1987 and 1991. The two major objectives of these studies have been to develop a data base for traffic operations data at stop-controlled intersections and to develop a set of models that can be used to forecast capacity and delay at stop-controlled intersections.

Empirical models have been developed for all-way stop-controlled (AWSC) intersections to estimate capacity and delay. The capacity model states that the capacity of a given approach is a function of the distribution of the volume among the intersection approaches, the number of lanes on each intersection approach, and the turning movement proportions on each approach. The delay model states that the average stopped delay on a given approach is a function of the volume/capacity ratio for that approach. A set of level of service ranges based on stopped delay has also been proposed. These two models form the basis for a new set of interim capacity materials recently adopted by the Committee on Highway Capacity and Quality of Service of the Transportation Research Board (TRB).

Empirical models have also been developed for two-way stop-controlled (TWSC) intersections to estimate capacity and delay. The capacity model states that the capacity of a minor street approach is a function of the flow rate on the major street, the flow rate on the opposing minor street approach, and the speed of traffic on the major street. The delay model includes two components, one for service time and one for queue time. Both are exponential in functional form. The service time on the minor street approach is a function of the flow rates on the major street and on the opposing minor street approach. The queue time on the minor street is a function of the flow rate on the subject minor street approach. These models are preliminary and further study is now underway to determine the effects on capacity and delay of turning movements on each approach and the degree of platooning on the major street.

1. Introduction

The most common type of intersection control in the United States is the stop sign. Stop signs are used to control access onto major arterials and highways from side streets or driveways and to assign rights-of-way at intersections in residential areas. Stop signs are also used as an intermediate level of intersection control in suburban areas until traffic signals can either be justified or afforded.

Unfortunately, aside from volume and accident warrants, U.S. traffic engineers are given only limited guidance in determining the appropriate kind of intersection control for a set of traffic and geometric conditions, or in deciding when a stop-controlled intersection should be upgraded to signal control [1].

Briefly, these are the relevant problems that confront the U.S. traffic engineer today:

1. The 1985 Highway Capacity Manual (HCM) [2] does not provide adequate methods for evaluating stop-controlled intersections. The method for evaluating TWSC intersections is based on a 1972 German method validated with only limited U.S. data [3]. The method for evaluating AWSC intersections is based on a study of three intersections; only limited capacity guidelines are provided [4].
2. There is no common measure of effectiveness (MOE) for that can be used to evaluate all types of intersection control. The 1985 HCM defines vehicle delay as the primary MOE for signalized intersections and reserve capacity for TWSC intersections. No MOE is provided for AWSC intersections.
3. There are simulation models available (notably NETSIM and TEXAS) that can

evaluate differing intersection control strategies for a given set of volume and geometric conditions. Unfortunately, these models were not validated with sufficient data for stop-controlled intersections and thus yield poor results when forecasting delay and capacity at stop-controlled intersections [5].

To alleviate these problems, TRB's Committee on Highway Capacity and Quality of Service has adopted a program plan to improve the methodologies for evaluating stop-controlled intersections [6]. This plan includes the following elements:

1. A research problem statement on unsignalized intersections.
2. A work plan to conduct new research activities.
3. A data collection methodology to encourage new research activities and to provide a consistent set of standards for data collection.

In response to this need to improve capacity analysis methods for stop-controlled intersections, several research projects have been undertaken in the U.S. in recent years. The purpose of this paper is to report on several recent efforts to develop models for estimating capacity and delay at stop-controlled intersections. The remainder of this paper is divided into two major sections. Section 2 describes the results of studies of traffic flow at AWSC intersections and presents a new interim procedure recently adopted by TRB's Committee on Highway Capacity and Quality of Service. Section 3 describes the results of studies of traffic flow at TWSC intersections and presents a set of preliminary models that have been developed. Both of these sections include a discussion of a theory of traffic flow, a summary of the data base that has been assembled, and the models developed for forecasting capacity and delay. The paper concludes with a summary and conclusions and a list of references.

2.0 All-Way Stop-Controlled Intersections

AWSC intersections are controlled by stop-signs on all approaches. No approach has priority or right-of-way above any other. While many jurisdictions in the United States mandate the right-of-way to the vehicle on the right, this rule is often not followed by drivers in the field.

The capacity guidelines contained in the 1985 HCM that are used for the evaluation of AWSC intersections are based on a study of three intersections by Hebert in 1963 [4]. Hebert found that the capacity of an AWSC intersection is a function primarily of the relative proportion of traffic on the two intersecting streets. Richardson developed an M/G/1 queueing model based on Hebert's headway data [7]. Zion, List, and Manning verified the basic form of Hebert's model using data collected in New York [8].

The capacity and delays models presented here were developed by Kyte and others in a series of studies from 1987 through 1990, primarily in the Northwest and Midwest United States [9, 10, 11]. The capacity model recognizes the essential importance of Hebert's work, namely that volume distribution is the major factor in determining the capacity of an AWSC intersection. The delay model proposes that delay increases exponentially with the volume/capacity ratio. These two models form the basis for a new set of interim capacity materials recently adopted by TRB's Committee on Highway Capacity and Quality of Service [12].

This section is divided into five parts. In section 2.1, the notation and terms are defined. In section 2.2, a theory of traffic flow at AWSC intersections is presented. The data base is described in section 2.3. The capacity and delay models are presented in sections 2.4 and 2.5

2.1 Notation and Terms

Designation of Intersection Approaches. The specific traffic stream under study travels on the *subject approach*. See Figure 1. The *opposing approach* and the *conflicting approaches* are also illustrated in Figure 1.

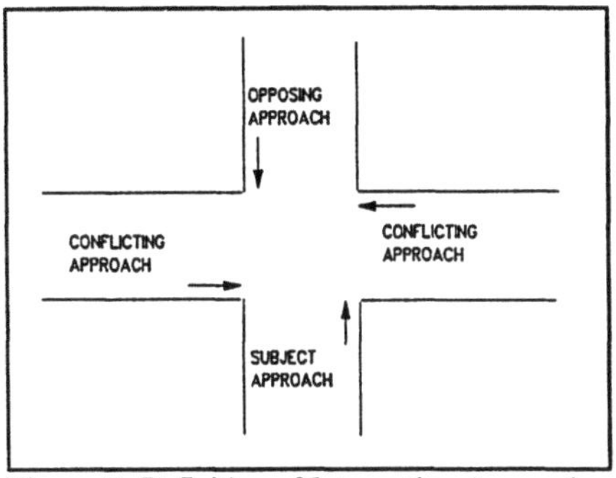

Figure 1. Definition of Intersection Approaches

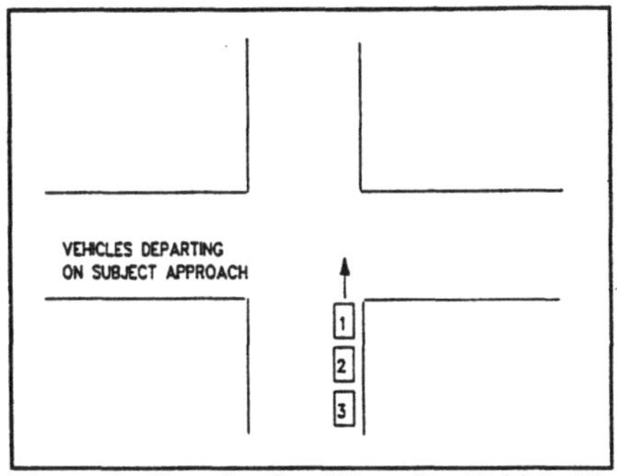

Figure 2. Capacity Conditions For Vehicle #2

Headways. The *departure headway* for a vehicle on the subject approach is defined as the difference between the successive times of departure of that vehicle and the previous departing vehicle on the subject approach. A departure headway is considered to be a *saturation headway* if, when a given vehicle arrives, there is already a vehicle ahead of it at the stop line (see Figure 2).

Delay. The total delay experienced by any vehicle traveling through a stop-controlled intersection includes the total time that a vehicle is traveling at less than its desired speed. Total delay includes two components, geometric delay and vehicle-interaction delay. Geometric delay is caused by acceleration and deceleration arriving at and departing from the stop line. Vehicle interaction delay results from vehicles waiting for their proper turn. There are a number of different methods that can be used to measure delay. In this paper, stopped delay is used. Stopped delay is the total elapsed time from when a vehicle becomes a part of the queue until the vehicle leaves the stop line.

2.2 Theory of Traffic Flow

2.2.1 Nature of Operations

AWSC intersections require that every vehicle stop at the intersection before proceeding. This requirement provides a framework for studying traffic operations at AWSC intersections. Since each driver must stop, the judgement as to whether to proceed into the intersection is a function of the traffic conditions on the other approaches. If there is no traffic present on the other approaches, a driver can proceed immediately after the stop is made. If there is traffic waiting on one or more of the other approaches, a driver proceeds only after determining that there are no vehicles currently in the intersection (i.e. it is safe) and that it is his or her turn.

While the "rules of the road" have traditionally suggested the driver-on-the-right right-of-way rule, the actual operation of AWSC intersections is somewhat more complex than this. The problem becomes one of determining, under capacity conditions for a given approach, the factors that influence the rate at which vehicles can successively depart from the stop line. The manner in which these factors influence the saturation headway for a given approach are discussed below.

If traffic is present on the subject approach only, vehicles depart as rapidly as individual drivers can safely accelerate into and clear the intersection. This is illustrated as case 1 in Figure 3. If traffic is present on the other approaches, as well as on the subject approach, the saturation headway on the subject approach will increase somewhat, depending on the degree of conflict that results between the subject approach vehicles and the vehicles on the other approaches. In case 2 (Figure 4), there is some uncertainty introduced with a vehicle on the opposing approach and thus the saturation headway will be greater than for case 1. In case 3 (Figure 5), vehicles on the conflicting approaches further restrict the departure rate of vehicles on the subject approach, and the saturation headway will be longer than for cases 1 or 2. When all approaches are loaded, case 4 (Figure 6), saturation headways are even longer than for the first three cases as the potential for conflict between vehicles is greatest.

This increasing degree of potential conflict translates directly into both longer driver decision times and thus longer saturation headways.

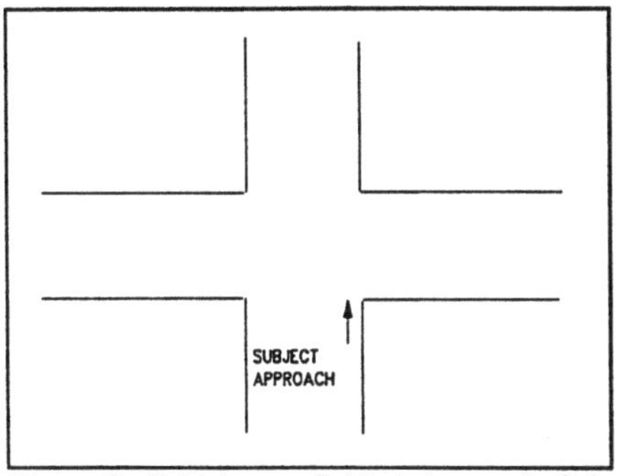

Figure 3. Vehicles on Subject Approach Only (Case 1)

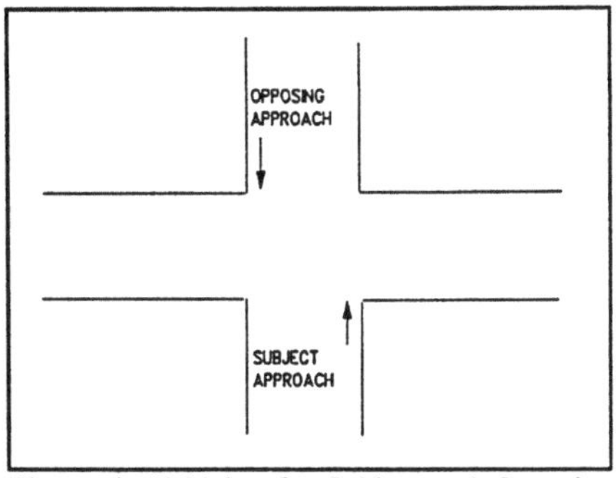

Figure 4. Vehicles On Subject and Opposing Approaches (Case 2)

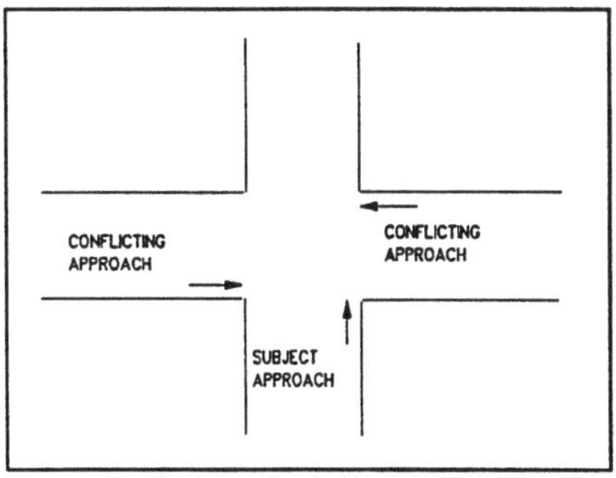

Figure 5. Vehicles On Subject And Conflicting Approaches (Case 3)

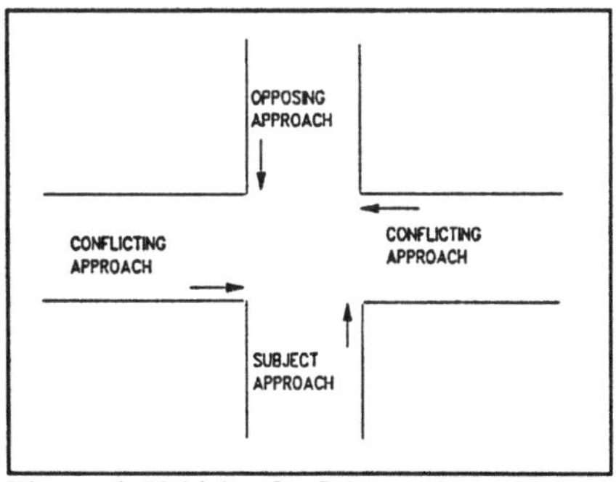

Figure 6. Vehicles On Subject, Opposing, and Conflicting Approaches (Case 4)

2.2.2 Implications For Model Form

This theory of traffic operations can be used as the basis for empirical model development.

Volume Distribution. The headway between vehicles departing from a stop line of an AWSC intersection is a function of the conditions present on the other intersection approaches. Minimum saturation headways are achieved if there is no traffic on any of the other intersection approaches. Maximum saturation headways result if traffic is present on all of the other approaches. This microscopic perspective has a direct analogue at the macroscopic level. The maximum flow rate on a given approach can be achieved if there is no traffic on any of the other approaches. The minimum flow rate on a given approach results if traffic is evenly distributed among all of the intersection approaches. Thus the key variable in the determination of approach capacity is the relative distribution of traffic volumes among the approaches. This variable is called *volume distribution.*

Number of Lanes. Unlike other highway facilities, increasing the number of lanes does not necessarily result in a corresponding increase in the capacity of an approach. This fact results from the nature of traffic flow at an AWSC intersection. For single-lane approach intersections operating at capacity, a two-phase operation results with traffic on opposing approaches flowing simultaneously. See Figure 7.

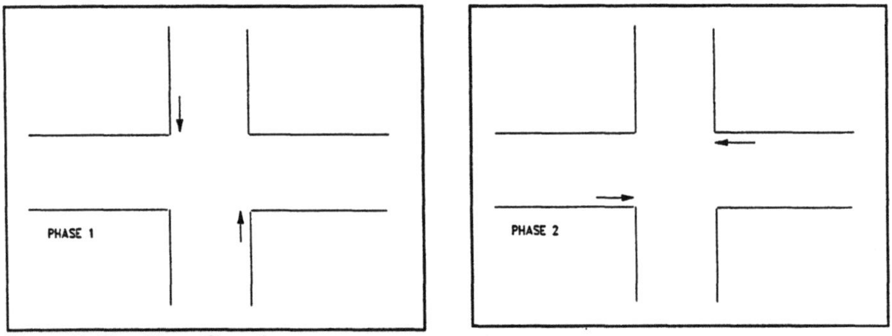

Figure 7. Two-Phase Operation

However, for multi-lane approaches, the nature of the operation is different. The addition of lanes, particularly on the opposing and conflicting approaches, introduces such a degree of uncertainty among drivers that a four-phase operation can result. That is, traffic on each approach flows as a group and saturation headways increase. See Figure 8.

Turning Movements. Left-turning movements increase the saturation headways on the subject approach and thus reduce the approach capacity. While two opposing through vehicles can travel through the intersection simultaneously without affecting each other, one vehicle is delayed if the other vehicle is turning left. Right-turning movements reduce the saturation headways on the subject approach as potential conflicts are decreased. More vehicles can travel through the intersection at shorter headways.

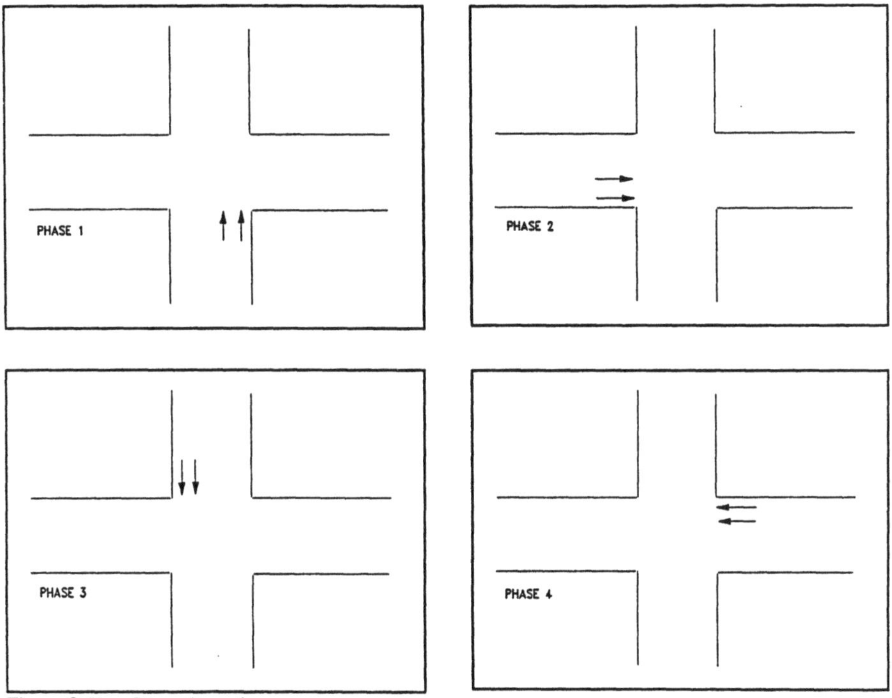

Figure 8. Four-Phase Operation

2.3 The Data Base

To provide a means to assess this theoretical framework for the capacity of an AWSC intersection, field data were collected at twenty sites encompassing thirty-three hours of vehicle operation. These sites represent a wide range of geometric and traffic conditions in the Pacific Northwest and Midwest regions of the United States.

Each site was videotaped with a video camera. The camera was placed in a position to observe vehicles on each approach as they passed through the intersection as well as the queue activity on one approach. Data collection software was used to record flow rate and delay data while observing the videotapes from each site. The time that each vehicle entered the intersection was recorded and its directional movement was noted. Flow rates were calculated based on these data. For one approach (referred to as the subject approach), the times that each vehicle entered the end of the queue and entered the intersection were also noted. Delays were calculated based on these data. The data collection and reduction method has been described elsewhere [13].

Table 1 provides a summary of the traffic flow and delay data collected at each site. The data that are summarized in Table 1 are 15-minute averages of traffic flow rates and average vehicle delays.

Table 1. AWSC Intersection Site Data

Variable	Mean	Minimum	Maximum	Standard Deviation
Delay (sec/veh)	15.1	4.2	59.2	10.6
Subject volume (vph)	372	136	676	102
Opposing Volume (vph)	235	0	556	159
Conflicting Volume (vph)	497	116	1224	240
Intersection Volume (vph)	1104	412	1892	354

2.4 Capacity Model

Capacity Flow and Departure Headways. Since one of the major objectives is to determine intersection capacity, it is important to identify those time periods during which the subject approach was *loaded* or operating at capacity conditions. The departure headway for each subject approach vehicle traveling during capacity conditions at each of the twenty sites was measured. This amounted to a total of 7,129 individual departure headways. The data were classified into each of the four cases described earlier. As expected from the proposed model, the observed departure headways for case 1 are the lowest among the four cases, while those for case 4 are the highest. These differences are stable between sites as illustrated in Figure 9.

From these departure headway data, the mean departure headway and capacity flow rate for each site were calculated. The mean departure headway is the weighted average of the departure headways for each case, weighted according to the number of observations for each case. The mean capacity flow rate is 3600 divided by the mean departure headway for the site. These data are summarized in Table 2.

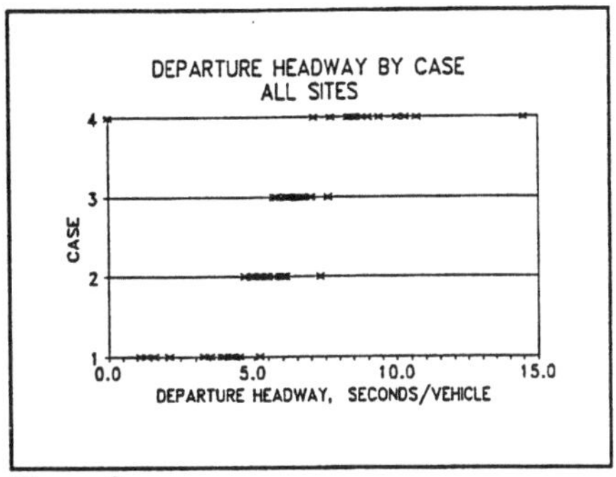

Figure 9

The capacity data were regressed on the traffic flow and geometric data for each site. The model for the capacity of an approach is given in equation (1).

$$C = 1000V_{ps} + 700V_{po} + 200L_s - 100L_o - 300LT_{po} + 200RT_{po} - 300LT_{pc} + 300RT_{pc} \quad (1)$$

where C is capacity of the subject approach, in veh/hr,
- V_{ps} is the proportion of the intersection volume on the subject approach,
- V_{po} is the proportion of the intersection volume on the opposing approach,
- L_s is the number of lanes on the subject approach,
- L_o is the number of lanes on the opposing approach,
- LT_{po} is the proportion of the volume on the opposing approach turning left,
- RT_{po} is the proportion of traffic on the opposing approach turning right,
- LT_{pc} is the proportion of traffic on the conflicting approaches turning left, and
- RT_{pc} is the proportion of traffic on the conflicting approaches turning right.

Effect of Volume Distribution. The equation clearly shows the importance of volume distribution in determining intersection capacity. When traffic is evenly distributed among the approaches, the capacity of a single-lane approach is approximately 500 vehicles per hour, and the capacity of a four-leg intersection is approximately 2000 vehicles per hour under ideal conditions. The capacity of a single-lane approach, when there is no traffic on any of the other approaches, is 1100 vehicles per hour under ideal conditions. Figure 10 shows a plot of approach capacity vs. the proportion of traffic on the subject approach.

Effect of Number of Lanes. Equation (1) also shows that the earlier discussion of two-phase and four-phase operation is verified by field observations. If vehicles are departing from the subject approach only, with no impedance from any vehicles on the opposing or

Table 2. Saturation Headway and Capacity Data

Site Number (4 legs, one lane per leg)	Mean Headway (sec/veh)	Mean Capacity (veh/hr)
10	5.7	637
11	7.0	512
13	6.2	580
21	6.3	574
22	6.8	531
23	7.1	505
24	7.3	492
26	5.8	619
27	5.7	631
29	5.1	709
30	5.5	660
32	5.8	624
33	5.6	645

Site Number (4 legs, multilanes per leg)	Mean Headway (sec/veh)	Mean Capacity (veh/hr)
17	6.4	564
25	7.2	499
34	5.0	715

Site Number (3 legs)	Mean Headway (sec/veh)	Mean Capacity (veh/hr)
15	7.3	494
16	7.0	514
19	4.9	729
20	5.8	622

conflicting approaches (Case 1), increasing the number of lanes does result in a corresponding increase in the approach capacity. The mean saturation headway for single-lane approach sites for Case 1 conditions is 3.9 seconds per vehicle, while for multi-lane sites it is 1.5 seconds. When traffic is present on all of the intersection approaches (Case 4), there is no difference in the saturation headways for the single and multi-lane sites. This means that the potential increase in capacity to be gained from additional lanes is offset by the increased uncertainty that results from each driver deciding whether it is his or her turn. These data are summarized in Table 3.

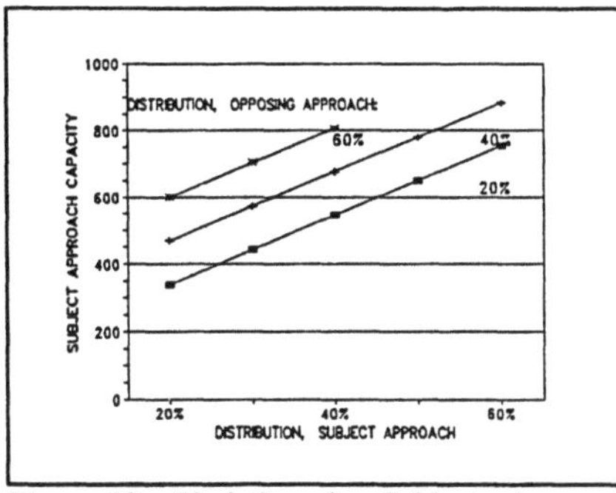

Figure 10. Variation in Subject Approach Capacity As A Function of Subject and Opposing Approach Volume Distribution

Table 3. Variation of Saturation Headway With Number Of Approach Lanes

	Mean Saturation Headway, Seconds/Vehicle			
Condition	Case 1	Case 2	Case 3	Case 4
All Data	3.5	5.5	6.5	9.0
Single-Lane Approach Sample Sites	3.9	5.6	6.5	9.0
Multi-Lane Approach Sample Sites	1.5	4.3	6.3	9.3

These data have an important implication for intersection capacity. When traffic is present on only one approach, the capacity of the approach directly increases with each increase in the number of lanes. That is, under ideal conditions, if one lane can carry 1100 vehicles per hour, then two lanes can carry 2200 vehicles per hour, and three lanes can carry 3300 vehicles per hour. However, when traffic is distributed among all of the approaches, the capacity of the intersection, at worst, is about the same no matter how many approach lanes there are. At best, the addition of approach lanes may slightly increase the capacity of the intersection.

2.5 Delay Model

The average stopped vehicle delay on each approach is a function of the volume on the approach and the capacity estimated for the approach. The delay model is given in equation (2).

$$D = e^{3.8V/C} \qquad (2)$$

where D is the average stopped delay on the subject approach, sec/veh,
V is the volume on the subject approach, veh/hr,
C is the calculated capacity of the subject approach, veh/hr, and
e is the base of natural logarithms.

Figure 11 shows variation of delay with the volume-capacity ratio. A new set of delay-based level of service ranges are included in the interim procedure. The ranges are somewhat different than the ranges currently used for signalized intersections, reflecting a difference in driver perception regarding the relative performance of these two kinds of intersections.

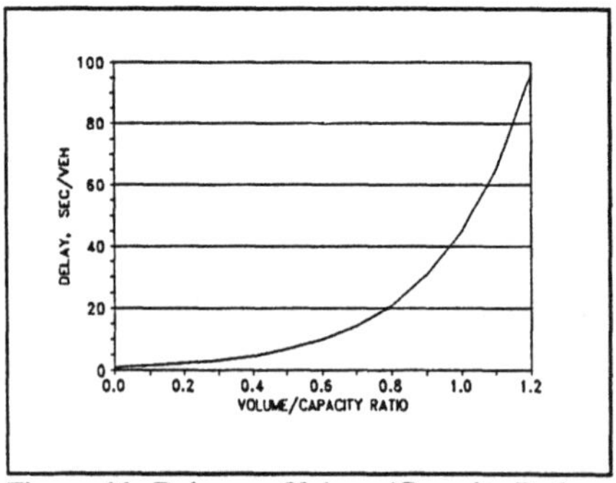

Figure 11. Delay vs. Volume/Capacity Ratio

Table 4. LOS Ranges For AWSC Intersections

LOS	Delay, sec/veh
A	< 5
B	5 - 10
C	10 - 20
D	20 - 30
E	30 - 45
F	> 45

3.0 Two-Way Stop-Controlled Intersections

TWSC intersections clearly assign the priority to the traffic on one street while requiring traffic on the other street to wait for a suitable gap in the priority traffic stream.

The capacity analysis procedure included in the 1985 HCM is based on a 1972 German procedure [3]. The procedure was validated with only a limited number of U.S. studies [14]. A number of questions have been raised about the procedure since it was released in 1985:

1. There are several procedural errors that have since been corrected in a new German procedure [15].
2. The procedure does not produce an estimate of vehicle delay.
3. There are several theoretical and procedural issues relating to the estimation of the critical gap [16]. It is important that these issues be resolved since the estimation of the critical gap forms the basis for the capacity estimation procedure.

The results presented in this section are based on studies that have been undertaken by Kyte, Lall, and others in the United States [17, 18]. The models described here are empirical, in the spirit and form of the models developed in the United Kingdom by Kimber, Taylor, and others [19, 20]. While Kimber has recently presented a compelling argument for the use of these flow-based empirical models [21], critical gap models are still much more widely used. See, for example, Brilon [22] and Troutbeck [23].

This section is divided into five parts. In section 3.1, the notation and terms are defined. In section 3.2, a theory of traffic flow at TWSC intersections is presented. The data base is described in section 3.3. The capacity and delay models are presented in sections 3.4 and 3.5.

3.1 Notation and Terms

Designation Of Intersection Approaches. The specific traffic stream under study travels on the subject approach. The other non-priority traffic stream travels on the opposing approach. The two priority traffic streams travel on the conflicting approaches. See Figure 12.

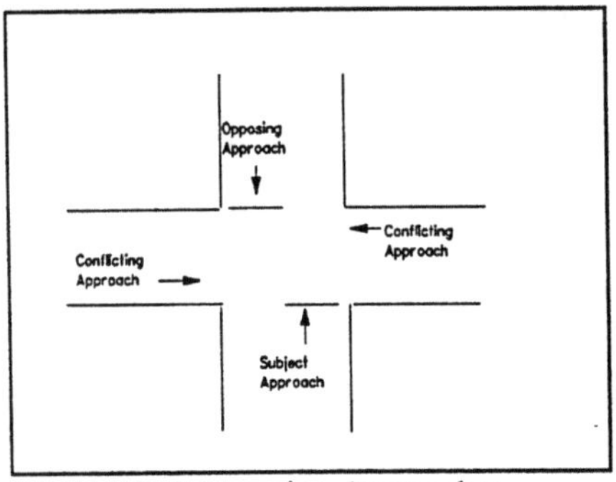

Figure 12. Intersection Approaches

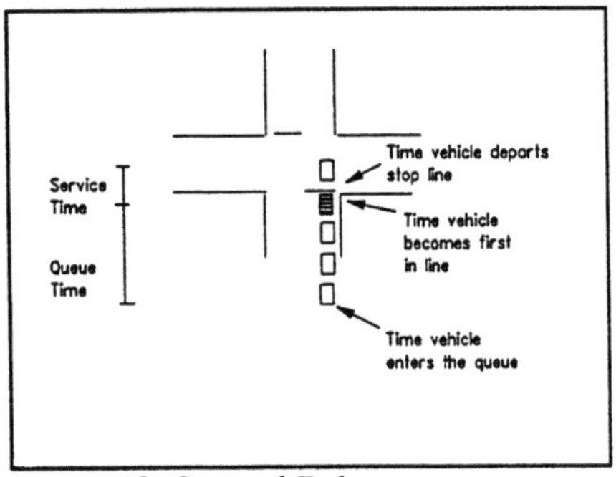

Figure 13. Stopped Delay

Delay. As described in section 2.1 of this paper, stopped delay (see Figure 13) is the total elapsed time from when a vehicle becomes a part of a queue until the vehicle leaves the stop line. For a TWSC intersection, stopped delay includes two components, queue time (d_q) and service time (d_s). Queue time is the elapsed time from when a vehicle becomes a part of a queue until it arrives first in line at the stop line. Service time is the elapsed time from when a vehicle arrives first in line at the stop line until it departs from the stop line into the intersection.

3.2 Theory of Traffic Flow

3.2.1 Nature of Operations

TWSC intersections require that each minor street vehicle stop at the intersection and wait for an acceptable gap in the major street traffic stream before proceeding. It is useful to study the sequence of events followed by a vehicle traveling through a TWSC intersection via the minor street, or non-priority, approach, as a framework for studying traffic operations at TWSC intersections.

First, consider the case of a minor street vehicle arriving with no other vehicle at the stop line ahead of it. The vehicle makes its stop at the stop line. When the driver decides that there is a safe gap in the major street traffic stream, he/she proceeds through the intersection. The time spent at the stop line (service time) is a function only of the nature of the traffic flow on the major street. The higher the flow rate and/or the more uniform the arrivals of the major street traffic, the longer will be the service time for the minor street vehicle. The presence of an opposing vehicle on the minor street will also tend to increase the service time since these vehicles would both be competing for the same available gap. Finally, since service time is directly related to the capacity or maximum rate of departure of the subject vehicle, these same factors also affect the capacity of the subject approach.

Next, consider the case when a minor street vehicle arrives with other vehicles already in queue at the stop line. If the flow on the subject approach is high enough to result in queue formation, additional dynamics must be considered. In this case, the subject approach vehicle must wait in line while vehicles in front of it wait at, and pass through, the intersection as safe or acceptable gaps become available. If a driver waits in a queue, the gap that he or she eventually accepts may be less than for a driver that does not have to wait in queue. This is due to the frustration experienced in waiting in queue; the frustration will likely increase as the length of the queue (and the queue time) increases. Similarly, the pressure on the driver that is first in line at the stop line from other vehicles behind it in queue will encourage the driver to accept a shorter gap than he or she might otherwise determine to be acceptable. Finally, the longer time that a driver spends in queue, the better he or she will be able to estimate the size of an acceptable gap, and thus the shorter the actual accepted gap may be. These three factors all combine into one effect: the longer that a driver waits in queue, the shorter the service time or the greater the minor street capacity will be.

3.2.2 Implications For Model Form

This theory of traffic operations can be used as a basis for empirical model development.

Conflicting Approach Volume. The characteristics of the flow on the conflicting approach are the primary influence on the rate at which vehicles can depart on the subject approach. The important characteristics are the arrival patterns on the conflicting approaches, the flow rates from the left and from the right, and the proportion of vehicles turning left and right from the conflicting approaches.
Opposing Approach Volume. The opposing flow rate reduces the capacity available to the subject approach vehicle as opposing vehicles use up some of the available gaps.
Subject Approach Volume. The subject approach flow rate affects the subject approach capacity: the higher the subject approach flow rate, the higher the capacity, and the lower the service time for the subject approach vehicle.

What are the implications for the empirical models for subject approach capacity, service time, and queue time, including the independent variables, the shape of the curve or surface, and the relevant boundary conditions?

Capacity Model Form. The capacity of the subject (minor street) approach decreases as conflicting and opposing flow rates increase, increases as subject approach flow rate increases, increases as the arrival pattern on the conflicting approaches goes from uniform arrivals to highly structured platoons, decreases as the proportion of turning movements increase, and decreases as the speed on the conflicting approaches increase. When the flow rates on the conflicting approaches are low or near zero, the rate of departure for subject approach vehicles can be estimated from AWSC intersection studies. If there are no vehicles on the opposing approach, the maximum rate of departure is 1100 vph or one vehicle every 3.3 seconds. If there are vehicles on the opposing approach, the capacity is 650 vph, or one vehicle every 5.5 seconds. As flow rates on the conflicting approaches increase, the capacity of the subject approach decreases. As the rate of flow on the conflicting approaches reaches some saturated level, the priority/nonpriority operations begin to break down and priority reversal begins to occur. No clear functional relationship should be apparent during these congested conditions.

Service Time Model Form. The service time model should show the inverse of relationships described above for the capacity model.

Queue Time Model Form. Queue time increases as flow rates increase on the subject, conflicting, and opposing approaches. Queue time is also affected by same variables as those that affect service time. Queue time is zero when subject flow rates are low and no queueing occurs.

3.3 The Data Base

3.3.1 Data Collection Method

To provide a means to assess this theoretical framework, field data were collected during twenty-three different days at twelve different TWSC intersections in the Northwest United States. The data set includes over twenty-eight hours of intersection operation.

Each site was videotaped with the camera in a position to observe vehicles on each approach as they passed through the intersection as well as the queue activity on one approach. Data collection software was used to record flow rate and delay data while observing the videotapes from each sample site. The time that each vehicle entered the intersection was recorded and its directional movement (left-turn, through, or right-turn) was noted. Flow rates were calculated based on these data. For one minor street approach (referred to as the subject approach), the times that each vehicle entered the end of the queue, arrived at the stop line, and entered the intersection were also noted. Delays were calculated based on these data.

Table 5 provides a summary of the important traffic flow and delay data collected at each site. The data that are summarized in Table 5 are 15-minute averages of traffic flow rates and vehicle delays.

Table 5. TWSC Intersection Site Data

Variable	Mean	Minimum	Maximum	Standard Deviation
Queue Time (sec/veh)	9.4	0.2	66.1	11
Service Time (sec/veh)	9.2	4.4	28.1	4
Subject Volume (vph)	187	4	496	133
Opposing Volume (vph)	109	0	360	85
Conflicting Volume (vph)	659	56	1412	316
Intersection Volume (vph)	954	96	1584	311
Subject Capacity (vph)	449	128	809	149

3.4 Capacity Model

The measured subject approach capacity data are plotted against conflicting approach flow rates, opposing approach flow rate, and the speed limit of the conflicting approaches, in Figures 14 through 16. These figures show several general trends:

1. Capacity decreases as conflicting approach flow rates and opposing approach flow rates increase.
2. Capacity decreases as speed on the conflicting approach increases.

Effect of Volume. A number of regression models for subject approach capacity, including various functional forms, were estimated. The most promising model is given in equation (3). A plot of this model and the field data are shown in Figure 17.

$$c_s = 683.43 - 0.31q_c - 0.25q_o \quad (3)$$

where c_s is the capacity of the subject approach, in veh/hr, q_c is the flow rate on the conflicting approaches, in veh/hr, and q_o is the flow rate on the opposing approach, in veh/hr.

Effect of Speed. The data were segregated according to speed limit on the conflicting approach. Separate capacity equations were estimated for two speed ranges, 25 to 35 miles per hour and 45 to 55 miles per hour. The results are given in equations (4) and (5).

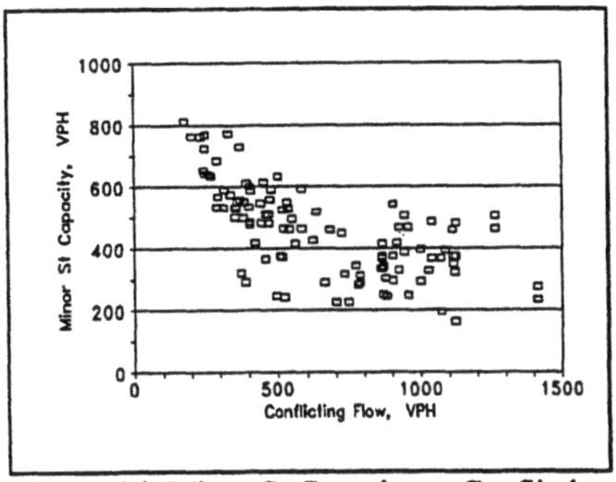

Figure 14. Minor St Capacity vs Conflicting Flow

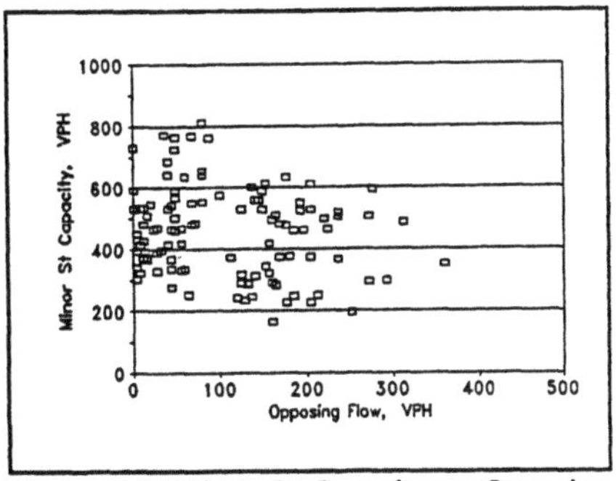

Figure 15. Minor St Capacity vs Opposing Flow

For 25 to 35 mph streets,

$$c_s = 770.49 - 0.48q_c \qquad (4)$$

For 45 to 55 mph streets,

$$c_s = 544.22 - 0.32q_c \qquad (5)$$

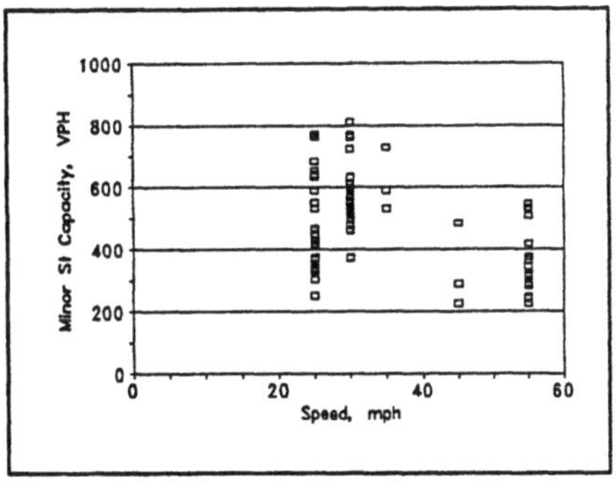

Figure 16. Minor St Capacity vs Speed

The two equations are plotted in Figure 18.

The difference between capacity estimates for the two speed ranges is clearly evident at the lower conflicting approach flow rates. The difference becomes less important as conflicting approach flows increase.

Conflicting Flow Regions. One of the interesting aspects of the plots of the subject approach capacity and the conflicting approach flow rates is the two apparently distinct regions of the plot (see Figure 19). For flow rates less than 600 vph, subject approach capacity declines linearly and fairly steeply. For higher flow rates, capacity remains fairly constant. The existence of these different regions might be explained by the difference in traffic flow behavior of the subject approach traffic. At mean gaps of greater than 6 seconds (equivalent to flow rates of less than 600 vph), drivers behave in a normal fashion and accept a safe gap

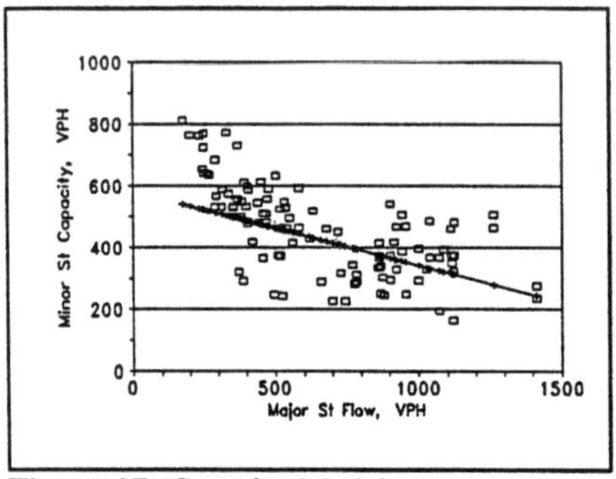

Figure 17. Capacity Model

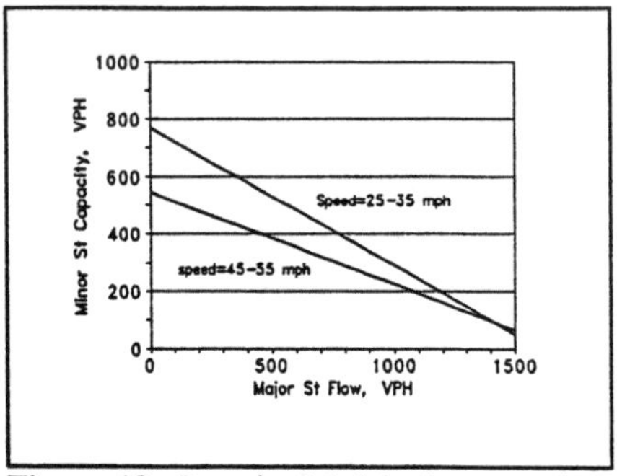

Figure 18. Capacity Models Segregated By Speed

when it appears. For lower mean gaps (which are less than the typical critical gap of 6 seconds), driver frustration begins to increase and a priority reversal begins. That is, subject approach drivers begin to accept smaller gaps, maybe even unsafe gaps. And at even higher conflicting approach flow rates, subject approach capacity does not appreciably decline.

Figure 20 shows idealized plots of the capacity-flow relationship for two different conflicting approach speed ranges.

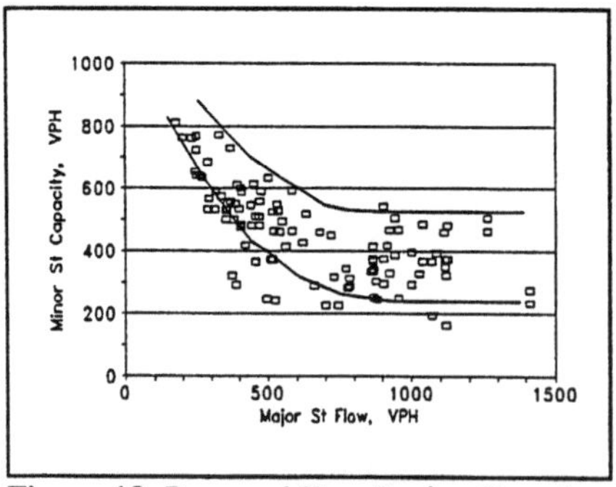

Figure 19. Proposed Two-Region Capacity-Flow Rate Curve

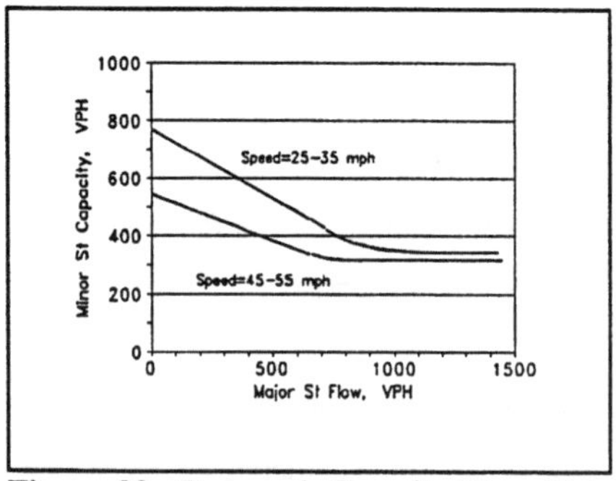

Figure 20. Proposed Capacity-Flow Rate Curve For Two Speed Ranges

3.5 Delay Model

3.5.1 Service Time Model

Plots of service time data versus conflicting approach flow rate and opposing approach flow rate are shown in Figures 21 and 22. Two general trends are evident from these figures, namely that service time increases as conflicting and opposing flow rates increase.

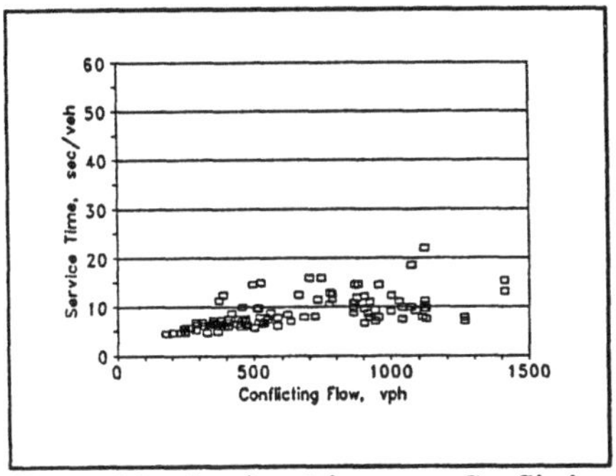

Figure 21. Service Time vs Conflicting Flow

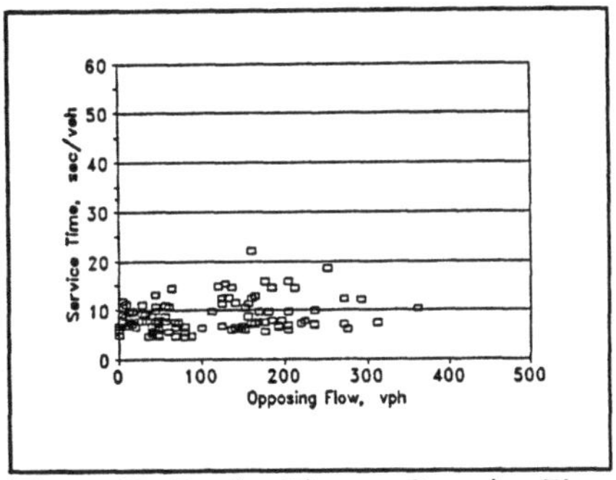

Figure 22. Service Time vs Opposing Flow

Equation (6) shows the model that was developed for estimating service time.

$$d_s = 4.98e^{0.00068q_c + 0.00064q_o} \qquad (6)$$

where d_s is the service time, in sec/veh, q_c is the flow rate on the conflicting approaches, in veh/hr, and q_o is the flow rate on the opposing approach, in veh/hr.

A plot of this model and the field data are shown in Figure 23.

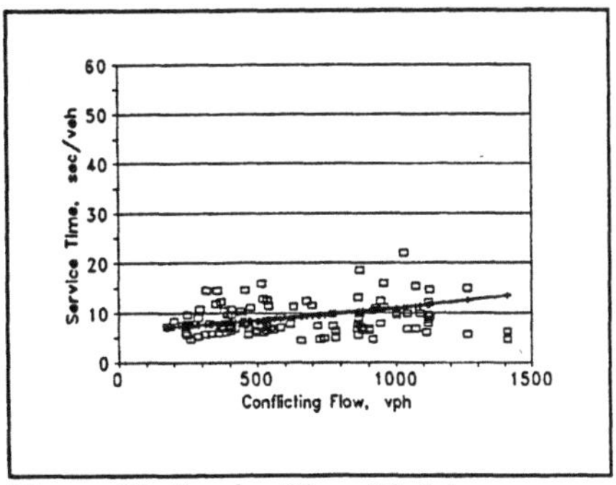

Figure 23. Service Time Model

3.5.2 Queue Time Model

Plots of queue time data are shown in Figures 24 and 25. Two general trends are evident from these plots, namely, that queue time increases as flow on the conflicting approaches increase and as flow on the subject approach increases.

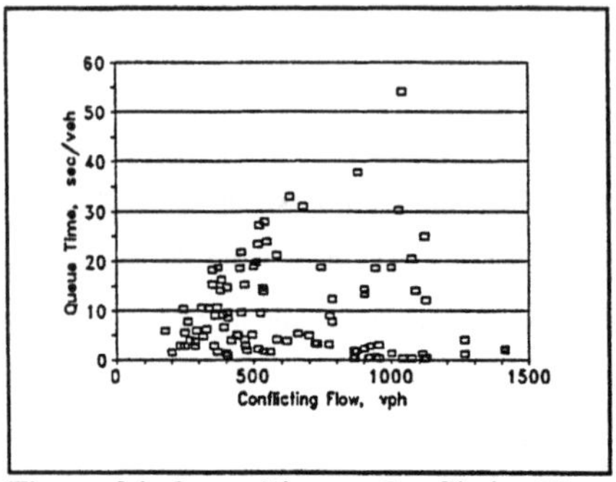

Figure 24. Queue Time vs Conflicting Flow

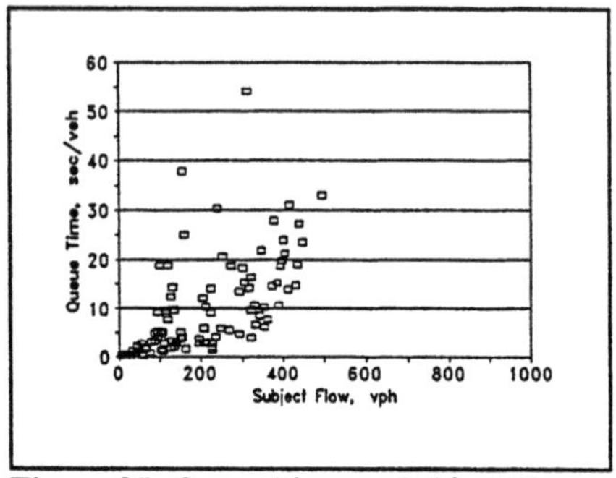

Figure 25. Queue Time vs Subject Flow

The equation for estimating queue time is given in equation (7):

$$d_q = .723e^{0.00035q_c + 0.00839q_s} \tag{7}$$

where d_q is the queue time, in sec/veh, q_c is the flow rate on the conflicting approaches, in veh/hr, and q_s is the flow rate on the subject approach, in veh/hr.

A plot of the model and the field data are shown in Figure 26.

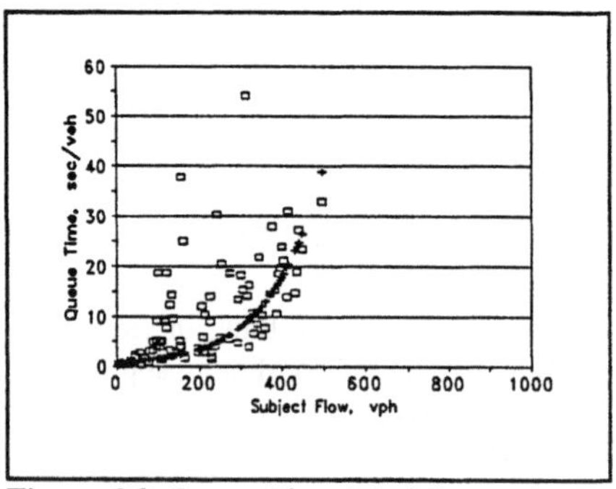

Figure 26. Queue Time Model

4.0 Summary and Conclusions

The purpose of this paper is to present the results of recent research activities in the United States on unsignalized intersections.

1. A new procedure for estimating capacity and delay for AWSC intersections has been

developed and approved for interim use by TRB's Committee on Highway Capacity and Quality of Service. Some of the important elements of the procedure are:

(1) Capacity is a linear function of the distribution of the volume on the intersection approaches, the number of lanes on each approach, and the proportion of turning movements on each approach.

(2) The addition of lanes on an approach does not significantly increase the capacity of the approach.

(3) Delay is an exponential function of the volume/capacity ratio.

2. A set of preliminary models for estimating capacity and delay at TWSC intersections has also been developed. Some of the findings relating to these models are listed below:

(1) Capacity is a linear function of the flow rates on the conflicting and opposing approaches, and the conflicting approach speed.

(2) There is some evidence that there are two different regions for this capacity curve. When the average gap on the conflicting approaches is greater than six seconds (the standard critical gap), capacity declines in a linear manner. When the average gap is less than six seconds, capacity is nearly independent of flow on the conflicting approaches. This may imply a different behavior pattern for minor street drivers when they are faced with longer queues, higher delays, and fewer acceptable gaps.

(3) Service time increases exponentially as flow rates on the conflicting and opposing approaches increase.

(4) Queue time increases exponentially as flow rates on the conflicting and subject approaches increase.

(5) Further study is now underway to determine the effects on capacity and delay of turning movements and the degree of platooning of the major street traffic stream.

5.0 References

1. Federal Highway Administration, Manual on Uniform Traffic Control Devices. Washington, D.C., 1988.

2. Transportation Research Board. 1985 Highway Capacity Manual, Special Report 209, Washington, D.C.

3. Merkblatt for Lichsignalanlagen an Landstrassen Ausgabe 1972. Forschungsgesellschaft fur das Strassenwesen, Koln, Germany, 1972.

4. Hebert, Jacques. "A Study of Four-Way Stop Intersection Capacities". Highway Research Record 27, 1963.

5. Clemow, Chris, John Ringert, Kathy Highly, Scott Lowe, and Ted Egerton. "Analysis of the Delay-Flow Relationship At Stop-Controlled Intersections Using The TEXAS Simulation Model. Unpublished research report, University of Idaho, May 1989.

6. Rouphail, Nagui, and Joel Leisch. "A Program of Research in Highway Capacity", Draft Report for the Committee on Highway Capacity and Quality of Service, Task Force on Research, December 1990.

7. Richardson, Anthony. "A Delay Model for Multiway Stop-Sign Intersections", Transportation Research Record 1112, 107-114.

8. Zion, Mark, George List, and Charles Manning. "Testing Delay Models with Field Data for Four-Way, Stop-Sign Controlled Intersections, Transportation Research Record 1225, 83-90.

9. Kyte, Michael and Joseph Marek. "Data Collection at Stop-Controlled Intersection". ITE Journal, April, 1989.

10. Kyte, Michael and Joseph Marek. "Estimating Capacity and Delay at a Single-Lane Approach All-Way Stop-Controlled Intersection". Transportation Research Record 1225, 73-82.

11. Kyte, Michael. "Estimating The Capacity of an All-Way Stop-Controlled Intersection", Transportation Research Record, 1990 (in-press).

12. Zegeer, John, Michael Kyte, and Wayne Kittelson, "Interim Materials on Unsignalized Intersection Capacity", Transportation Research Circular, Transportation Research Board, 1991. (in press)

13. Kyte, Michael and Anthony Boesen. Traffic Data Input Program (TDIP) User's Manual. Version 3.0. 1990.

14. Bakare and Jovanis. "Analysis of Unsignalized Intersection Capacity", Transportation Research Record 971.

15. Brilon, Werner and Michael Grossmann. "Recommendation For a New Calculation Method For Unsignalized Intersections", Lehrstuhl fur Verkehrswesen, Ruhr-Universitat, Bochum, July 1990.

16. Kittelson, Wayne and Mark Vandehey. "Delay Effects on Driver Gap Acceptance Characteristics at Two-Way Stop-Controlled Intersections". Presented at 1991 annual meeting, Transportation Research Board.

17. Kyte, Michael, Chris Clemow, Naseer Mahfood, B. Kent Lall, C. Jotin Khisty. "Capacity And Delay Characteristics of Two-Way Stop-Controlled Intersections." Presented at 1991 Annual Meeting, Transportation Research Board, Washington, D.C.

18. Kyte, Michael, Kent Lall, Naseer Mahfood, Gopal Panchavati, and Chris Clemow. "Development of Empirical Models to Forecast Delay and Capacity at Two-Way Stop-Controlled Intersections". Presented at 1991 Annual Meeting, special session on two-way stop-controlled intersections, Transportation Research Board.

19. Kimber, R. M. and R. D. Coombe. The Traffic Capacity of Major/Minor Priority Junctions. TRRL Supplementary Report 582. Transportation and Road Research Laboratory, 1980.

20. Semmens, Marie C. "The Capacity of Non-Signalised Road Junctions With High-Speed Controlling Flows". Traffic Engineering and Control, 1988, 280-283.

21. Kimber, R. M. "Gap-Acceptance and Empiricism in Capacity Prediction". Transportation Science. Volume 23, Number 2, May 1989, 100-111.

22. Brilon, Werner. "Recent Development in Calculation Methods for Unsignalized Intersections in West Germany". In Intersections Without Traffic Signals, Werner Brilon, editor, Springer-Verlag, Berlin, 1988.

23. Troutbeck, R. J. "Current and Future Australian Practices For The Design Of Unsignalized Intersections". In Intersections Without Traffic Signals, Werner Brilon, editor, Springer-Verlag, Berlin, 1988.

MIX
Papier aus verantwortungsvollen Quellen
Paper from responsible sources
FSC® C105338

If you have any concerns about our products,
you can contact us on
ProductSafety@springernature.com

In case Publisher is established outside the EU,
the EU authorized representative is:
Springer Nature Customer Service Center GmbH
Europaplatz 3, 69115 Heidelberg, Germany

Printed by Libri Plureos GmbH
in Hamburg, Germany